Third Generation Mobile Communication Systems

For a listing of recent titles in the *Artech House Universal Personal Communications Series,* turn to the back of this book.

Third Generation Mobile Communication Systems

Ramjee Prasad
Werner Mohr
Walter Konhäuser

Editors

Artech House
Boston • London

Library of Congress Cataloging-in-Publication Data
Prasad, Ramjee.
  Third generation mobile communication systems / Ramjee Prasad, Werner Mohr,
Walter Konhäuser.
      p. cm. — (Artech House universal personal communications library)
  Includes bibliographical references and index.
  ISBN 1-58053-082-6 (alk. paper)
  1. Wireless communication systems. 2. Mobile communication systems —
Standards. 3. Code division multiple access. 4. Time division multiple access.
  I. Mohr, Werner, 1955- II. Konhäuser, Walter. III. Title. IV. Series.

TK5103.2. P72 2000   99-089511
621.3845—dc21   CIP

British Library Cataloguing in Publication Data
  Prasad, Ramjee
  Third generation mobile communication systems. — (Artech House
  universal personal communications library)
  1. Mobile communication systems
  I. Title II. Mohr, Werner III. Konhäuser, Walter
  621.3'845

  ISBN 1-58053-082-6

Cover design by Dutton and Sherman

International Standard Book Number: 1-58053-082-6
Library of Congress Catalog Card Number: 99-089511

10 9 8 7 6 5 4 3 2 1

Dedication

To
my wife Jyoti, to our daughter Neeli,
and to our sons Anand and Rajeev
Ramjee Prasad

To
my parents
Werner Mohr

To
my wife Tina and to our son Philipp
Walter Konhäuser

Contents

Chapter 6 UTRA Transport Control Function   165

J. Lundsjö, M. Rinne

J. Arponen, J. Eldståhl, A. Näsman

W. Mohr

Preface

कर्मण्येवाधिकारस्ते मा फलेषु कदाचन ।
मा कर्मफलहेतुर्भूर् मा ते संगोऽस्त्वकर्मणि ॥

karmaṇy evādhikāras te
mā phaleṣu kadācana
mā karma-phala-hetur bhūr
mā te saṅgo 'stv akarmaṇi

You have a right to perform your prescribed duty, but you are not entitled to the fruits of action. Never consider yourself the cause of the result of your activities, and never be attached to not doing your duty.

The Bhagavad Gita (2.47)

This book is the output of the research and development contributions of several major players from industries, a network operator, research laboratories and universities, which were carried out during the European Advanced Communication Technology and Services (ACTS) Future Radio Wideband Multiple Access System (FRAMES) project. The main objective of the FRAMES project was to develop a radio interface proposal, which fulfills the requirements on terrestrial third generation mobile radio systems, and to contribute to the international standardization process. The FRAMES project was the only ACTS project dealing with the terrestrial communications of the Universal Mobile Telecommunications System (UMTS) radio interface.

Third Generation Mobile Communication Systems is the first book to take a comprehensive look at UMTS, providing an in depth description of all the elements required to understand and develop the third generation mobile radio systems and networks. Figure 1 illustrates the coverage of the book. Chapter 1 presents an overview of International Mobile Telecommunications – 2000 (IMT-2000) / UMTS. Chapter 2 explains the basic principles of Time Division – Code Division Multiple Access (TD-CDMA). The basic concept of Wideband CDMA

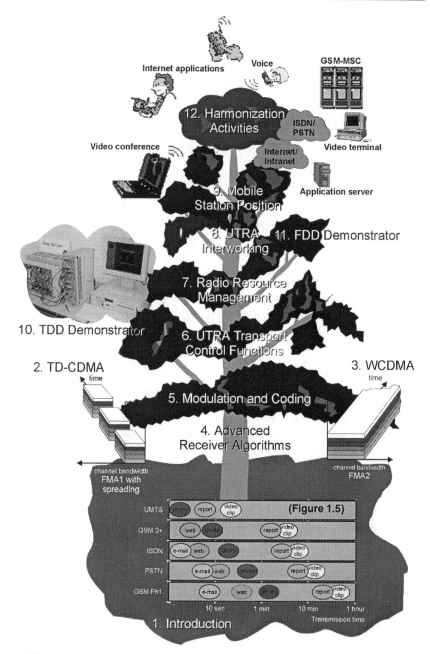

Figure 1 Illustration of the coverage of the book.

(WCDMA) is introduced in Chapter 3. Advanced receiver algorithms are covered in Chapter 4. Chapter 5 discusses modulation and coding techniques. UMTS Terrestrial Radio Access (UTRA) transport control functions are presented in detail in Chapter 6. Radio resource management is introduced in Chapter 7. Chapter 8 presents UTRA networking and mobile station positioning is presented in Chapter 9. Chapters 10 and 11 cover the Time Division Duplexing (TDD) and Frequency Division Duplexing (FDD) demonstrators developed during the FRAMES project, respectively. Finally the international harmonization activities are discussed in Chapter 12.

Thus this book delivers the basic principles and the analytical models for the UTRA TDD mode using TD-CDMA and for the UTRA FDD mode using WCDMA, allowing everyone to understand how these multiple access systems fulfill the UMTS requirements. Plus, several interesting topics are presented, viz. very advanced receiver algorithms, coding and modulation techniques, layer 2 issues including the UTRA architecture, protocol architectures, signaling protocols and Automatic Repeat Request (ARQ) schemes. Included is an examination of TDD and FDD mode compatibility with Global System for Mobile Communications (GSM) and the methods used for calculating mobile station location within the coverage area. With this new book one gets an integrated resource that examines the fundamentals and applications of today's most important mobile communication technologies.

FRAMES started in September 1995 before the detailed standardization activities in European Telecommunications Standardization Institute Special Mobile Group 2 (ETSI SMG2) began for the UMTS radio interface. During the first phase of the project we had serious discussions on the technical approach for the radio interface. This approach was also changed several times in this period. These technical discussions took place in the starting phase of a big project, where all partners were working to understand each other and their positions. Experience showed that this needs time when colleagues from different countries all over Europe, namely

- Austria

- Finland

- France

- Germany

- Portugal

- Spain

- Sweden

- Switzerland

- The Netherlands

- United Kingdom

and from different organizations as

- Manufacturers

  - Ericsson Radio Systems AB

  - Nokia Corporation

  - Siemens AG and Roke Manor Research

- Network operator

  - France Télécom – CNET

- SME

  - Integracion y Sistema de Medida

- Research Center and Academia

  - CSEM – Centre Suisse d'Electronique et de Microtechnique SA

  - Eidgenössische Technische Hochschule Zurich

  - Chalmers University of Technology AB

  - Delft University of Technology

  - Instituto Superior Técnico

  - Oulu Technical University

  - Royal Insitute of Technology

  - University of Kaiserslautern

started to cooperate.

The technical discussions in the first phase have been determined by the different interest of partners. However, FRAMES defined in that period the FMA scheme, which combined TDMA and CDMA based techniques. In December 1996 FRAMES participated in the ETSI SMG2 Workshop on UMTS in Sophia Antipolis with two presentations to present the first time publicly the FMA scheme. This was the starting point of the standardization process in ETSI SMG. In January 1997 a first Long-term Research Workshop was organized in Gothenburg. During 1997 we presented a lot of joint contributions to the international standardization. In 1997 we again had serious technical discussions in the standardization bodies. This is understandable due to the different interests of the different partners, which cannot be solved by a research project. However, the project contributed significantly to the international consensus building process. Despite all discussions FRAMES presented the FMA scheme in the ITU IMT-2000 Workshop in September 1997 in Toronto again with two deeply technical presentations.

January 1998 was a very important period for FRAMES. The ETSI decision on the UTRA concept on January 29, 1999 with WCDMA in the paired bands based on FMA2, and with TD-CDMA in the unpaired bands based on FMA1 with spreading was finally a big success for FRAMES. FMA1 without spreading was adopted in the U.S. for the high speed mode in UWC-136. From that point of view FRAMES had a big impact on the UMTS and IMT-2000 standardization. In 1998 FRAMES partners participated extensively in the standardization process with many contributions, which have been prepared in the framework of the project. FRAMES adopted the ETSI decision and focused its work on the optimization of the UTRA concept. UTRA TDD was mainly developed in the project. During 1998 FRAMES performed two successful workshops, one in Beijing, China, and one in Yokosuka, Japan. In addition, during the project's lifetime we did a large number of presentations and wrote several publications. Therefore, FRAMES is well-known internationally.

In January 1999 an open workshop was organized in Delft to present our results. We invited other ACTS projects as well as the European Commission. FRAMES was invited to workshops organized by other ACTS projects namely OnTheMove in Singapore in September 1997 and RAINBOW in December 1998 in Torino. In the last phase of the project, mainly during 1999, we concentrated our effort on the demonstrator, which is finally integrated. Joint trials with RAINBOW took place and in September 1999 and CNET performed trials and measurements. Beginning in November 1999 an Open Day was organized at France Télécom - CNET to present the demonstrator.

With respect to the difficult environment as different partner interests and the ongoing international standardization process, FRAMES was a very successful project with a significant contribution to and impact on the Third Generation Mobile Radio Systems.

All this was only possible because all colleagues worked together, respecting the interests of their organizations. It was a very good experience to cooperate in such an environment. We always could talk to each other and we were able to find reasonable solutions. The international workshops and the book, which was prepared by the FRAMES project, showed our ability to cooperate.

The relation to the European Commission was very trustful and fair. Discussions and negotiations have been needed in difficult situations. We were always able to find suitable solutions for the European Commission and the project.

We have tried our best to make each chapter quite complete in itself. This book will help in finding the solution in deploying the Third Generation Mobile Communications Systems IMT-2000 / UMTS. Any remarks to improve the text and correct any errors would be highly appreciated.

Acknowledgments

The material in this book originates from the FRAMES project and contributions to the international standardization process. Therefore, we would like to thank all the colleagues involved in the project for their support and cooperation that made success possible. This success is not only in completing the project successfully, but also finalizing the book as an additional part of the project. We hope that our personal relations remain and possibly we will cooperate in other projects or international bodies in the future.

FRAMES was partly funded by the European Union. We thank especially Dr. Joao Schwarz Dasilva and Mr. Bartolomé Arroyo-Fernandez from the European Commission for their continuous support. We would like to acknowledge the contributions of our colleagues from Siemens AG, Roke Manor Research Limited, Ericsson Radio Systems AB, Nokia Corporation, Technical University of Delft, University of Oulu, France Télécom CNET, CSEM – Centre Suisse d'Electronique et de Microtechnique SA, Eidgenössische Technische Hochschule Zürich, University of Kaiserslautern, Chalmers University of Technology AB, the Royal Institute of Technology, Instituto Superior Técnico and Integracion y Sistema de Medida.

Ljupco Jorguseski from Delft University of Technology / KPN Research helped to prepare the complete manuscript, freeing us from the enormous editorial burden. He was supported by Albena Mihovska and Martijn Kuipers from the Center for PersonKommunikation, Aalborg University, Denmark.

Last but not least, Per-Olof Anderson from Ericsson Radio Systems AB is greatly acknowledged for giving his support to finish this book as the technical manager of the FRAMES project.

Ramjee Prasad
Werner Mohr
Walter Konhäuser

December 1999

Chapter 1

Introduction

The International Telecommunications Union (ITU) has launched one of its most ambitious projects ever: a federation of systems for third generation mobile telecommunications that will provide wireless access to the global telecommunication infrastructure at anytime and anywhere. This new framework of standards is known under the generic name of International Mobile Telecommunications-2000 (IMT-2000) and represents the culmination of ten years of study and design work.

The most exciting development in mobile communications since the advent of digital systems back in the early 1990s, IMT-2000 also represents one of the ITU's most important achievements in the last decade of the 20th century. IMT-2000 identifies the following key factors as essential for the success of the next generation of mobile communications [1-33].

- High-speed access, supporting broadband services such as fast Internet access or multimedia-type applications. Demand for such services is already growing fast;

- Flexibility, supporting new kinds of services such as universal personal numbering and satellite telephony. Features will greatly extend the reach of mobile systems, benefiting consumers and operators alike;

- Affordable, as those of today, if not more so;

- Compatible, perhaps most vital, offering an effective evolutionary path for existing wireless networks.

Universal mobile telecommunications system (UMTS) is one of the most important third generation mobile communications systems being developed within the IMT-2000 framework. It has the support of many major telecommunications operators and manufacturers because it represents a unique opportunity to create a mass market for highly personalized and user-friendly mobile access to tomorrow's information society.

UMTS will deliver wideband information as well as voice, data, and multimedia, direct to people who can be on the move. UMTS will build on and extend the capability of today's mobile technologies (like digital cellular and cordless) by providing increased capacity, data capability, and a far greater range

1

of services using an innovative radio access scheme and an enhanced, evolving core network.

This introductory chapter will provide a general overview of IMT-2000/ UMTS and outline some of the research and standardization work devoted to those concepts worldwide. This will provide a background for the following chapters on the technical concepts.

1.1 OVERVIEW OF IMT-2000 / UMTS

Market expectations for mobile radio systems show an increasing demand for a wide range of services from voice and low rate data services up to high rate data services, and including advanced services such as mobile multimedia. This leads to technical requirements, which can only be met with a new generation of mobile radio systems, IMT-2000/UMTS, which are currently being standardized worldwide. Circuit and packet-oriented services will be supported. These systems will operate in all radio environments to provide service to anyone, anytime and anywhere.

ITU has identified spectrum for the allocation of IMT-2000. However, these frequency bands are currently not available worldwide. In different regions research activities of IMT-2000 have been initiated to support the international consensus-building process and standardization activities. Based on these activities system proposals have been submitted to ITU TG 8/1.

The code division multiple access (CDMA) proposals from Europe, Japan, Korea, and the United States are very similar. This enables the chance for a worldwide standard of a third generation radio interface. For this reason, the newly formed Third Generation Partnership Projects 3GPP and 3GPP2 have the objective to harmonize proposals and to define detailed standards.

Some proposals focus mainly on the new IMT-2000 spectrum, while others such as time division multiple access/136 (TDMA/136) (EDGE) are more directed to the introduction of IMT-2000 services in spectrum already used by second generation systems. The evolution and migration of second generation systems to the third generation takes into account the deployed investment to save today's investment where useful and necessary.

1.1.1 Market Requirements and Services for Third Generation Mobile Radio Systems

Third generation mobile radio systems are one of the key communication technologies in research, development, and international standardization bodies, due to the high market demand for advanced wireless communication.

First generation mobile radio systems, such as advanced mobile phone system (AMPS), total access communication system (TACS), C-net, and Nordic Mobile Telephones (NMT), were analog voice telephone systems. They were first deployed in the early 1980s, and provided reasonably good quality and capacity. In some cases, they were quite successful. Second generation mobile radio systems, which use digital technology in contrast to the first generation systems, provided low rate data services and other auxiliary services, in addition to voice. Second generation systems are very successful worldwide in providing service to users. The customer base is increasing much faster than initially expected. The UMTS Forum is expecting around 400 million mobile subscribers worldwide in the year 2000 and nearly 1800 million subscribers in the year 2010 (Figure 1.1, [5, 32, 33]). Especially in Asia, tremendous growth is expected with a dominant subscriber base in 2010.

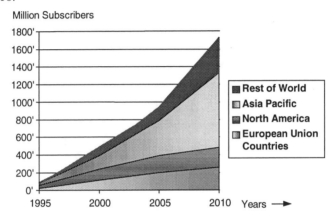

Figure 1.1 World mobile subscribers.

These second generation systems are dominated by the global system for mobile communications (GSM), IS-136, IS-95 CDMA and in Japan by personal digital system (PDC) and personal handyphone system (PHS) technology (Figure 1.2). GSM is the mobile radio standard with today's highest penetration worldwide.

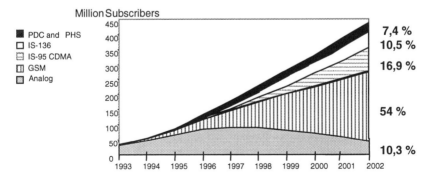

Figure 1.2 Mobile market segmentation by technology.

The penetration of analog systems (first generation) is starting to decrease. However, these second generation systems are limited in the maximum data rate. On the other hand, the percentage of mobile multimedia users will increase significantly after the year 2000. According to the UMTS Forum, in 2010 about 60% of the traffic in Europe will be created by mobile multimedia applications ([5, 32, 33]). A similar growth of mobile data traffic is expected worldwide, with an expected growth rate per year in the order of 70% in the next 5 years, starting from about 3 million data users in 1998 up to about 77 million data users in 2005 (Figure 1.3).

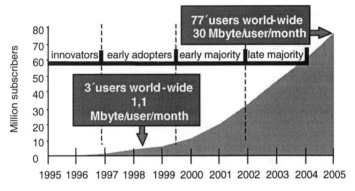

Figure 1.3   Mobile data market worldwide, exceptional growth of subscribers and usage
(Source: Merrill Lynch).

The market expectations are stemming from visions for the future communication needs. More advanced services than current voice and low data rate services are foreseen and will bring together the three worlds according to the three basic categories (Figure 1.4):

- Computer-data with Internet access, electronic mail, real-time image transfer, multimedia document transfer, mobile computing;

- Telecommunications with mobility, video conferencing, GSM-based and integrated switched digital network (ISDN)-based services, video telephony, wideband data services;

- Audio/video-content with video on demand, interactive video services, infotainment, electronic newspaper, teleshopping, value-added Internet services, TV and radio contribution.

Future services will range from low up to high user data rate (maximum 2 Mbps for IMT-2000/UMTS systems). The transmission of, for example, a large presentation of a size of 2 Mbyte would need 8 sec with a 2 Mbps service compared to current data transmission in GSM with 9.6 Kbps, where 28 minutes would be necessary (Figure 1.5).

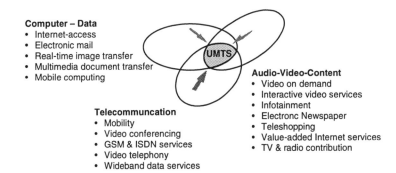

Computer – Data
- Internet-access
- Electronic mail
- Real-time image transfer
- Multimedia document transfer
- Mobile computing

Telecommuncation
- Mobility
- Video conferencing
- GSM & ISDN services
- Video telephony
- Wideband data services

Audio-Video-Content
- Video on demand
- Interactive video services
- Infotainment
- Electronc Newspaper
- Teleshopping
- Value-added Internet services
- TV & radio contribution

Figure 1.4 Converging applications offer new business opportunities.

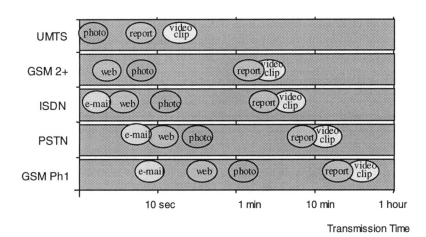

Transmission Time

Figure 1.5 Transmission times for different services and systems.

Multimedia applications use several services such as voice, audio/video, graphics, data and e-mail, and circuit switched and packet-oriented service types in parallel, which have to be supported by the radio interface and the network subsystem. One main driver for data services is the exponential growth of the Internet in the last few years.

The GSM Association is expecting a high grade of asymmetry between uplink and downlink for data transmission for services such as Internet access with much higher necessary capacity on the downlink (Figure 1.6). Similar predictions are obtained by the UMTS Forum (Figure 1.7).

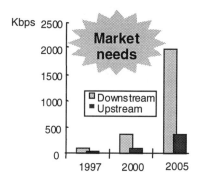

Figure 1.6   Asymmetrical data traffic.

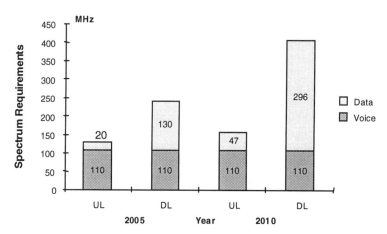

Figure 1.7   Terrestrial spectrum requirements for mobile services in the years 2005/2010.

Within the third generation system (IMT-2000/UMTS), these different service needs are supported in a spectrum-efficient way by a combination of frequency and time division duplex (FDD and TDD). The FDD component supports wide area coverage mainly for symmetrical services, whereas TDD is suited especially for asymmetrical services. Only the opportunity of combining both modes will provide maximal efficiency and flexibility for third generation networks.

The cell capacity does strongly depend on the user behavior in terms of required data rate and throughput and the applied services. Therefore, service demands for higher data rates require sufficient available frequency spectrum to support a mass market with sufficient grade and quality of service. These basic market expectations are also supported by the European Commission (e.g., in [6]).

1.1.2 Technical Requirements and Radio Environments

These market requirements and service needs result in international technical requirements for the ongoing definition of third generation mobile radio systems (IMT-2000 / UMTS) in the different regions America, Asia, and Europe ([7-9]). These systems are required to support a wide range of bearer services from voice and low rate data services up to high rate data services. High data rate requirements are up to at least 144 Kbps in vehicular, up to at least 384 Kbps in outdoor to indoor, and up to 2 Mbps in indoor and picocell environment (Figure 1.8).

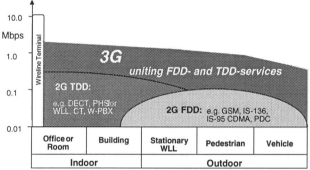

Figure 1.8 Application area of third generation systems.

Circuit-switched and packet-oriented services for symmetric and asymmetric traffic need to be supported. Third generation systems will operate in all radio environments like urban and suburban areas, hilly and mountainous areas, microcell, picocell, and indoor environments to provide service to anyone, anytime, and anywhere. These requirements are quite well aligned in the different regions of America, Asia, Europe, and in the ITU. This enables a much wider application range of third generation systems compared to the second generation (Figure 1.9). In addition, the ability for global roaming has to be supported in the system design.

In the following technical chapters, a Europe-centered approach is taken with respect to the activities in the European research project ACTS FRAMES.

1.2 EUROPEAN RESEARCH ACTIVITIES

In Europe, the European Commission has partly funded research activities in the ACTS (advanced communication technologies and services) framework related to third generation mobile radio systems, to support consensus-building and international standardization activities ([14-16]). Preceding framework research programs have been RACE I and II (research of advanced communication technology in Europe), especially in the ATDMA and CODIT projects to investigate advanced TDMA and CDMA-based radio interfaces ([17], p. 500).

The European Commission has launched a 5th framework program with the theme "Creating a User Friendly Information Society," which is also open for non-European countries [6].

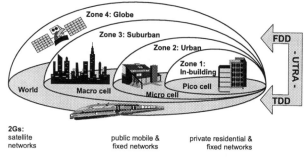

Figure 1.9 User environments and related UTRA modes.

The ACTS program started in 1995 with about 250 projects in total. More than 20 projects are related to the mobile domain with the following areas and projects [6]:

- Enabling technologies: FIRST, SORT, SUCOMS, SUNBEAM, TSUNAMI II;

- Services and applications: CAMELEON, MEMO, MICC, MOMENTS, MONTAGE, MOVE, OnTheMove, UMPTIDUMPTI, USECA;

- UMTS: Cobuco, Exodus, FRAMES, Momusys, RAINBOW, STORMS;

- WLANs, MBS: Awacs, CABSINET, Median, Samba, Wand;

- Satellites: Asset, CRABS, Insured, Newtest, Secoms, Sinus, SUMO, Tomas, WISDOM.

The FRAMES project (future radio wideband multiple access system) was the only ACTS project dealing with the terrestrial component of the UMTS radio interface. It is a consortium comprised of partners from manufacturers, operators, small medium enterprises (SME), and the research community [18].

It has the main goal of developing a radio interface proposal, which fulfills the requirements on terrestrial third generation mobile radio systems. The definition of the original FRAMES multiple access scheme (FMA) has fulfilled these requirements, and has taken into account the big worldwide footprint of the GSM family with respect to harmonized radio parameters and the GSM core network.

Based on the UMTS requirements and the activities and decisions in different regions of the world in the beginning of the project, the FMA scheme was developed in 1996. It combined TDMA and CDMA techniques into one harmonized platform to cope with all possible UMTS scenarios and to address possible technical solutions from a global perspective (Figure 1.10) [19]:

- Mode 1 (FMA 1): wideband TDMA with and without spreading;

- Mode 2 (FMA 2): wideband CDMA.

Technical details are presented in [19-31].

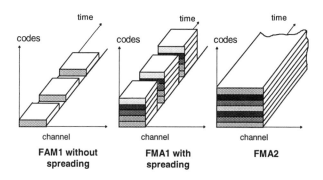

Figure 1.10 Basic FMA multiple access schemes.

The FRAMES project assumed that UMTS (and partly IMT-2000) will build on an evolved GSM platform with several hundred million subscribers at the time of the UMTS launch. Under the assumption that UMTS and GSM will operate in parallel for a long time, it is important that dual-mode terminals be easy to build. Therefore, FMA was harmonized with the GSM network in terms of radio parameters and the protocol stack as far as possible. Table 1.1 shows the basic parameters of the original FMA scheme.

With respect to the status of the worldwide standardization activities at the time of the definition of the FMA concept at the end of 1996, the main goal of harmonization was to meet different system needs of different possible access schemes in different regions, and to ensure compatibility to second generation systems. Unnecessary differences from a technical point of view between a TDMA- and CDMA-based multiple access scheme are avoided without sacrificing the good aspects of each technology. The harmonized approach between a TDMA- and CDMA-based mode and the GSM family minimizes the terminal hardware complexity.

At the beginning of 1998, the FRAMES project adopted the European Telecommunications Standards Institute (ETSI) consensus decision on the UTRA concept for its further investigations (c.f. section 1.3). These investigations have been focused in more detail on channel coding and modulation, receiver algorithms, layer 1 aspects and network aspects for the FDD and the TDD component to optimize the selected scheme even further. In addition, the TDD component is presented in trials and service demonstrations.

Table 1.1

Basic Parameters of FMA 1 and FMA2 (status January 1997)

	FMA1 (TDMA with and without spreading)		FMA2 (CDMA)
	Without spreading	With spreading	
Multiple access method	TDMA	TDMA/CDMA	WB-CDMA
Carrier Chip/bit rate	2.6 Mbps	2.167 Mcps	4.096 Mcps
Bandwidth	1.6 MHz		4.4 - 5 MHz
Duplex method	FDD and TDD		FDD TDD for further study
Interference reduction	Joint detection supported	Joint detection	Multi user detection supported
Spreading codes	N/A	Orthogonal of length 16 chips	Spreading factor 4 to 256, short codes
Multirate concept	Multislot	Multislot and multicode	Variable spreading and multicode
Detection	Coherent, based on midamble		Coherent (reference symbol or pilot)
Handover	Mobile assisted hard handover		Mobile controlled soft handover
Interfrequency handover	Supported		Supported
Frequency hopping	Frame by frame / slot by slot		N/A

The results and proposals of FRAMES have been used successfully as input to the ETSI standardization and evaluation process for the UMTS terrestrial radio access (UTRA) concept. In addition, FRAMES partners have contributed to the standardization process in ARIB in Japan, to TIA and T1P1 in the United States, and to TTA in Korea. The high-speed mode of TDMA/136 is based on FMA1 without spreading.

In parallel with the work on the physical layer, the FRAMES project set out to define the protocol structure and architecture of the radio interface.

The generic radio access network (GRAN) concept has been supported in the project. The fundamental ideas of this concept were to separate the core network from the radio access network, and to keep radio-dependent functionalities within the GRAN [19].

In conjunction with this work, the project considered how to describe radio bearers in a generic, parametric manner. Such a description would be very useful at the interface between the core network and the GRAN; the core network would be able to request a radio bearer based on a set of parameters, and leave it to the GRAN to realize the bearer on the radio interface. This work has inspired the work in standardization bodies on the UMTS terrestrial radio access network (UTRAN).

The project developed protocol structures, and partly protocols as well, for the FMA concept, both FMA1 and FMA2. Within the FMA2 protocol work, the concept of transport channels was introduced and developed. The usual split of physical and logical channels, and the "traditional" way to use these concepts, was found to be inconsistent and insufficient. Therefore, the transport channels were introduced as a significant intermediate step in the mapping from physical to logical channels. The same concept was incorporated early in the UTRA FDD description, and soon after that also in UTRA TDD.

The fact that the early analysis came up with a bimodal concept, FMA, lead obviously to considerations on harmonization between the modes. The harmonization aspects relate to the physical layer as well as the protocol stack. At the physical layer, issues like chip (or symbol) rate and clock generation, as well as carrier spacing and bandwidth, were seen as the most relevant to solve. On the protocol side, the protocols were separated into mode-specific and nonspecific parts, see Figure 1.11. This work on harmonization was later put to use during the refinement of the UTRA decision.

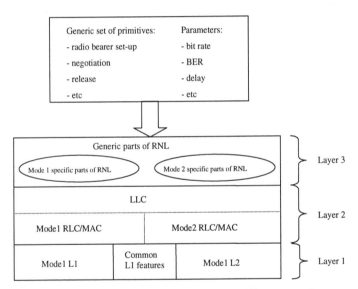

Figure 1.11 Protocol stack, showing where the two modes imply different protocols.

1.3 EUROPEAN STANDARDIZATION ACTIVITIES

ETSI SMG (European Telecommunications Standards Institute) was responsible for the standardization process for IMT-2000/UMTS in Europe. These activities are now moved to the newly established body 3GPP (Third Generation Partnership Project) according to section 1.4.1.

The ETSI process started at the end of 1996 and was subdivided in a grouping phase of proposals, a refinement and synthesis phase, and a definition phase. This open process resulted in five concept groups, which are related to the basic technical proposals based on:

- WCDMA (wideband code division multiple access);
- OFDMA (orthogonal frequency division multiple access);
- TDMA (time division multiple access);
- Hybrid schemes by combining TDMA and CDMA as TD/CDMA;
- ODMA (Opportunity Driven Multiple Access), a radio relay system using mobile terminals as relay stations for network organization.

ETSI SMG decided on the basic access scheme (UTRA concept - UMTS terrestrial radio access) in January 1998 for WCDMA in paired bands (FDD) and for TD/CDMA in unpaired bands (TDD). This concept with the FDD mode WCDMA is partly based on FMA2, and the TDD mode TD/CDMA is based on FMA1 with spreading. In addition, the SMG decision also emphasized the following objectives when selecting the technical parameters ([10, 11]):

- Low-cost terminals;
- Harmonization with GSM;
- FDD/TDD dual-mode operation;
- Fit into 2×5 MHz spectrum allocation.

The harmonization between both UTRA components results from the manufacturers', operators', and customers' viewpoint with the main benefit to implement cost-effective, dual-mode terminals and base station equipment. The parameter harmonization of the FDD and TDD component has been achieved with respect to the implementation of terminals and to achieve a worldwide standard. ETSI special mobile group (SMG) has submitted the UTRA concept to ITU-R for IMT-2000 [12].

Some key characteristics of the UTRA concept are shown in Table 1.2. ETSI SMG submitted this basic concept to the IMT-2000 radio transmission technology development process in ITU.

The motivation for having different schemes for the FDD and TDD components of the UTRA concept is to provide the scheme that is best suited for FDD and TDD, respectively. In order to facilitate manufacturing of dual-mode (FDD-TDD) terminals, care is taken (as seen in Table 1.2) to harmonize key parameters, such as chip rate, pulse shape, and frame length. This harmonization is done without sacrificing performance in either mode. The lower layer protocol structures are harmonized as far as possible.

Table 1.2

Key Characteristics of the UTRA Concept (status June 1998)

	UTRA FDD	*UTRA TDD*
Multiple access	WCDMA	TD-CDMA
Duplex method	FDD	TDD
Carrier spacing	5/10/20 MHz	5 MHz
Chip rate	4.096 Mcps	4.096 Mcps
Slot structure	16 slots per frame	16 slots per frame
Spreading	Spreading factor 4-256, short codes for DL and UL; long optional for UL	Orthogonal, spreading factor 1, 2, 4, 8 and 16 chips/symbol
Frame length	10 ms	10 ms
Multiple rates	Multicode, variable spreading factor	Multislot, multicode
Modulation	DL: QPSK, UL: (Dual channel) BPSK	QPSK
Pulse shaping	Root Raised Cosine, roll-off = 0.22	Root raised cosine, roll-off = 0.22
Handover	Mobile-controlled soft handover	Mobile-assisted hard handover
IF handover	Mobile-assisted hard handover	Mobile-assisted hard handover

From Table 1.1 and Table 1.2 it is clear that large parts of the original FMA concept are also found in the UTRA concept. FRAMES adapted during its lifetime to actual developments in the international standardization bodies. There are, however, some important differences:

- The UTRA FDD (WCDMA) is based on FMA2 for the uplink, and on the ARIB proposal for the downlink. The reason is that the FMA2 proposal employs code multiplexing of data and physical control information, which fits best in the uplink, where power amplifier limitations are very important. On the other hand, the time multiplexing in the ARIB proposal fits better in the downlink;

- The UTRA TDD (TD-CDMA) is based on FMA1 with spreading, but with parameters modified for application to TDD, and harmonized with UTRA FDD according to the decision in ETSI SMG.

1.4 RESEARCH AND STANDARDIZATION ACTIVITIES IN OTHER REGIONS

1.4.1 Global Activities

The international standardization activities for IMT-2000/UMTS have been mainly concentrated in different regions in:

- ETSI SMG (European Telecommunications Standards Institute) in Europe (c.f. section 1.3);

- RITT (Research Institute of Telecommunications Transmission) in China;

- ARIB (Association of Radio Industry and Business) and TTC (Telecommunication Technology Committee) in Japan;

- TTA (Telecommunications Technologies Association) in Korea;

- TIA (Telecommunications Industry Association) and T1P1 in the United States.

According to Figure 1.2, the second generation systems are mostly dominated by GSM, followed by IS-136, IS-95 CDMA, and PDC. These systems have been taken into account by the related regional bodies in the design of third generation system proposals for backward compatibility reasons as UTRA, cdma2000, TDMA/136 and EDGE, where useful and necessary, to save today's investment of already deployed systems (Figure 1.12). These radio interfaces are connected to the two dominating types of core networks MAP and ANSI-41 and their evolutions to third generation.

ITU-R TG 8/1 (Task Group 8/1) called for RTT (radio transmission technology) proposals by the end of June 1998 and for the system evaluation by the end of September 1998 [9]. The system evaluation for the development of radio interface recommendations is an open process under participation of ITU members. A similar approach was applied in regional bodies.

ETSI SMG submitted the UTRA concept to ITU-R for IMT-2000 (c.f. section 1.3).

China has presented to ITU-R a TD-SCDMA proposal based on a synchronous TD-CDMA scheme for TDD and WLL (wireless local loop) applications.

The Japanese standardization body ARIB proposed WCDMA. The Japanese and the European WCDMA proposal for FDD were practically aligned. The ARIB WCDMA concept was submitted to ITU-R. Japan intends to start service in 2001.

TTA in Korea has prepared two proposals for ITU-R; one is close to the ARIB WCDMA scheme and the other proposal is similar to the TIA cdma2000 approach.

In the United States, TIA has prepared several proposals for the third generation with TDMA/136 as an evolution of IS-136, cdma2000 as evolution of

IS-95 and a wideband CDMA system - so-called WIMS W-CDMA. T1P1 is supporting WCDMA-NA, which corresponds to UTRA FDD. WCDMA-NA and WIMS W-CDMA have merged into WP-CDMA (wideband packet CDMA), which is well aligned with UTRA FDD. All these concepts have been submitted to ITU-R.

For the satellite component of IMT-2000, four proposals have been submitted to ITU-R from ESA, ICO, and INMARSAT.

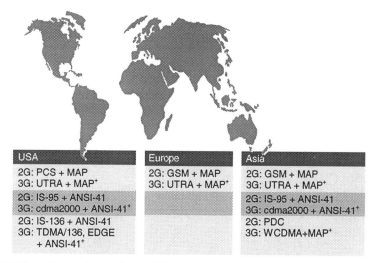

Figure 1.12 Second generation standards in different regions and related third generation systems.

Details of all proposals for IMT-2000 are available in [13]. The international consensus-building and harmonization activities between different regions and bodies started immediately after the contributions were submitted. A harmonization between regions would lead to a quasi-world standard, which would allow economical advantages for customers, network operators, and manufacturers. By avoiding market fragmentation, economy of scale can be obtained. Therefore, two international bodies are established: 3GPP to harmonize and to standardize in detail the similar ETSI, ARIB, TTA, and T1P1 WCDMA and the related TDD proposals on one hand, and 3GPP2 for the cdma2000-based proposals from TIA and TTA on the other hand. These bodies are working on a harmonized solution initiated by major international operators (c.f. chapter 12).

1.4.2 Japanese Activities

With the PDC standard as the dominant standard and PHS as a competing cordless system, Japan has currently almost 40 million subscribers. Furthermore, two operators started with IS-95 at the end of 1998, which is an American standard, based on CDMA.

The fast growth of the number of subscribers puts great pressure on the current systems in Japan as they run out of capacity. Therefore, a fast introduction is needed for the third generation system in new spectrum bands with a starting date around 2001.

To overcome the problems of the fast introduction on one side and the need for wide adoption on the other side, an IMT-2000 study committee was established in 1993 under the Association of Radio Industries and Business (ARIB). The main tasks for this committee are as follows:

- Study of the technical aspects of IMT-2000;

- Preparation of standard proposals;

- Fitting the proposals within the IMT-2000 framework;

- Cooperation with other standardization bodies.

Several companies have developed IMT-2000 air interface proposals:

- Thirteen CDMA-based FDD proposals;

- Three CDMA-based TDD proposals;

- Eight TDMA-based proposals.

From the 16 originally proposed CDMA based solutions in the beginning of 1995, the following proposals were generated after an evaluation, grouping, and merging phase:

- Three CDMA/FDD proposals (Core A, B, and C);

- A CDMA/TDD proposal based on core A;

- The TDMA-based proposals have not been investigated further in more detail.

At the end of 1996, these schemes were combined into one single proposal, of which the main parameters are from core A.

ARIB cooperated with ETSI (Europe) and TIA (United States) to come up with harmonized standards during the evaluation phase in 1997 and 1998. These activities were moved to the new bodies 3GPP and 3GPP2. ARIB, TTC, ETSI, TTA, and T1P1 have succeeded in harmonizing proposals in the 3GPP process. The harmonization with TIA was mainly solved within the activities of the Operators Harmonization Group, which have been initiated by major international operators (c.f. chapter 12).

The Ministry of Post and Telecommunications and ARIB itself have set out the following general guidelines for IMT-2000:

- Meet IMT-2000 requirements;

- Introduction around year 2001;

- High-quality services comparable to fixed network;

- Multimedia service capability up to about 2 Mbps;

- Dual-mode terminals to solve backward-compatibility issues;

- Coordination with other countries for a global standard.

ARIB has chosen WCDMA as the technology for IMT-2000.

1.4.3   North American Activities

Three recently introduced (1995) wireless systems in the U.S. market are IS-136 and PCS-1900 (GSM 1900) that are based on TDMA and cdmaOne (IS-95) based on CDMA. However, with 49 million subscribers analog AMPS is still the dominant system in the United States.

The Federal Communication Commission (FCC) is a governmental body that regulates telecommunications in the United States. It is guided by the principle of three Cs: competition, community, and common sense. The FCC wants competition among providers, serving everybody, and in meanwhile avoiding regulations that could limit the development of the first two principles.

The third generation objectives are difficult to determine since the U.S. policy toward the wireless market is that the introduction of innovations and new technologies should be market driven. As well, very huge investments on second generation systems have recently been made that first need to bring returns before moving toward replacement with the third generation. However, the FCC established the general framework for the third generation technologies:

- Superior voice quality;

- Ubiquitous, seamless coverage;

- Enhanced messaging;

- Significantly higher data rates.

The challenge for third generation standardization should further be as open and flexible as possible to gain acceptance in the market.

When the introduction will take place is unclear, since the main philosophy is still that the market decides when and how to act. The FCC expects an introduction between the years 2003 and 2005.

1.5   INTERNATIONAL FREQUENCY ALLOCATION

Available spectra are a prerequisite for the economical success of a new system. Therefore, WARC 1992 (World Administrative Radio Conference) has identified spectra for third generation mobile radio systems (Figure 1.13).

Europe and Japan have basically followed these recommendations for FDD systems. In the lower band, parts of the spectrum are currently used for digital enhanced cordless telephones (DECT) and PHS, respectively. The Federal Communication Commission (FCC) in the United States has allocated a significant part of the WARC IMT-2000 spectrum in the lower band to second generation PCS systems. Most of the American countries are following the FCC frequency allocation. In China, big parts of the WARC spectrum are currently allocated to wireless local loop applications. Currently, no common spectrum is available worldwide for third generation mobile radio systems.

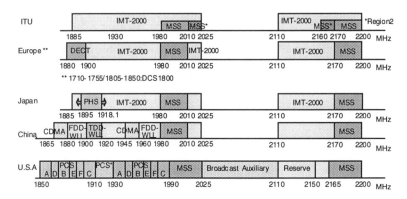

Figure 1.13 Worldwide frequency allocation.

1.6   EVOLUTION AND MIGRATION FROM SECOND TO THIRD GENERATION SYSTEMS

In different regions, different second generation mobile radio systems have been deployed according to Figure 1.14, which are the starting points for the evolution and migration to third generation systems. In the fixed network, an evolutionary path is envisaged, whereas in the radio interface a revolutionary approach is needed to support high data rate services. This enables a stepwise introduction of third generation features from increasing capacity, higher data rates with GSM2+ services in the core network to wireless multimedia applications.

The goal of the evolution and migration from second to third generation systems is to secure the second generation investment by an evolutionary upgrade either of the GSM or ANSI 41 core network, depending on the available infrastructure, to ensure roaming from second to third generation systems and handover for dual-mode terminals.

It is assumed that UMTS will start in islands as business areas, where more capacity and advanced services are needed. Figure 1.14 shows a possible evolutionary scenario for the introduction of third generation systems in a quite complete picture of current networks with different transport technologies and their related interfaces. The two different starting points of evolution from a mobile operator (e.g., GSM core network based) and a fixed network operator (ATM, IP-network based, N-ISDN) have to be taken into account.

Third generation mobile radio systems may be connected to public third generation radio access networks (RAN) as well as private networks (SOHO - small office/home office). To maintain full coverage for second generation services, dual-mode terminals can also be connected to second generation RANs as GSM. These radio access networks are linked to the related fixed network:

- Via the A interface and the mobile switching center (MSC) for circuit switched services;

- Via the Gb interface, the serving GPRS support node (SGSN), and the gateway GPRS support node (GGSN) for packet-oriented services.

The fixed network can comprise N-ISDN and/or public switching telephone networks (PSTNs) and Internet networks. ATM (asynchronous transfer mode) may be used in the backbone network for transport.

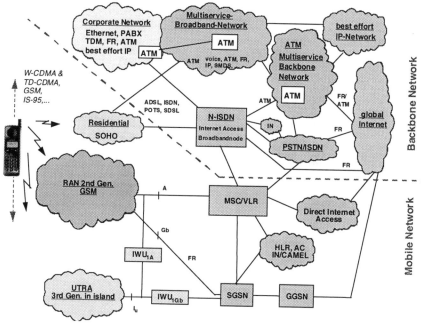

Figure 1.14 Second generation scenario: introduction of third generation mobile radio system.

For an evolutionary introduction of UMTS in islands (e.g., in hot spots), an evolved GSM core network comprising MSC, home location register (HLR), visiting location register (VLR), SGSN, and GGSN functions together with future proof innovations from GPRS phase 2 is needed. ATM transport and enhancements with IP (Internet protocol) technology is very attractive for current mobile operators to reuse investment. ATM can be applied as layer 2 transport technology and IP for routing in layer 3.

Advanced services are mobile multimedia that are parallel services as voice, audio, and video (e.g., H.320, H.324M, and real-time streaming), animated text and graphics as Web page hypertext markup language (HTML) and Java animation, graphics animation as real-time streaming over IP, and data and e-mail for file transfer (Figure 1.15). Therefore, Internet-type traffic will dominate multimedia. In the case of the receiving and sending of independent parallel media streams, independent transport paths are possible. For receiving and sending of parallel-dependent media streams, a single network path is preferred. This has to be supported by the core network.

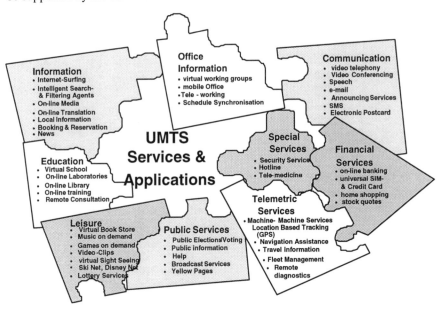

Figure 1.15 UMTS services and applications.

Therefore, key requirements on the GSM evolution are the support of multimedia, high data rates, virtual home environment, software download to mobiles, and global roaming. The main goals are attractive new features in line with a new generation of mobile terminals as personal digital assistants (PDAs), wireless laptops, communicators, and new radio-technology based, for example, on the UTRA concept (WCDMA for FDD and TD/CDMA for TDD). The network should be as flexible as possible and open to fixed mobile convergence via IN services. The same services to fixed (e.g., cordless) and cellular users should be supported (similar degree of authentication, one-number reachability, dual-mode or single-mode, third generation terminals).

Access to Internet has to be optimized by broadband backbone networks for high quality of service, especially for service providers who support large amounts of data. The World Wide Web (WWW) will be one key application for third generation mobile systems. The network architecture should be flexible for the introduction of new functionality and should allow for medium term substitution of existing functions. This flexibility is needed because services will not be standardized.

The service creation is dependent on the operator to achieve differentiation. This would guarantee a smooth evolution as for example, exercised with GSM CAMEL (customized applications for mobile network enhanced logic). Modular and scalable architecture and equipment allows an evolution in steps, secures investment, and fits diverse operators' strategies.

The development and deployment of UMTS is an evolutionary process, especially in the core network. A possible evolutionary path for a core network is shown in Figure 1.16, which harmonizes new technology with the needs of existing operators.

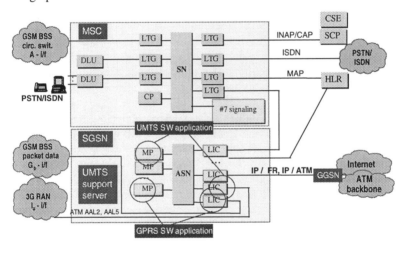

Figure 1.16 Possible evolution scenario from GPRS to third generation core network
(Source: Siemens).

In the first phase of UMTS, the new third generation RAN is connected via the standardized Iu interface to an UMTS support server (also known as IWU - interworking unit), which is connected to the GSM MSC and SGSN to reuse several GSM functions as the GSM mobility management. This server has the same ATM platform as the SGSN.

Thus, an SGSN can be upgraded to provide UMTS functionality. During this phase-circuit switched services will be supported via the IWU and GSM MSC, whereas high data rate packet-oriented services (Internet access) are supported via the IWU and third generation GPRS functions.

The mobility management is envisaged to be based on MAP (mobile application part) by reusing the HLR and VLR databases. Control of circuit switched mode and packet mode connections will be possible by the evolution of

INAP/CAP (intelligent network / CAMEL application part). The Iu interface between the RNC (radio network controller) and the MSC/SGSN will be completely standardized.

However, in different regions different evolution paths may be reasonable according to the available network infrastructure and its potential for evolution (Figure 1.12).

1.7 PREVIEW OF THE BOOK

This book consists of 12 chapters. The main idea behind this book is to present the crucial research results from the ACTS project FRAMES that paved the way for standardization of the third generation of mobile communication systems. Therefore, in the following chapters, the scientific output generated from the FRAMES research activities and the actual developments in the international standardization are presented together with basic principles from multiple access techniques, advanced receiver techniques, coding, modulation, and so forth.

Chapter 2 is focussed on the UMTS UTRA, TDD, TD/CDMA mode. First the basic concept of TD/CDMA is discussed by explaining the physical channel, spreading and modulation, training sequences, channel allocation, and adaptive antenna. Then the system model is analyzed and performance results are presented and discussed.

UTRA frequency division duplex (WCDMA) is presented in Chapter 3. The basic concept of WCDMA is introduced by presenting carrier spacing and deployment scenarios, logical and physical channels, spreading multirate, packet data, handover and interoperability between GSM and WCDMA.

Advanced receiver algorithms are introduced in Chapter 4. First, the limitations of conventional RAKE receivers are discussed. Then, the joint detection for TD/CDMA, uplink multiuser detection for CDMA, and improved downlink receiver for CDMA are presented.

Modulation and coding are discussed in Chapter 5. Research has been carried out during FRAMES to improve modulation and coding. Some of these investigations are presented in this chapter: multi-amplitude linearized Gaussian minimum shift keying (MALGMSK) modulation scheme, multicode CDMA, turbo codes with TD/CDMA, optimum distance spectrum convolutional codes, convolutional codes for rate matching, sequential decoding of convolutional codes, code spread CDMA, and coding for packet data transmission.

Chapter 6 discusses the UTRA transport control functions and the principle in the research work carried out to define the model. Radio interface protocol architecture, urban architecture, RRC connection, and mobility and physical layer interface are briefly outlined. Medium access control, radio link control, and radio resource control are also presented.

Radio resource management is the subject of Chapter 7. First, its fundamental limits and properties are discussed. Then, the resource management for the TDD and FDD mode are introduced. Scheduling is also discussed briefly.

In Chapter 8, we explain the UTRA interworking. Coexistence and compatibility, GSM, UMTS handover measurements, and adjacent channel interference in a UTRA FDD system are discussed.

Mobile station (MS) positioning refers to the determination of the coordinates of a desired mobile user. This interesting topic is introduced in Chapter 9. After discussing MS positioning applications, criteria to evaluate MS location methods, MS position estimation methods and problems in MS positioning uplink and downlink MS positioning methods, are presented.

Chapter 10 outlines the development of TD/CDMA demonstrator and describes the key system concept and technologies used in the demonstrator. It provides a brief history of the project, particularly related to the demonstrator. Chapter 11 presents the WCDMA evaluation system developed by Ericsson and TDMA system by Nokia.

Finally, the outlook of the international standardization bodies 3GPP, 3GPP2, and ITU/R is discussed in Chapter 12.

References

[1]   R. Prasad, CDMA for Wireless Personal Communications, Norwood MA: Artech House, 1996.

[2]   R. Prasad, Universal Wireless Personal Communications, Norwood MA: Artech House, 1998.

[3]   T. Ojanperä and R. Prasad, Wideband CDMA for Third Generation Mobile Communications, Norwood MA: Artech House 1998.

[4]   R. van Nee and R. Prasad, OFDM for Mobile Multimedia Communications, Norwood MA: Artech House, 1999.

[5]   UMTS Forum, A Regulatory Framework for UMTS, Report no. 1 from the UMTS Forum, 26 June 1997.

[6]   Infowin, "Mobile Communications," The ACTS Multimedia Information Window, Publisher INTRACOM S.A., Athens, 1998.

[7]   ETSI, Requirements for the UMTS Terrestrial Radio Access System, UMTS 04-01 UTRA, June 1997.

[8]   ITU TG8/1, Guidelines for Evaluation of Radio Transmission Technologies for FPLMTS, 10th Meeting of TG8/1, Mainz, Germany, 15 - 26 April 1996.

[9]   ITU-R, Request for submission of Candidate Transmission Technologies (RTTs) for IMT-2000/FPLMTS Radio Interface, Circular letter 8/LCCE/47, 1997 and further Corrigendums and Addendums.

[10]   ETSI SMG, Proposal for a Consensus Decision on UTRA, ETSI SMG Tdoc 032/98.

[11]   ntz, Einigung über UMTS-Standard. Nachrichten-technische Zeitschrift, No. 4, April 1998, p. 11.

[12]   ETSI, The ETSI UMTS Terrestrial Radio Access (UTRA) ITU-R RTT Candidate Submission, ETSI SMG 2, Tdoc 260/98, June 1998.

[13]   ITU, ITU WWW page http://www.itu.int/imt/2-radio-dev/rtt/index.html, July 1998.

[14]   J.S. DaSilva and B.E. Fernandes, "The European Research Program for Advanced Mobile Systems," IEEE Personal Communications Magazine, February 1995, Vol. 2, No. 1, pp. 14 – 19.

[15]   H. Erben, J.S. DaSilva, B. Arroyo, B. Barani and D. Ikonomou, "European R & D Towards Third Generation Mobile Communication Systems," International Conference on Communications (ICT '96), Istanbul, Turkey, April 14 - 17, 1996, pp. 661 – 666.

24

[16] J.S. DaSilva, B. Arroyo, B. Barani and D. Ikonomou, "European Third generation Mobile Systems," IEEE Communications Magazine, October 1996, Vol. 35, No. 10, pp. 68 – 83.

[17] J.D. Gibson, The Mobile Communications Handbook, CRC Press in cooperation with IEEE Press, 1996.

[18] FRAMES, ACTS FRAMES Workshop on the First International Symposium on Wireless Personal Multimedia Communications, Yokosuka, Japan, November 4 - 6, 1998, pp. 12 – 63.

[19] E. Berruto, M. Gudmundson, R. Menolascino, W. Mohr and M. Pizarroso, "Research Activities on UMTS Radio Interface, Network Architectures, and Planning," IEEE Communications Magazine, February 1998, Vol. 36, No. 2, pp. 82 – 95.

[20] T. Ojanperä, A. Klein and P.-O. Anderson, "FRAMES Multiple Access for UMTS," IEEE Vehicular Technology Conference, May 1997, Phoenix, USA, pp. 490 – 494.

[21] T. Ojanperä, J. Sköld, J. Castro, L. Girard and A. Klein, "Comparison of Multiple Access Schemes for UMTS," Fourth IEE Colloquium on CDMA Techniques and Applications for Third Generation Mobile Systems. May 19, 1997, London.

[22] T. Ojanperä, M. Gudmundson, P. Jung, J. Sköld, R. Pirhonen, G. Kramer and A. Toskala, "FRAMES - Hybrid Multiple Access Technology," International Symposium on Spread Spectrum Technology and Applications (ISSSTA), Mainz, Germany, September 22 - 25, 1996, pp. 320 – 324.

[23] T. Ojanperä, A. Klein and P.-O. Anderson, "FRAMES Multiple Access for UMTS," IEEE Colloquium on CDMA Techniques and Applications for 3rd Generation Mobile Systems, London, May 19, 1997.

[24] A. Klein, R. Pirhonen, J. Sköld and R. Suoranta, "FRAMES Multiple Access Mode 1 - Wideband TDMA With and Without Spreading," Personal, Indoor, and Mobile Radio Conference (PIMRC), Helsinki, Finland, 1997.

[25] F. Ovesjö, E. Dahlman, T. Ojanpera, A. Toskala and A. Klein, "FRAMES Multiple Access Mode 2 - Wideband CDMA," Personal, Indoor, and Mobile Radio Conference (PIMRC), Helsinki, Finland, 1997.

[26] W. Mohr, "FRAMES Multiple Access, Harmonized Radio Interface," 1997 IEEE International Conference on Personal Wireless Communications, Mumbai, India, December 17 - 19, 1997, pp. 252 – 256.

[27] P.-O. Anderson, "FRAMES Contributions to the UMTS Standardization Process," The First International Symposium on Wireless Personal Multimedia Communications, Yokosuka, Japan, November 4 to 6, 1998, pp. 18 – 23.

[28] W. Mohr, "UTRA FDD and TDD, a harmonized proposal for IMT-2000," International Conference on Communication Technology ICCT '98, Beijing, China, October 22 - 24, 1998, pp. S22-03-1 - S22-03-5.

[29] W. Mohr, "Research and Development of IMT-2000 / UMTS Mobile Radio Systems," The First International Symposium on Wireless Personal Multimedia Communications. Yokosuka, Japan, 4 - 6 November 1998, pp. 12.

[30] W. Mohr, "ACTS FRAMES Project Towards IMT-2000 / UMTS", 1999 IEEE International Conference on Personal Wireless Communications (ICPWC), February 17 - 19, 1999, Jaipur, India, pp. 44 – 49.

[31] P.-O. Anderson, C. Boisseau, P. Croft, M. Dillinger, H. Erben, L. Girard, W. Mohr, E. Nikula, K. Richardson and F. Schweifer, "Investigations in the FRAMES Project on the UMTS Radio Interface," 4th ACTS Mobile Communications Summit, Sorrento, Italy, June 8 - 11, 1999, pp. 39 – 44.

[32] W. Konhäuser, "Applications and Services From 2nd to 3rd Generation," ICCT 1998, October 22 – 24, 1998, pp. S22-08-1 – S22-08-4.

[33] W. Konhäuser: Keynote speech on IEEE Vehicular Technology Conference Fall 99, Amsterdam, September 19 – 22, 1999.

Chapter 2

TD-CDMA

As mentioned in Chapter 1, in January 1998 the Special Mobile Group (SMG) of ETSI recommended the use of TD/CDMA in the unpaired portion and WCDMA in the paired portion of the radio frequency spectrum of the UTRA proposal. In other words, TD/CDMA is utilized in the TDD mode and WCDMA in the FDD mode. Based on the recommendation of ETSI SMG and FRAMES, a specification of UMTS becomes possible which allows low-cost terminals, a harmonization with GSM, and the implementation of FDD/TDD dual mode terminals. Moreover, UTRA should support operation already in an allocated spectrum as small as 2×5 MHz and enable the deployment of the TDD component within a minimum spectrum of 5 MHz. The subject of this chapter is the system concept TD-CDMA [1-7].

The component TD-CDMA of UTRA should be usable from low to high data rates concerning symmetrical and especially asymmetrical services, based on the UMTS and international system for mobile telecommunications 2000 (IMT 2000) requirements. The design goals in the maximum data rates achievable in the TDD mode are at least 384 Kbps for bidirectional transmission in microcells and 2 Mbps for unidirectional transmission, combined with a low data rate transmission in the other direction, in picocells. The support of a 2 Mbps circuit switched service in a symmetrical duplex mode is not required. The TD/CDMA concept provides the performance as well as the flexibility to fulfill these requirements. Pooling of CDMA codes and TDMA timeslots can easily provide the capacity demanded by a certain user (i.e., by the allocation of several codes, timeslots, or both to a single user). Furthermore, the frequency spectrum can be efficiently utilized, since the total available capacity can be adaptively partitioned and distributed to the uplink (UL) and downlink (DL) by just changing the allocation of the timeslots within a frame to the two links. Therefore, the need for higher overall data rate on the one link can be satisfied by the taking of a part of releasable resources of the other link, which is a major advantage of a TDD system.

This chapter is structured as follows: First, the basic concept of TD-CDMA is explained in Section 2.1. After the description of the traffic channels and logical channels is referenced, the physical channels are introduced, where the multiframe, frameslot and timeslot structure, the switching point configuration, as well as the burst types and the synchronization channel concept are considered. Then follow subsections on spreading and modulation, on the training sequences, and channel allocation. In Section 2.2, the analytical system model of TD-CDMA applying joint detection is presented, which is the basis of the performance evaluation. The obtained performance results are contained in Section 2.3. The conclusions on the chapter are drawn in Section 2.4.

2.1 BASIC CONCEPT

Section 2.1 presents the physical channels, spreading and modulation, training sequences, channel allocation and adaptive antennas. The description of the transport channels and the logical channels is contained in Sections 6.5.1 and 6.6.1, respectively.

2.1.1 Physical Channels

Frame and timeslot structuring, switching point configuration, multiframe structures, burst types, and synchronization channel concept are discussed below.

2.1.1.1 Frame and Timeslot Structure

The TDMA frame shown in Figure 2.1 has a length of 10 ms and is subdivided into 16 timeslots (TS) of 625 μs each. A timeslot corresponds to 2,560 chips. The physical content of a timeslot is the burst of corresponding length as described in Section 2.1.1.4. Each of the 16 timeslots of a frame can be allocated either to the uplink or the downlink. Thanks to this flexibility, the TDD mode can be adapted to different environments and deployment scenarios.

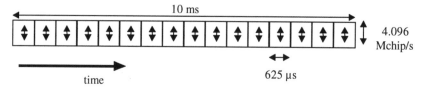

Figure 2.1 TD-CDMA frame structure.

2.1.1.2 Switching Point Configurations

In Figure 2.2, examples of possible switching point configurations are given. Single or multiple-switching-point configurations as well as symmetric or asymmetric downlink/uplink allocations are considered. In any configuration, at least one timeslot has to be allocated to the downlink and at least one timeslot has to be allocated to the uplink.

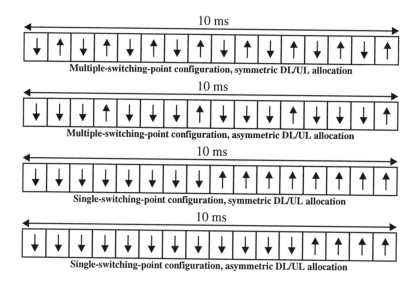

10 ms

Multiple-switching-point configuration, symmetric DL/UL allocation

10 ms

Multiple-switching-point configuration, asymmetric DL/UL allocation

10 ms

Single-switching-point configuration, symmetric DL/UL allocation

10 ms

Single-switching-point configuration, asymmetric DL/UL allocation

Figure 2.2   Examples of switching point configurations.

2.1.1.3   Multiframe Structure

A strong requirement concerning the multiframe structure originates in the realization of low-cost dual mode FDD/TDD terminals and the intended compatibility with GSM of the UTRA proposal. In this respect, the superframe and multiframe structure of the FDD and TDD mode have to be compatible and harmonized with GSM. Therefore, in the proposed structure a multiframe is composed of 24 frames of 10 ms length each. Consequently, the multiframe period is 240 ms, which is twice the length of the GSM TCH-F multiframe.

All frames in the traffic channel multiframes are used to carry both user data and dedicated signaling for the following reasons:

- The use of a signaling frame like the slow associated control channel (SACCH) frame in GSM is avoided by the use of in-band dedicated signaling, or the allocation of a standalone dedicated control channel (SDCCH).

- There is no need for an idle slot to read broadcast control channels (BCCHs) of adjacent cells as in GSM. Adjacent cells in the TDD network are frame synchronized.

- The bursty nature of TD-CDMA transmission and reception allows the mobile station to make measurements of signals of GSM and FDD networks in idle timeslots. This is valid also for high-bit-rate users. BCCH and RACH timeslots could also be used for this purpose.

Therefore, the multiframe length is given by the common channel with the lowest bit rate. This channel is in the present case the synchronization channel (SCH), if its multiframe structure is compatible with the GSM TCH-F multiframe. This leads to a multiframe length of 240 ms. Three TDD multiframes exactly fit into an FDD multiframe, which ensures the compatibility of both components.

2.1.1.4   Burst Types

Traffic bursts and access bursts are explained in the following paragraph.

Traffic Bursts

Several burst types can be defined according to the different spreading options and spreading factors, see Section 2.1.2.2. As an example, consider the traffic bursts depicted in Figure 2.3 and Figure 2.4, which correspond to multicode transmission with fixed spreading. Both traffic bursts consist of two fields of data symbols, a midamble and a guard period. The length of the midamble of traffic burst 1 and 2 is 512 chips or 256 chips, respectively.

Sample sets of midamble codes are given in Section 2.1.3.1. The longer midamble of traffic burst 1 with $W = 57$ is suited for the uplink, capable to estimate up to eight different channel impulse responses of length WT_c=13.9 ms. Here, Tc denotes the chip length and W the maximum length of the channel impulse response in chips. Traffic burst 2, which allows the estimation of channel impulse responses with $W = 64$ taps, can be used for the downlink and, if the bursts within a timeslot are allocated to less than four users, also for the uplink. In this case, up to three different channel impulse responses of length $WT_c = 15.6$ ms. can be estimated. Obviously, more channel impulse responses can be estimated if the individual channel impulse responses have a shorter maximum length.

With this parameterization, traffic burst 1 can be used for the

- Uplink independently of the number of active users in one timeslot;

- Downlink independently of the number of active users in one timeslot.

Traffic burst 2 can be used for the

- Uplink, if the bursts within a timeslot are allocated to less than four users;

- Downlink independently of the number of active users in one timeslot.

The data fields of the traffic burst 1 (Figure 2.3) contain 61 symbols, spread to 976 chips, whereas the traffic burst 2 (Figure 2.4) contains 69 symbols corresponding to 1,104 chips. The guard period of traffic bursts 1 and 2 is 96 chip periods long.

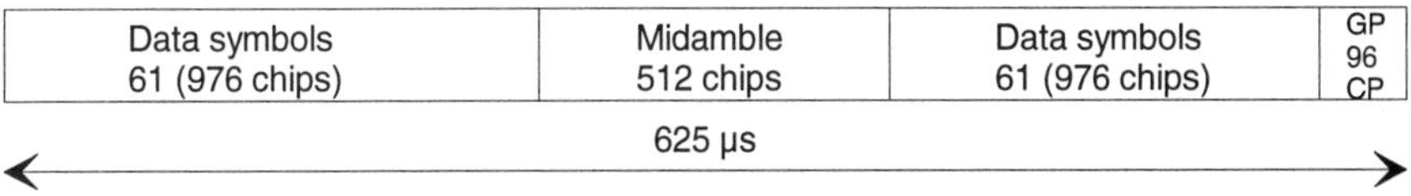

Data symbols 61 (976 chips)	Midamble 512 chips	Data symbols 61 (976 chips)	GP 96 CP

←———————————————— 625 µs ————————————————→

Figure 2.3 Structure of traffic burst 1.

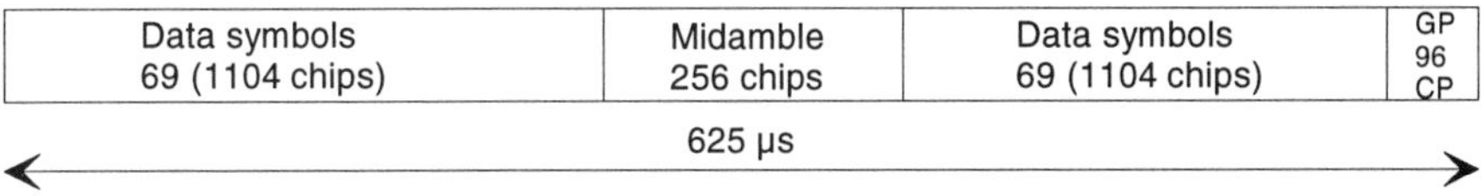

Data symbols 69 (1104 chips)	Midamble 256 chips	Data symbols 69 (1104 chips)	GP 96 CP

←———————————————— 625 µs ————————————————→

Figure 2.4 Structure of traffic burst 2.

The contents of the traffic burst fields are described in Table 2.1 and Table 2.2. The two traffic bursts 1 and 2 defined here are well suited for the different applications mentioned above. It may be possible to further optimize the structure of the traffic bursts for specific applications, for instance for unlicensed operation.

Table 2.1

Contents of Traffic Burst 1 Fields

Chip number	Field length (in chips)	Field length (in symbols)	Field length (in µs)	Field content
0-975	976	61	238.3	Data symbols
976-1487	512	-	125.0	Midamble
1488-2463	976	61	238.3	Data symbols
2464-2559	96	-	23.4	Guard period

Table 2.2

Contents of Traffic Burst 2 Fields

Chip number	Field length (in chips)	Field length (in symbols)	Field length (in µs)	Field content
0-1103	1104	69	269.55	Data symbols
1104-1359	256	-	62.5	Midamble
1360-2463	1104	69	269.55	Data symbols
2464-2559	96	-	23.4	Guard period

Access Bursts

The mobiles transmit the access bursts randomly in the random access channel (RACH). This leads to time-divided collision groups. The usage of up to 8 orthogonal codes per timeslot increases the amount of collision groups and RACH throughput, respectively. A further improvement is achieved by using two distinct access bursts, which can both be transmitted within one timeslot without collision. Access burst 1 uses only the first half of a timeslot; access burst 2 the second half. The access bursts are depicted in Figure 2.5 and Figure 2.6, respectively. The contents of the access burst fields are listed in Table 2.3 and Table 2.4.

Table 2.3

Contents of Access Burst 1 Fields

Chip number	Field length (in chips)	Field length (in symbols)	Field length (in μs)	Field content
0-335	336	21	82.0	Data symbols
336-847	512	-	125.0	Midamble
848-1183	336	21	82.0	Data symbols
1184-1279	96	-	23.4	Guard period
1279-2559	1280	-	312.5	Extended guard period

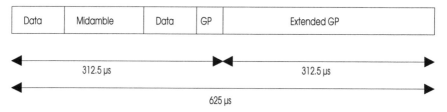

Figure 2.5   Structure of access burst 1.

Table 2.4

Contents of Access Burst 2 Fields

Chip number	Field length (in chips)	Field length (in symbols)	Field length (in μs)	Field content
0-1279	1280	-	312.5	Extended guard period
1280-1615	336	21	82.0	Data symbols
1616-2127	512	-	125.0	Midamble
2128-2463	336	21	82.0	Data symbols
2464-2559	96	-	23.4	Guard period

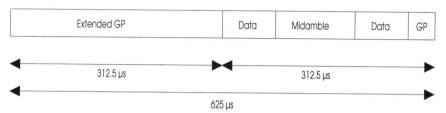

Figure 2.6 Structure of access burst 2.

2.1.1.5 Synchronization Channel Concept

Overview

In order to achieve harmonization between the FDD and the TDD mode, the described approach of the TDD synchronization channel concept uses the synchronization scheme of the UTRA FDD mode with as few changes as possible. In one step the terminal can acquire synchronization and the coding scheme of the cell for the BCCH. Thus, it can detect cell messaging instantly. The monitoring of other cells while in traffic (e.g., for the preparation of a handover) is also simplified. A second goal is to put as few restrictions as possible on the system coming from this SCH/BCCH scheme, especially not to limit the uplink/downlink asymmetry, which is a main feature of any TDD system.

From the requirements above, we propose to map the SCH to only two downlink slots per frame, namely timeslot 0 and timeslot 8, as shown Figure 2.7.

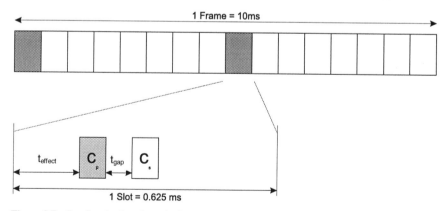

Figure 2.7 Synchronization channel scheme.

As depicted in Figure 2.7, there are primary and secondary synchronization sequences. The sequences used are the same unmodulated orthogonal gold codes of length, 256 chips as in the case of the UTRA FDD mode. The difference is on the one hand the mapping of the SCH to two slots per frame only, and on the other hand the introduction of a time offset t_{off}. Furthermore, we propose to disperse the synchronization sequences over time, and for interference reasons not to add their transmission power simultaneously.

The advantages of this SCH concept are the following:

- The mapping on two slots per frame provides only minimum restrictions on the UL/DL asymmetry.

- The introduction of the time offset t_{off} enables the system to overcome the capture effect, which arises due to network synchronization and for which no solution has been proposed up to now. This will be further explained in the later paragraphs.

- The time offset t_{gap} is introduced to give enough time for calculations. The exact value is yet to be determined.

In the following paragraphs, the different synchronization steps and the reasoning for determining the time offset are explained in detail.

Primary Synchronization

In a first step (e.g., after switching on) the mobile permanently monitors the primary synchronization code, which is the same unmodulated orthogonal gold code as used in the FDD mode. As stated in the FDD system description, there will be a dedicated filter, for example a matched filter, for this purpose, which can be reused here. Thus, the mobile can obtain slot synchronization and, up to a remaining factor of 2, frame synchronization.

Secondary Synchronization and BCCH Pointing

In the second step, the mobile not only can acquire the exact frame synchronization but also the time offset t_{off}, the coding, and the position of the BCCH of the cell. For this purpose, the secondary synchronization codes are chosen from the same set of 17 different unmodulated orthogonal gold codes as in the FDD mode, thus reusing the capabilities of the terminal. After four frames the mobile can acquire a sequence of eight of these codes. With this sequence, it is possible to code the following information:

- The exact factor of frame synchronization, remaining after primary synchronization;

- The time offset t_{off};

- The midamble and spreading code set of the base station;

- The spreading code(s) used for the BCCH.

The number of possible combinations of the parameters above is low, in comparison to the given coding possibilities of 17^8. Additionally, high coding gain and thus a high security in detecting these cell parameters are achieved.

Reasoning for Determining the Time Offset t_{off}

Due to interference between MSs, it is mandatory for TDD systems to keep synchronization between base stations. When searching for synchronization by engaging the primary SCH, a situation as outlined in Figure 2.8 may occur. The correlations that are shown separately in the figure superimpose at the receiver of the mobile. In this example there is no possibility of detecting cell 3 as it is captured by the correlation of cell 1.

The introduction of the time offset t_{off} enables the receiver to detect even cells with low correlation peaks, as there is additional separation in the time domain. Information on the specific offset of the cell can be transmitted to the mobile in the secondary SCH with no additional resources required, and thus, an appropriate mechanism is obtained to overcome the capture effect.

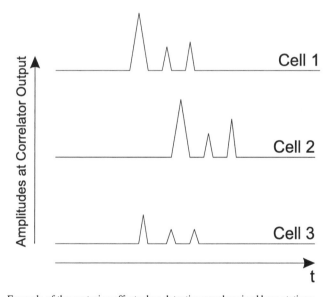

Figure 2.8 Example of the capturing effect when detecting synchronized base stations.

2.1.2 Spreading and Modulation

This section explains the data modulation, different spreading options, spreading codes, pulse shape filtering, and spread data symbol and data block signal.

2.1.2.1 Data Modulation

Each user burst has two data carrying parts $\underline{d}^{(k,i)} = \left(\underline{d}_1^{(k,i)}, \underline{d}_2^{(k,i)} \dots \underline{d}_N^{(k,i)} \right)^{\mathrm{T}}$, $i = 1,2$, $k = 1 \dots K$, termed data blocks. Data block $\underline{d}^{(k,1)}$ is transmitted prior to the midamble and data block $\underline{d}^{(k,2)}$ after the midamble. Each of the N data symbols $\underline{d}_n^{(k,i)}$, $i = 1,2$, $k = 1 \dots K$, $n = 1 \dots N$, is QPSK modulated. Each data symbol $\underline{d}_n^{(k,i)}$ originates in two interleaved and encoded data bits

$$\underline{b}_{l,n}^{(k,i)} \in \{0,1\}, \quad l = 1,2, \ n = 1 \dots N, \ i = 1,2, \ k = 1 \dots K \tag{2.1}$$

Where,

$$\mathbf{Re}\left\{ \underline{d}_n^{(k,i)} \right\} = \frac{1}{\sqrt{2}} \left(2\underline{b}_{1,n}^{(k,i)} - 1 \right), \ n = 1 \dots N, \ i = 1,2, \ k = 1 \dots K$$

$$\tag{2.2}$$

$$\mathbf{Im}\left\{ \underline{d}_n^{(k,i)} \right\} = \frac{1}{\sqrt{2}} \left(2\underline{b}_{2,n}^{(k,i)} - 1 \right), \ n = 1 \dots N, \ i = 1,2, \ k = 1 \dots K$$

2.1.2.2 Spreading Options

Overview

Two spreading options are considered in UTRA TDD: either a multicode transmission with fixed spreading and a spreading factor of $Q = 16$, or a single code transmission with variable spreading factors of $Q = 1, 2, 4, 8, 16$. Both options allow a high degree of bit rate granularity and flexibility. Therefore, the implementation of the whole service range from low to high bit rates can be supported.

Multicode Transmission With Fixed Spreading

Within each timeslot of length 625 µs, several spreading codes of constant length $Q = 16$ may be used. These multiple resources within the same timeslot can be allocated to different users, but also partly or completely to a single user. Multiple resources within the same timeslot can be distinguished by different spreading codes.

Single Code Transmission With Variable Spreading

Additionally to the multicode transmission, single code transmission with variable spreading using a spreading factor of $Q = 1, 2, 4, 8$ or 16 can be applied.

Uplink   In the uplink, the mobile terminal always uses single code transmission by adapting the spreading factor as a function of the data rate within each timeslot of length 625 μs. In this way, the peak-to-average ratio requirements on the power amplifiers can be reduced significantly. This is particularly important at the mobile terminal for high rate services. At the base station, several mobiles can still be received in the same timeslot. They are separated by their codes and the individual decoding is improved by joint detection [1].

Downlink   Also in the downlink, the base station can adapt the spreading factor as a function of the data rate. High-rate data transmission requiring more than one resource per mobile can be supported by terminals having the processing power for joint detection only on a single slot. The required throughput occupies an integer number of slots plus a fraction of an extra slot. Single-burst transmission occurs in all slots except the extra slot, in which the extra bursts of several mobiles superimpose. Here, joint detection is only needed for this extra timeslot.

2.1.2.3   Spreading Codes

Each data symbol $\underline{d}_n^{(k,i)}$ is spread by a CDMA code $\underline{c}^{(k)}$, $k = 1...K$, of length $Q = 1, 2, 4, 8, 16$. The elements $\underline{c}_q^{(k)}$, $q = 1...Q$, $k = 1...K$, are taken from the complex set

$$\underline{V}_c = \left\{ 1, j, -1, -j \right\} \tag{2.3}$$

In (2.3), j denotes the imaginary unit. The complex CDMA codes $\underline{c}^{(k)}$ are generated from binary CDMA codes $a^{(k)}$ with the elements $a_q^{(k)}$, $q = 1...Q$, $k = 1...K$, by relation

$$\underline{c}_q^{(k)} = (j)^q \cdot a_q^{(k)}, \quad a_q^{(k)} \in \left\{ 1, -1 \right\}, \quad q = 1...Q, \quad k = 1...K \tag{2.4}$$

Hence, the elements $\underline{c}_q^{(k)}$ of the CDMA codes $\underline{c}^{(k)}$ are alternatingly real and imaginary.

Table 2.5 shows binary CDMA codes, which can be used as the quantities $a^{(k)}$, $k = 1...K$, in (2.4).

Table 2.5

Exemplary Binary CDMA Code Set

Code 1	$(-1 \ -1 \ \ 1 \ \ 1 \ \ 1 \ -1 \ \ 1 \ -1 \ \ 1 \ -1 \ \ 1 \ -1 \ \ 1 \ -1 \ -1 \ -1)^{T}$
Code 2	$(-1 \ -1 \ \ 1 \ \ 1 \ \ 1 \ -1 \ \ 1 \ -1 \ -1 \ \ 1 \ -1 \ \ 1 \ -1 \ \ 1 \ \ 1 \ \ 1)^{T}$
Code 3	$(-1 \ -1 \ \ 1 \ \ 1 \ -1 \ \ 1 \ -1 \ \ 1 \ -1 \ \ 1 \ -1 \ \ 1 \ \ 1 \ -1 \ -1 \ -1)^{T}$
Code 4	$(-1 \ -1 \ \ 1 \ \ 1 \ -1 \ \ 1 \ -1 \ \ 1 \ \ 1 \ -1 \ \ 1 \ -1 \ -1 \ \ 1 \ \ 1 \ \ 1)^{T}$
Code 5	$(-1 \ -1 \ -1 \ -1 \ -1 \ \ 1 \ \ 1 \ -1 \ \ 1 \ -1 \ -1 \ \ 1 \ -1 \ \ 1 \ -1 \ -1)^{T}$
Code 6	$(-1 \ -1 \ -1 \ -1 \ -1 \ \ 1 \ \ 1 \ -1 \ -1 \ \ 1 \ \ 1 \ -1 \ \ 1 \ -1 \ \ 1 \ \ 1)^{T}$
Code 7	$(-1 \ -1 \ -1 \ -1 \ \ 1 \ -1 \ -1 \ \ 1 \ -1 \ \ 1 \ \ 1 \ -1 \ -1 \ \ 1 \ -1 \ -1)^{T}$
Code 8	$(-1 \ -1 \ -1 \ -1 \ \ 1 \ -1 \ -1 \ \ 1 \ \ 1 \ -1 \ -1 \ \ 1 \ \ 1 \ -1 \ \ 1 \ \ 1)^{T}$
Code 9	$(-1 \ \ 1 \ -1 \ \ 1 \ \ 1 \ \ 1 \ -1 \ -1 \ \ 1 \ \ 1 \ -1 \ -1 \ \ 1 \ \ 1 \ \ 1 \ -1)^{T}$
Code 10	$(-1 \ \ 1 \ -1 \ \ 1 \ \ 1 \ \ 1 \ -1 \ -1 \ -1 \ -1 \ \ 1 \ \ 1 \ -1 \ -1 \ -1 \ \ 1)^{T}$
Code 11	$(-1 \ \ 1 \ -1 \ \ 1 \ -1 \ -1 \ \ 1 \ \ 1 \ -1 \ -1 \ \ 1 \ \ 1 \ \ 1 \ \ 1 \ \ 1 \ -1)^{T}$
Code 12	$(-1 \ \ 1 \ -1 \ \ 1 \ -1 \ -1 \ \ 1 \ \ 1 \ \ 1 \ \ 1 \ -1 \ -1 \ -1 \ -1 \ -1 \ \ 1)^{T}$
Code 13	$(-1 \ \ 1 \ \ 1 \ -1 \ -1 \ -1 \ -1 \ -1 \ \ 1 \ \ 1 \ \ 1 \ \ 1 \ -1 \ -1 \ \ 1 \ -1)^{T}$
Code 14	$(-1 \ \ 1 \ \ 1 \ -1 \ -1 \ -1 \ -1 \ -1 \ -1 \ -1 \ -1 \ \ 1 \ \ 1 \ -1 \ \ 1)^{T}$
Code 15	$(-1 \ \ 1 \ \ 1 \ -1 \ \ 1 \ \ 1 \ \ 1 \ \ 1 \ -1 \ -1 \ -1 \ -1 \ -1 \ -1 \ \ 1 \ -1)^{T}$
Code 16	$(-1 \ \ 1 \ \ 1 \ -1 \ \ 1 \ \ 1 \ \ 1 \ \ 1 \ \ 1 \ \ 1 \ \ 1 \ \ 1 \ \ 1 \ \ 1 \ -1 \ \ 1)^{T}$

These 16 orthogonal binary CDMA codes are generated based on Walsh-Hadamard codes for $Q = 16$ followed by a multiplication with a binary pseudo-random (PN) sequence. Typically, K is smaller than 16 and, therefore, in (2.4) less than 16 binary CDMA codes are needed. The CDMA codes given in Table 2.5 are one example. Other sets of 16 CDMA codes can be generated by multiplying the 16 orthogonal binary Walsh-Hadamard CDMA codes with other PN sequences. In this way, different sets of binary CDMA codes are generated for use in different cells.

2.1.2.4   Pulse Shape Filtering

Pulse shape filtering is applied to each chip at the transmitter. In this context, the term chip represents a single element $c_q^{(k)}$ with $q = 1 \ldots Q$, $k = 1 \ldots K$, of a CDMA code $\underline{c}^{(k)}$, $k = 1 \ldots K$, see also Section 2.1.2.3.

The impulse response of the above-mentioned chip impulse filter is a root-raised cosine. The corresponding raised cosine impulse is

$$C_0(t) = \frac{\sin\left(\pi \dfrac{t}{T_c}\right)}{\pi \dfrac{t}{T_c}} \cdot \frac{\cos\left(\alpha\pi \dfrac{t}{T_c}\right)}{1 - 4\alpha^2 \dfrac{t^2}{T_c^2}} \tag{2.5}$$

The roll-off factor shall be $\alpha = 0.22$, and the chip duration is

$$T_c = \frac{1}{\text{chip rate}} = 0.24414 \ \mu s \qquad (2.6)$$

The impulse response $C_0(t)$ according to (2.5) and the energy density spectrum $\Phi_0(f)$ of $C_0(t)$ are depicted in Figure 2.9 and Figure 2.10, respectively.

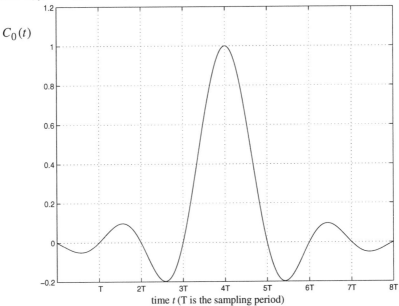

Figure 2.9 Impulse response C0(t) of chip impulse filter.

2.1.2.5   Spread Data Symbol and Data Block Signal

With the root-raised cosine chip impulse $C_0(t)$, the spread signal belonging to an arbitrary data symbol $\underline{d}_n^{(k,i)}$ can be expressed as

$$\underline{d}_n^{(k,i)}(t - T_0) = \underline{d}_n^{(k,i)} \sum_{q=1}^{Q} \underline{c}_q^{(k)} \cdot C_0\!\left(t - (q-1)T_c\right)$$

$$= \underline{d}_n^{(k,i)} \sum_{q=1}^{Q} (j)^q \cdot a_q^{(k)} \cdot C_0\!\left(t - (q-1)T_c\right) \qquad (2.7)$$

$$i = 1,2, \ k = 1...K, \ n = 1...N$$

In (2.7), T_0 denotes an arbitrary time shift. The transmitted signal belonging to the data block $\underline{d}^{(k,1)}$ transmitted prior to the midamble is

$$\underline{d}^{(k,1)}(t) = \sum_{n=1}^{N} \underline{d}_n^{(k,1)} \sum_{q=1}^{Q} \underline{c}_q^{(k)} \cdot C_0\big(t - (q-1)T_c - nT_c\big), k = 1...K \qquad (2.8)$$

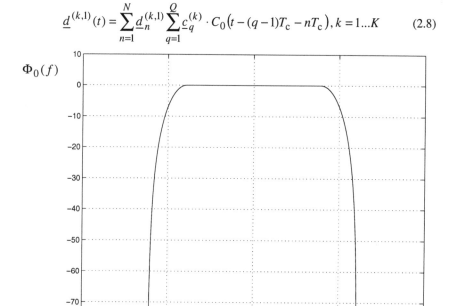

Figure 2.10 Energy density spectrum $\Phi_D(f)$ of $C_D(t)$.

and for the data block $\underline{d}^{(k,2)}$ transmitted after the midamble

$$\underline{d}^{(k,2)}(t) = \sum_{n=1}^{N} \underline{d}_n^{(k,2)} \sum_{q=1}^{Q} \underline{c}_q^{(k)} \cdot C_0\big(t - (q-1)T_c - nT_c - NQT_c - L_m T_c\big) \qquad (2.9)$$

$$k = 1...K$$

where L_m is the number of midamble chips.

2.1.3   Training Sequences

The midambles of different users active in the same timeslot are time-shifted versions of one single periodic basic code. Different cells use different periodic basic codes (i.e., different midamble sets). Joint channel estimation of the channel impulse responses of all active users within one timeslot can be achieved by one single cyclic correlator. Such an estimator is termed the Steiner estimator, because it was invented by B. Steiner [3]. The different user-specific channel impulse response estimates are obtained sequentially at the output of the correlator.

The midambles of the uplink have a length of 125 µs and a cyclic shift of 15.625 µs, enabling eight different channels to be estimated simultaneously. Hence, a midamble set comprises eight different midambles.

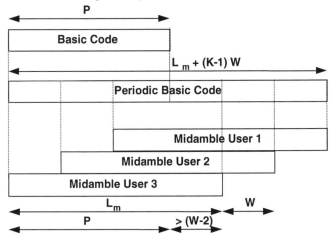

Figure 2.11 TD-CDMA midamble structure in case of 3 to be estimated channels.

Section 2.1.1.4 contains a description of the spread speech/data bursts. These traffic bursts contain L_m midamble chips, which are also termed midamble elements. The L_m elements $\underline{m}_i^{(k)}$, $i = 1 \ldots L_m$, $k = 1 \ldots K$, of the midamble codes $\underline{\boldsymbol{m}}^{(k)}$, $k = 1 \ldots K$, of the K users are taken from the complex set

$$\underline{V}_m = \left\{ 1, j, -1, -j \right\} \tag{2.10}$$

The elements $\underline{m}_i^{(k)}$ of the complex midamble codes $\underline{\boldsymbol{m}}^{(k)}$ fulfill the relation

$$\underline{m}_i^{(k)} = (j)^i \cdot m_i^{(k)}, \quad m_i^{(k)} \in \left\{ 1, -1 \right\}, \quad i = 1 \ldots L_m, \quad k = 1 \ldots K \tag{2.11}$$

Hence, the elements $\underline{m}_i^{(k)}$ of the complex midamble codes $\underline{\boldsymbol{m}}^{(k)}$ of the K users are alternatingly real and imaginary. With W, the number of taps of the impulse response of the mobile radio channels, the L_m binary elements $\underline{m}_i^{(k)}$, $i = 1...L_m$, $k = 1...K$, of (2.11) for the complex midambles $\underline{m}^{(k)}$, $k = 1...K$, of the K users are generated according to [3] from a single periodic basic code

$$
\begin{aligned}
\mathbf{m} &= \left(m_1, m_2 \; \mathrm{K} \, m_{L_m + (K-1)W} \right)^{\mathrm{T}} \\
m_i &\in \{1,-1\}, \, i = 1...(L_m + (K-1)W)
\end{aligned}
\tag{2.12}
$$

The elements m_i, $i = 1...(L_m + (K-1)W)$ of (2.12) fulfill the relation

$$
m_i = m_{i-P}
\tag{2.13}
$$

for the subset $i = (P+1)...(L_m + (K-1)W)$ as illustrated in Figure 2.11. The P elements m_i, $i = 1...P$, of one period of $\boldsymbol{m}$ according to (2.12) are contained in the vector

$$
\boldsymbol{m}_P = \left(m_1, m_2 ... m_P \right)^{\mathrm{T}}
\tag{2.14}
$$

With $\boldsymbol{m}$ according to (2.12) the L_m binary elements $\underline{m}_i^{(k)}$, $i = 1...L_m$, $k = 1...K$, of (2.11) of the midambles of the K users are generated based on

$$
m_i^{(k)} = m_{i+(K-k)W}, \quad i = 1...L_m, \; k = 1...K
\tag{2.15}
$$

Table 2.6

Mapping of Four Binary Elements m_i on a Single Hexadecimal Digit

Four binary Elements m_i	Mapped on hexadecimal digit	Four binary Elements m_i	Mapped on hexadecimal digit
-1 -1 -1 -1	0	1 -1 -1 -1	8
-1 -1 -1 1	1	1 -1 -1 1	9
-1 -1 1 -1	2	1 -1 1 -1	A
-1 -1 1 1	3	1 -1 1 1	B
-1 1 -1 -1	4	1 1 -1 -1	C
-1 1 -1 1	5	1 1 -1 1	D
-1 1 1 -1	6	1 1 1 -1	E
-1 1 1 1	7	1 1 1 1	F

In the following, the term midamble code set or midamble code family denotes K-specific midamble codes $\underline{m}^{(k)}$, $k = 1...K$. Different midamble code sets m(k) are in the following specified based on different periods mp according to (2.14). In adjacent cells of the cellular mobile radio system, different midamble code sets m(k) are used to guarantee a proper channel estimation. As mentioned above, a single midamble code set $\underline{m}^{(k)}$ consisting of K midamble codes is based on a single period $\underline{m}_p$, according to (2.14).

In the following, several periods mp according to (2.14), which are used to guarantee different midamble code sets $\underline{m}^{(k)}$, will be listed in tables in a hexadecimal representation. As shown in Table 2.6, always 4 binary elements m_i are mapped on a single hexadecimal digit. The mean degradations, see equation (38) in [3], which serve as a quality measure of the periods mp according to (2.14) and, hence, of the midamble code sets $\underline{m}^{(k)}$, $k = 1...K$, will also be given.

2.1.3.1   Examples of Midamble Code Sets

In the case of burst type 1 (see Section 2.1.1.4), the midamble length is $L_m = 512$, which corresponds to $K = 8$, $W = 57$, $P = 456$.

Table 2.7

Examples of Periods $\underline{m}_p$ According to (2.14) in the Case of Burst Type 1

Periods $\underline{m}_p$ of length $P = 456$	Degradation (in dB)
C482462CA7846266060D21688BA00B72E1EC84A3D5B7194C8DA39E 21A3CE12BF512C8AAB6A7079F73C0D3E4F40AC555A4BCC453F1D FE3F6C82	0.649471
56F3ACE0A65B96FC326A30B91665BD4380907C2B08DEC98C16A0B 0339AEA855C3D8BDD016E4C3E0F3DA5DF5C0891C851BA30A6C19 ABE6C3ED4	0.695320

In the case of burst type 2, see (Section 2.1.1.4), the midamble length is $L_m = 256$, which corresponds to $K = 3$, $W = 3$, $P = 192$.

Table 2.8

Examples of Periods $\underline{m}_p$ According to (2.14) in the Case of Burst Type 2

Periods of length $P = 192$	Degradation (in dB)
D4A124FE4D11BC14C258546A18C5DE0E3AA3F0617245DBFE	0.615566
48D76A687E21D22321C5201977F620D7A4CB5945F5693A1C	0.638404

2.1.3.2   Performance Aspects

With an uplink midamble set of size eight, the number of users within one slot and one cell is limited to eight. Basic differences between the UTRA TDD and FDD modes complicate an assessment of system characteristics. In the following, we see that the limit of eight midambles is not a restriction to TDD TD-CDMA system resources. Where it is useful, the midamble concept is compared with FDD as a reference.

It is a characteristic of the midamble concept that the spreading code is completely independent of the midamble. Hence, the limit of eight midambles only limits the number of users in one timeslot and one cell to eight, but does not limit the number of spreading codes they use. In other words, a mobile station can use different spreading codes together with the same midamble. Even if eight users are active within one timeslot and one cell more than eigth codes may be used (e.g., in the case of multicode spreading for one service).

The pilot concept of FDD with potentially large spreading factors enables many different codes, which can be discriminated, due to their processing gain. However, FDD pilots are allocated for a whole frame, whereas TDD midambles are only allocated for a slot. Though the number of midambles within one slot is limited to eight, the number of midambles within one frame is $16 \times 8 = 128$. Therefore, the possible number of midambles in TDD is equal to 128, which is equivalent to the number of pilots in a FDD frame, if they are all spread with 128, which is the spreading factor in case the of the speech service.

Due to the CDMA technique used, both FDD and TDD are generally interference-limited systems. This means that far fewer than the available number of codes can be used. Furthermore, it depends solely on the interference level whether a new link can be established.

2.1.3.3   Midamble Transmit Power

In the case of the downlink, $2K$ data blocks are simultaneously transmitted in a burst. Also in the uplink, if K' greater than one CDMA codes assigned to a single user, $2K'$ data blocks are simultaneously transmitted in a burst by this user. This is the so-called multicode uplink situation. In the downlink and in the multicode uplink, the mean powers used to transmit the midambles on the one hand and the $2K$ (or $2K'$) data blocks on the other hand are equal. This is achieved by multiplying the midamble codes $\underline{m}^{(k)}$, $k = 1...K$, with a proper real factor to achieve an attenuation or an amplification.

2.1.3.4   Summary

The midamble concept enables easy joint channel estimation in every receiver, yielding a channel impulse response estimate of every active user in the actual timeslot. The number of available midambles per timeslot and cell is 8, which is equivalent to 128 possible users per cell. The 128 users can be allocated more than one spreading code. This amount of resources is proven to be more than sufficient.

2.1.4   Channel Allocation

For the UTRA TDD mode, a physical channel, which is also termed *resource unit* (RU), is characterized by a combination of its carrier frequency, timeslot, and spreading code as explained in Section 2.1.1. The allocation of these resources to cells, which is called slow dynamic channel allocation (DCA), and the allocation of resource units to bearer services, called fast DCA, are discussed below.

2.1.4.1   Resource Allocation to Cells (Slow DCA)

The channel allocation to cells follows the subsequent rules:

- A cluster size of 1 is used in the frequency domain. In terms of an interference-free DCA strategy, a timeslot-to-cell assignment is performed, resulting in a timeslot clustering. A reuse 1 cluster in the frequency domain does not require frequency planning. If there is also more than one carrier available for a single operator, other frequency reuse patterns with cluster sizes greater than 1 are possible.

- Any specific timeslot within the TDD frame is available either for uplink or downlink transmission. Thus, UL/DL resource allocation can be adapted to time-varying asymmetric traffic.

- In order to accommodate the traffic load in the various cells, the assignment of the timeslots of UL and DL to the cells is rearranged dynamically on a coarse time scale (slow DCA), with the aim that strongly interfering cells use different timeslots. Thus, resources allocated to adjacent cells may also overlap depending on the interference situation.

- Due to idle periods between successively received and transmitted bursts, the user equipment (UE) can provide the network with interference measurements in timeslots different from the one currently used. The availability of such information enables the operator to implement DCA algorithms suited for the network. For instance, the prioritized assignment of timeslots based on interference measurements results in a clustering in the time domain, and in parallel takes into account the demands on locally different traffic loads within the network.

2.1.4.2   Resource Allocation to Bearer Services (Fast DCA)

Fast channel allocation refers to the allocation of one or multiple physical channels to any bearer service. Resource units are acquired and released according to a cell related preference list derived from the slow DCA scheme.

The following principles hold for fast channel allocation:

- The basic resource unit used for channel allocation is one code per timeslot per frequency band.

- Multirate services are achieved by pooling of resource units. This can be done both in the code domain by pooling of multiple codes within one timeslot, termed multicode operation, and in the time domain by pooling of multiple timeslots within one frame, termed multislot operation. Additionally, any combination of both approaches is possible.

- The maximum number of codes per timeslot in the uplink or the downlink depends on several physical circumstances (e.g., channel characteristics, propagation environments). Additional techniques to further enhance capacity, for example smart antennas, are applied as well. Therefore, the DCA algorithm has to be independent of the number of actually used codes. Additionally, time-hopping can be used to average over intercell interference in the case of users requiring low to medium bit rates.

- Channel allocation distinguishes between real time (RT) and nonreal time (NRT) bearer services:

  - RT services: Channels remain allocated for the whole duration for which the bearer service is established. The allocated resources may change because of a channel reallocation procedure, for example in the case of a VBR.

  - NRT services: Channels are allocated for the period of the transmission of a dedicated data packet only. Channel allocation for unconstrained delay data (UDD) services is performed using the *best effort strategy*. This means that resources available for NRT services are distributed to all admitted NRT services with pending transmission requests. The number of channels allocated to any NRT service is variable, and depends at least on the number of currently available resources and the number of NRT services attempting to perform packet transmission simultaneously. Additionally, prioritization of admitted NRT services is possible.

- Channel reallocation procedures (intracell handovers) are triggered for many reasons:

  - To cope with varying interference conditions.

  - In the case of high rate RT services (i.e., services requiring multiple resource units), a *channel reshuffling procedure* is required to prevent a fragmentation of the allocated codes over too many timeslots. This is achieved by freeing the least loaded timeslots (timeslots with minimum used codes) by a channel reallocation procedure.

  - When using smart antennas, channel reallocation is useful to keep the different users active in the same timeslot spatially separated.

2.1.5 Adaptive Antennas

Since the uplink and the downlink operate on the same frequency, adaptive antennas can easily be implemented in UTRA TDD. Due to the reciprocity of the uplink and the downlink spatial characteristics at the base station antenna array, the spatial covariance matrices obtained on the uplink can be used for efficient downlink beamforming. Moreover, the frame-synchronized base stations facilitate an efficient suppression of cochannel interference from adjacent cells via adaptive antennas.

If adaptive antennas are used for downlink beamforming, the same connection-dedicated midamble sequences used on the uplink are also transmitted on the downlink instead of using the same midamble for all downlink connections. This enables user-specific, space-selective beamforming techniques on the downlink to provide a significant capacity increase through interference reduction.

Although the UTRA TDD component does not require the use of smart antennas, the resulting signal-to-interference-plus-noise-ratio (SINR) can significantly improve by incorporating various smart antenna concepts at the base station on the uplink as well as the downlink.

These SINR gains may be exploited

- To increase the capacity, (e.g., by reducing the amount of interference suffered (BS receiver) and created (BS transmitter) in the system);

- To increase the quality;

- To decrease the delay spread;

- To reduce the transmission powers;

- To reduce the electromagnetic pollution and user health hazards;

- To enhance spatial user location due to the estimation of the dominant directions of arrivals;

or a combination thereof.

Three different smart antenna concepts, namely

- Diversity antennas;

- Sector antennas;

- Adaptive antenna arrays;

can be incorporated into the UTRA TDD mode.

2.2 ANALYTICAL MODEL FOR DATA ESTIMATION ALGORITHM USING JOINT DETECTION

This section describes the TD/CDMA system model. Only the uplink is considered, and receiver antenna diversity is taken into account. The described model can also be easily applied to the downlink. The basic structure of the uplink is shown in Figure 2.12. Within the same cell in the same frequency band, a number of K users (i.e., mobile stations) are simultaneously active, which can be separated based on their user-specific CDMA codes. Each MS is assumed to have a single transmitter antenna. The transmitted signals are received at the uplink receiver of the BS over K_a receiver antennas. Therefore, the transmission of the K user signals takes place over $K \cdot K_a$ different radio channels with the time-variant complex impulse responses $\underline{h}^{(k,k_a)}(\tau,t)$, $k = 1...K$, $k_a = 1...K_a$, where the radio channel with the impulse response $\underline{h}^{(k,k_a)}(\tau,t)$ refers to the connection of mobile

k with receiver antenna k_a. The parameter τ denotes the delay time referring to time spreading of the transmitted signals due to multipath reception, and t denotes the real time referring to the time variance of the channel.

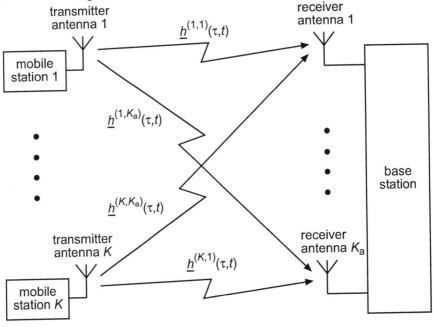

Figure 2.12 Uplink system model.

In Figure 2.13, the block structure of transmitter k is shown. A bit stream representing voice or data information is encoded and interleaved to avoid burst errors. The encoded binary data stream is mapped onto a symbol stream consisting, in general, of complex data symbols. Each data symbol is spread by a user-specific CDMA code. Bursts as described in Section 2.1.1.4 are generated by linking together two data blocks and a midamble. After linear modulation and D/A-conversion, the signal is passed through a transmitter filter and an amplifier.

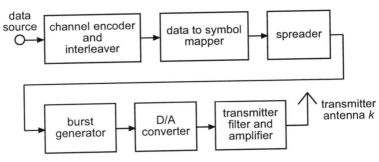

Figure 2.13 Block structure of transmitter k.

The block structure of the uplink receiver is illustrated in Figure 2.14. The received signal at the receiver antenna k_a is the sum of K user signals. This sum is filtered for band limitation and noise suppression, and D/A-converted. The sampling frequency is the inverse of the chip duration T_c. A synchronization unit in the receiver has to compensate for the slightly different delay times of the K_a received signals resulting from the different propagation paths to the K_a antennas. The set of samples resulting from the K user bursts is separated into a subset of samples corresponding to the K midambles of the K users, and two subsets of samples corresponding to the K · N data symbols of the K user bursts transmitted before and after the midamble, respectively. From the subset of samples corresponding to the K midambles, estimates of the K channel impulse responses $\underline{h}^{(k,k_a)}(\tau,t)$, k = 1...K bandlimited to the user bandwidth, are determined by channel estimator k_a. A data estimator applying a joint detection algorithm determines continuous valued estimates of the K · N data symbols transmitted before and after the midamble, respectively. The user-specific CDMA codes are known at the uplink receiver, as they are allocated to each mobile by the base station during first access or handover. The 2 · N complex, continuous valued estimates of the data symbols transmitted by user k, k = 1...K, are mapped onto a real-valued data stream. This data stream is deinterleaved and decoded in a soft-input decoder.

The description of the JD data estimation algorithm using ZF equalization is based upon the discrete-time lowpass representation of the uplink of the TD-CDMA mobile radio system (see Figure 2.15) representing the blocks between the data-to-symbol mapper in Figure 2.13, and the symbol-to-data mapper in Figure 2.14. The sample interval for the discrete-time lowpass representation is the chip duration T_c. In the following, sequences, vectors, and matrices are in boldface, complex values are underlined, and the symbols $(\cdot)^*$ and $(\cdot)^T$ designate the complex conjugation and transpositon, respectively. The transmission of one single data block before or after the midamble is considered, and the influence of the midamble on the data blocks due to multipath reception is assumed to be perfectly eliminated. The K users, which are simultaneously active in the same frequency band, are transmitting finite data symbol sequences

$$\underline{d}^{(k)} = \left(\underline{d}_1^{(k)},\underline{d}_2^{(k)}...\underline{d}_N^{(k)}\right)^T, \quad \underline{d}_n^{(k)} \in \underline{V}, \ k = 1...K, \ n = 1...N \quad (2.16)$$

of N M-ary complex data symbols $\underline{d}_n^{(k)}$ with duration T_s, which are taken from the complex set:

$$\underline{V} = \left\{\underline{v}_1,\underline{v}_2...\underline{v}_M\right\} \quad (2.17)$$

The size M of the set $\underline{V}$, and thus the actual data rate, may be varied due to the service provided or due to the desired transmission quality. Each of the data symbols $\underline{d}_n^{(k)}$, n = 1...N, of mobile k is spread by the user-specific CDMA codes

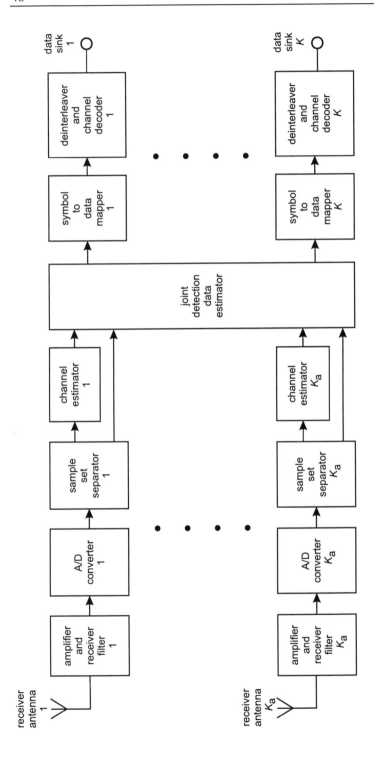

Figure 2.14 Block structure of uplink receiver.

$$\underline{c}^{(k)} = \left(\underline{c}_1^{(k)}, \underline{c}_2^{(k)} \ldots \underline{c}_Q^{(k)}\right)^{\mathrm{T}}, \quad k = 1 \ldots K \tag{2.18}$$

consisting of Q complex chips $\underline{c}_q^{(k)}$ of duration T_c. The influences of the modulator in Figure 2.13, the analog components in Figure 2.13 and Figure 2.14, and the mobile radio channel with the impulse response $\underline{h}^{(k,k_a)}(\tau,t)$, are represented by the discrete-time channel impulse response

$$\underline{h}^{(k,k_a)} = \left(\underline{h}_1^{(k,k_a)}, \underline{h}_2^{(k,k_a)} \ldots \underline{h}_W^{(k,k_a)}\right)^{\mathrm{T}}, \quad k = 1 \ldots K, \; k_a = 1 \ldots K_a \tag{2.19}$$

consisting of W complex samples $\underline{h}_w^{(k,k_a)}$ taken at the chip rate $1/T_c$. The combined channel impulse responses are defined as

$$\underline{b}^{(k,k_a)} = \left(\underline{b}_1^{(k,k_a)}, \underline{b}_2^{(k,k_a)} \ldots \underline{b}_{Q+W-1}^{(k,k_a)}\right)^{\mathrm{T}} = \underline{h}^{(k,k_a)} * \underline{c}^{(k)}$$
$$k = 1 \ldots K, \; k_a = 1 \ldots K_a \tag{2.20}$$

That is, as the discrete-time convolution of $\underline{h}^{(k,k_a)}$ introduced in (2.19) with the signature sequences $\underline{c}^{(k)}$ defined by (2.18).

First, the case of a single receiver antenna k_a is considered [7]. According to Figure 2.15, the received sequence $\underline{e}^{(k_a)}$ of length $(N \cdot Q + W\text{-}1)$ occurs at antenna k_a. Equivalently to the single antenna receiver discussed in [7], each received sequence $\underline{e}^{(k_a)}$ consists of a sum of K sequences, each of length $(N \cdot Q + W\text{-}1)$, which are perturbed by an additive stationary noise sequence

$$\underline{n}^{(k_a)} = \left(\underline{n}_1^{(k_a)}, \underline{n}_2^{(k_a)} \ldots \underline{n}_{N \cdot Q+W-1}^{(k_a)}\right)^{\mathrm{T}}, \quad k_a = 1 \ldots K_a \tag{2.21}$$

with zero mean and covariance matrix

$$\underline{R}_{\mathrm{n}}^{(k_a)} = \mathrm{E}\left\{\underline{n}^{(k_a)} \underline{n}^{(k_a)*\mathrm{T}}\right\}, \quad k_a = 1 \ldots K_a \tag{2.22}$$

With the total data vector

$$\underline{d} = \left(\underline{d}^{(1)\mathrm{T}}, \underline{d}^{(2)\mathrm{T}} \ldots \underline{d}^{(K)\mathrm{T}}\right)^{\mathrm{T}} = \left(\underline{d}_1, \underline{d}_2 \ldots \underline{d}_{K \cdot N}\right)^{\mathrm{T}} \tag{2.23}$$

in which the components of $\underline{d}$ are given by

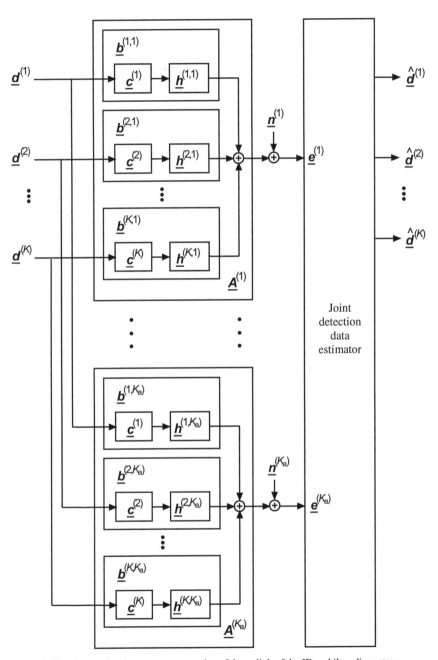

Figure 2.15 Discrete-time lowpass representation of the uplink of the JD mobile radio system.

$$\underline{d}_{N\cdot(k-1)+n} \overset{\text{def}}{=} \underline{d}_n^{(k)}, \quad k=1...K, \; n=1...N \tag{2.24}$$

and the $(N\cdot Q+W-1)\times K\cdot N$ matrix

$$\underline{A}^{(k_{\mathrm{a}})} = \left(\underline{A}_{i,j}^{(k_{\mathrm{a}})}\right), \quad i=1...N\cdot Q+W-1, \; j=1...K\cdot N, \; k_{\mathrm{a}}=1...K_{\mathrm{a}} \tag{2.25}$$

$$\underline{A}_{Q\cdot(n-1)+l,N\cdot(k-1)+n}^{(k_{\mathrm{a}})} = \begin{cases} \underline{b}_l^{(k,k_{\mathrm{a}})} & \text{for} \quad k=1...K, k_{\mathrm{a}}=1...K_{\mathrm{a}} \\ & \phantom{\text{for}} \quad l=1...Q+W-1, n=1...N \\[2mm] 0 & \text{otherwise} \end{cases} \tag{2.26}$$

the received sequence can be represented by

$$\underline{e}^{(k_{\mathrm{a}})} = \left(\underline{e}_1^{(k_{\mathrm{a}})}, \underline{e}_2^{(k_{\mathrm{a}})} \dots \underline{e}_{N\cdot Q+W-1}^{(k_{\mathrm{a}})}\right)^{\mathrm{T}} = \underline{A}^{(k_{\mathrm{a}})}\underline{d} + \underline{n}^{(k_{\mathrm{a}})}, \quad k_{\mathrm{a}}=1...K_{\mathrm{a}} \tag{2.27}$$

Now the unified mathematical representation for the case of K_a receiver antennas is presented. First the $K_{\mathrm{a}}\cdot(N\cdot Q+W-1)\times K\cdot N$ matrix

$$\underline{A} = \left(\underline{A}^{(1)^{\mathrm{T}}}, \underline{A}^{(2)^{\mathrm{T}}} \dots \underline{A}^{(K_{\mathrm{a}})^{\mathrm{T}}}\right)^{\mathrm{T}} \tag{2.28}$$

and the combined noise vector

$$\underline{n} = \left(\underline{n}^{(1)^{\mathrm{T}}}, \underline{n}^{(2)^{\mathrm{T}}} \dots \underline{n}^{(K_{\mathrm{a}})^{\mathrm{T}}}\right)^{\mathrm{T}} = \left(\underline{n}_1, \underline{n}_2 \dots \underline{n}_{K_{\mathrm{a}}\cdot(N\cdot Q+W-1)}\right)^{\mathrm{T}} \tag{2.29}$$

in which

$$\underline{n}_{(N\cdot Q+W-1)\cdot(k_{\mathrm{a}}-1)+n} \overset{\text{def}}{=} \underline{n}_n^{(k_{\mathrm{a}})}, \quad k_{\mathrm{a}}=1...K_{\mathrm{a}}, \; n=1...N\cdot Q+W-1 \tag{2.30}$$

holds, are introduced and $\underline{n}$ has the covariance matrix

$$\underline{R}_n = \mathrm{E}\left\{\underline{nn}^{*\mathrm{T}}\right\} = \begin{bmatrix} \underline{R}_n^{(1,1)} & \underline{R}_n^{(1,2)} & \cdots & \underline{R}_n^{(1,K_a)} \\ \underline{R}_n^{(2,1)} & \underline{R}_n^{(2,2)} & \cdots & \underline{R}_n^{(2,K_a)} \\ \vdots & \vdots & & \vdots \\ \underline{R}_n^{(K_a,1)} & \underline{R}_n^{(K_a,2)} & \cdots & \underline{R}_n^{(K_a,K_a)} \end{bmatrix} \tag{2.31}$$

$$\underline{R}_n^{(i,j)} = \mathrm{E}\left\{\underline{n}^{(i)}\,\underline{n}^{(j)^{*\mathrm{T}}}\right\}, \quad i,j = 1...K_a \tag{2.32}$$

With $\underline{A}$, $\underline{d}$ and $\underline{n}$ the combined received vector becomes

$$\underline{e} = \left(\underline{e}^{(1)^{\mathrm{T}}}, \underline{e}^{(2)^{\mathrm{T}}} \cdots \underline{e}^{(K_a)^{\mathrm{T}}}\right)^{\mathrm{T}} = \left(\underline{e}_1, \underline{e}_2 \cdots \underline{e}_{K_a\cdot(N\cdot Q+W-1)}\right)^{\mathrm{T}} = \underline{A}\,\underline{d} + \underline{n} \tag{2.33}$$

where

$$\underline{e}_{(N\cdot Q+W-1)\cdot(k_a-1)+n} \overset{\mathrm{def}}{=} \underline{e}_n^{(k_a)}, \quad k_a = 1...K_a, \; n = 1...N\cdot Q+W-1 \tag{2.34}$$

holds. The combined received vector $\underline{e}$ of (2.33) is processed by a JD data estimator in order to determine continuous valued estimates

$$\underline{\hat{d}} = \left(\underline{\hat{d}}^{(1)^{\mathrm{T}}}, \underline{\hat{d}}^{(2)^{\mathrm{T}}} \cdots \underline{\hat{d}}^{(K)^{\mathrm{T}}}\right)^{\mathrm{T}} = \left(\underline{\hat{d}}_1, \underline{\hat{d}}_2 \cdots \underline{\hat{d}}_{K\cdot N}\right)^{\mathrm{T}} \tag{2.35}$$

of $\underline{d}$ defined by (2.33). The basic concept of the JD technique is given by the set of equations

$$\underline{\hat{d}} = \underline{M}\,\underline{e} \tag{2.36}$$

where

$$\underline{M} = (\underline{M}_{i,j}), \quad i = 1...K\cdot N, \; j = 1...K_a\cdot(N\cdot Q+W-1) \tag{2.37}$$

is a $K\cdot N \times K_a(N\cdot Q+W-1)$ matrix. The choice of the matrix $\underline{M}$ determines the equalizer type. The mathematical derivation and explanation of JD data estimators is not the subject of this chapter but is contained in the subsection *Joint Detection* of Chapter 4.

2.3 PERFORMANCE EVALUATION

The performance of the considered radio transmission technology TD-CDMA, applied in the UTRA TDD mode, the basic parameters of which were described in Section 2.1, was evaluated by means of computer simulations using the analytical model already described in Section 2.2. This evaluation is carried out based on the methods and conditions described in [14], Annex B. The evaluated test cases specified in Table 2.9 were taken from [5], Attachment 7. In that document, it is recommended to evaluate the performance of the considered RTT for the lowest and highest possible bit rate. The following test data rates were chosen from the comprehensive list of test cases described in [5]:

- For the speech service the 8 Kbps bearer was selected.

- Long constrained delay (LCD) data services were evaluated. The low delay data (LDD) services are excluded, because there is no specific delay requirement defined in [5].

- The data packet services are called unconstrained delay data (UDD) in Table 2.9 and are modeled as a packet service with automatic repeat request (ARQ) protection, and no loss of data is expected over the radio link. Hence, there is no need to specify a bit error ratio (BER) requirement.

Table 2.9

Simulation Test Cases

Cell type	Indoor (A), 3km/h	Pedestrian (A), 3 km/h	Vehicular (A), 120 km/h
	Pico	Micro	Macro
Speech $BER = 10^{-3}$	8 Kbps	8 Kbps	8 Kbps
LCD $BER = 10^{-6}$	64 Kbps 2048 Kbps	64 Kbps 384 Kbps	64 Kbps 144 Kbps
UDD	64 Kbps 2048 Kbps	64 Kbps 384 Kbps	64 Kbps 144 Kbps

The performance evaluation consists of two stages: the link-level simulation and the system-level simulation. This section describes the detailed conditions and assumptions for each stage. Note that the results provided may be subject to further refinement and also to the ongoing development of the radio interface specification.

The subsequent section presents TD-CDMA link level simulation results for the services enumerated in Table 2.9.

2.3.1   Link-Level Simulations

In the following, the results of link-level simulation for the TD-CDMA mode are presented. The circuit switched services, which are the speech service and the LCD services (see Table 2.9), are implemented with forward error correction (FEC) and the UDD packet services use ARQ in combination with FEC. The basic assumptions and technical choices for the link-level simulations are summarized in Table 2.10.

Table 2.10

Basic Assumptions and Technical Choices for the Link-Level Simulations

Carrier frequency	2 GHz
Carrier spacing	5 MHz
Chip rate	4.096 Mchip/s
Duration of a TDMA frame	10 ms
Duration of a timeslot	625 μs
Data modulation	QPSK
Chip modulation	Root raised cosine roll-off α=0.22
Spreading characteristics	Orthogonal Q chips/symbol
Number of chips per symbol	16
Chip duration	0.24414 μs
Channel coding	Convolutional coding + puncturing for rate matching
Outer coding	Reed-Solomon coding
Interleaving	Block interleaving
Data detection	Joint detector: minimum mean square error block linear equalizer
Channel estimation	Joint channel estimator according to [3] based on correlation
Power control	Frame based C-level power control

In the simulations, all intracell interferers are completely modeled with their entire transmission and reception chains. Intercell interference is modeled as white Gaussian noise. In the following, BERs are given as a function of the average E_b/N_o in decibels (E_b is the energy per bit and N_0 is the one-sided spectral noise density) with the intracell interference, (i.e., the number K of active users per timeslot as a parameter). The relation between the E_b/N_o and the carrier to interference ratio C/I, with C denoting the carrier power per CDMA code and I denoting the intercell interference power, is given by

$$\frac{C}{I} = \frac{E_b}{N_0} \cdot \frac{R_c \cdot \log_2(M)}{B \cdot Q \cdot T_c} \qquad (2.38)$$

with

- R_c: the service depending rate of the channel encoder;

- $M = 4$: the size of the data symbol alphabet;

- $B = 4.096$ MHz: the user bandwidth;

- $Q = 16$: the number of chips per symbol;

- $T_c = 0.24414$ μs: the chip duration.

The $\log_2(M)$ is the number of bits per data symbol and $(Q \cdot T_c)/\log_2(M)$ is the bit duration at the output of the encoder. One net information bit is transmitted in the time period $(Q \cdot T_c)/(R_c \cdot \log_2(M))$. Therefore, (2.38) is equivalent to

$$\frac{C}{I} = \frac{E_b/T_b}{N_0 \cdot B} \tag{2.39}$$

This means that C equals E_b/T_b, and I equals $N_0 \cdot B$, with T_b the duration of a net information bit. The carrier-to-interference ratio per user is K_c times the carrier-to-interference ratio per CDMA code, with K_c denoting the number of CDMA codes per timeslot per user.

Speech Service

In this section, the results of link-level simulation for the speech service are given. The system parameters for implementing the speech service are summarized in Table 2.11.

The values of Eb/N0 and C/I required to not exceed a BER of 10^{-3} as defined for the speech service are summarized in Table 2.12. The values of C/I are obtained from the values of Eb/N0 according to (2.38) by subtracting 13.9 dB for the spread speech/data burst 1.

Table 2.11

System Parameters of the Speech Service

Service	Speech, 8 Kbps, 20 ms delay
User bit rate	8000 bps
Number of timeslots per frame per user	1
Number of codes per timeslot per user	1
Burst type	Spread speech/data burst 1 for the uplink and downlink
Data modulation	QPSK
Convolutional code rate	0.31 for the spread speech/data burst 1
Interleaving depth	2 frames = 2 bursts
User block size	160 bits
Antenna diversity	Uplink: yes (2 branches), downlink: no

Table 2.12

Required Values of E_b / N_0 and C / I for the Speech Service

Speech, 8 Kbps	$10 \log_{10}(E_b/N_0)$ *in dB @* BER $= 10^{-3}$			$10 \log_{10}(C/I)$ *in dB @* BER $= 10^{-3}$		
$K_c = 1$	$K = 1$ UL / DL	$K = 4$ UL / DL	$K = 8$ UL / DL	$K = 1$ UL / DL	$K = 4$ UL / DL	$K = 8$ UL / DL
Vehicular A, 120 km/h, without power control	5.3/-	5.8/8.3	6.1/8.5	-8.6/-	-8.1/-5.6	-7.8/-5.4
Outdoor to indoor and pedestrian A, 3 km/h, with power control	3.8/-	3.7/6.1	4.0/6.1	-10.1/-	-10.2/-7.8	-9.9/-7.8
Indoor A, 3 km/h, with power control	3.2/-	3.6/6.0	4.0/5.7	-10.7/-	-10.3/-7.9	-9.9/-8.2

K_c is the number of codes per timeslot per user and K is the number of users per timeslot. The values E_b/N_0 required for speech 8 Kbps in order to not exceed a BER of 10^{-3} are given as a function of K for the propagation environments indoor A, pedestrian A, and vehicular A in the downlink and uplink. Values of K between 1 and 8 are taken into account. There occurs a slight degradation with increasing number K of active users per timeslot. This is due to the increase of intracell interference with increasing K. The degradation is less severe for the indoor and pedestrian channels, which do not have as many propagation paths as the vehicular channel.

LCD Services

In this section, the results of link-level simulations for the LCD services are presented. The system parameters for implementing the LCD 64 Kbps service are summarized in Table 2.13, the system parameters for implementing the LCD 144 Kbps service are summarized in Table 2.14, and for implementing the LCD 384 Kbps service in Table 2.15. Furthermore, an LCD 2048 Kbps service is investigated, for which the system parameters are given in Table 2.16. Antenna diversity in the uplink and downlink are also included in the results.

For the LCD 64 Kbps service four codes in one of the 16 timeslots are allocated to a user (LCD 64) with outer coding. For the LCD 144 Kbps service, nine codes in one of the 16 timeslots are allocated to a user (LCD 144) with outer coding. For the LCD 384 Kbps service, nine codes are allocated in three of the 16 timeslots to a user (LCD 384) with outer coding. For the LCD 2084 Kbps service, nine codes are allocated in 13 of the 16 timeslots to a user (LCD 2048) with outer coding.

Table 2.13

System Parameters of the LCD 64 Kbps Service

Service	LCD, 64 Kbps, 300 ms delay LCD 64; 2 users
User bit rate	64 Kbps
Number of timeslots per frame per user	1
Number of codes per timeslot per user	4
Burst type	Burst type 2
Data modulation	QPSK
Convolutional code rate (inner code)	0.61
Reed Solomon code rate (outer code)	120/127
Total code rate	0.58
Interleaving depth	30 frames = 30 bursts
User block size	4,800 bits

Table 2.14

System Parameters of the LCD 144 Kbps Service

Service	LCD, 144 Kbps, 300 ms delay LCD 144
User bit rate	144 Kbps
Number of timeslots per frame per user	1
Number of codes per timeslot per user	9
Burst type	Burst type 2
Data modulation	QPSK
Convolutional code rate (inner code)	0.61
Reed Solomon code rate (outer code)	120/127
Total code rate	0.58
Interleaving depth	30 frames = 30 bursts
User block size	4,800 bits

Table 2.15

System Parameters of the LCD 384 Kbps Service

Service	*LCD, 384 Kbps, 300 ms delay*
	LCD 384
User bit rate	388.9 Kbps
Number of timeslots per frame per user	3
Number of codes per timeslot per user	9
Burst type	Burst type 2
Data modulation	QPSK
Convolutional code rate (inner code)	0.57
Reed Solomon code rate (outer code)	108/118
Total code rate	0.52
Interleaving depth	30 frames = 30 bursts
User block size	4,320 bits

Table 2.16

System Parameters of the LCD 2048 Kbps Service

Service	*LCD, 2048 Kbps, 300 ms delay*
	LCD 2048
User bit rate	2059.0 Kbps
Number of timeslots per frame per user	13
Number of codes per timeslot per user	9
Burst type	Burst type 2
Data modulation	QPSK
Convolutional code rate (inner code)	0.72
Reed Solomon code rate (outer code)	66/75
Total code rate	0.64
Interleaving depth	30 frames = 30 bursts
User block size	5,280 bits

To reach the required BER of 10^{-6}, LCD services use a concatenated coding scheme with an inner convolutional code and an outer Reed Solomon code. The required values of E_b/N_0 and C/I are summarized in Table 2.17 for LCD 64 Kbps, in Table 2.18 for LCD 144 Kbps, in Table 2.19 for LCD 384 Kbps and in Table 2.20 for LCD 2048 Kbps. The values of C/I are obtained from the values of E_b/N_0 according to (2.38) by subtracting 11.4 dB for LCD 64, 11.4 dB for LCD 144, 11.9 dB for LCD 384, and 11.0 dB for LCD 2048. The required E_b/N_0 and C/I values for the downlink when using antenna diversity are identical to those given in the uplink, except for the case LCD 64 Kbps with $K = 2$ users. Antenna diversity would be a reasonable assumption for applications that are, for instance, executed on a notebook.

Table 2.17

Required Values of E_b/N_0 and C/I for the LCD 64 Kbps Service

LCD 64	$10 \log_{10}(E_b/N_0)$ in dB @ BER $= 10^{-6}$ RS	$10 \log_{10}(C/I)$ in dB @ BER $= 10^{-6}$ RS
$K_c = 4$	$K = 2$ UL / DL	$K = 2$ UL / DL
Indoor (A), 3km/h, with power control	3.2/3.1	-8.2/-8.3
Outdoor to indoor and pedestrian (A), 3 km/h, with power control	3.3/3.1	-8.1/-8.3
Vehicular (A) 120 km/h; without power control	3.9/3.7	-7.5/-7.7

Table 2.18

Required Values of E_b/N_0 and C/I for the LCD 144 Kbps Service

LCD 144	$10 \log_{10}(E_b/N_0)$ in dB @ BER $= 10^{-6}$ RS	$10 \log_{10}(C/I)$ in dB @ BER $= 10^{-6}$ RS
$K_c = 9$		
Vehicular (A), 120 km/h, without power control	4.1	-7.3

Table 2.19

Required Values of E_b/N_0 and C/I for the LCD 384 Kbps Service

LCD 384	$10 \log_{10}(E_b/N_0)$ in dB @ BER $= 10^{-6}$ RS	$10 \log_{10}(C/I)$ in dB @ BER $= 10^{-6}$ RS
$K_c = 9$		
Outdoor to indoor and pedestrian (A), 3 km/h, with power control	1.4	-10.5

Table 2.20

Required Values of E_b/N_0 and C/I for the LCD 2048 Kbps Service

LCD 2048	$10 \log_{10}(E_b/N_0)$ *in dB* @ $BER = 10^{-6}$ RS	$10 \log_{10}(C/I)$ *in dB* @ $BER = 10^{-6}$ *RS*
$K_c = 9$		
Indoor (A), 3km/h, with power control	2.8	-8.2

The results for LCD were achieved without the application of turbo codes; however, turbo coding is likely to be used for LCD services. According to the first simulations, the implementation of turbo codes would lead to an improvement of about 2 dB over the used convolutional and RS codes.

UDD Services

In this section, the results of link-level simulations for UDD services are given. UDD services are implemented by using a type II hybrid ARQ scheme. This ARQ scheme is applied for improving the code rate from one transmission to the next from 1 to 1/2 and to 1/3, and is explained in the following. Some of the code rates can also be omitted, for instance the 1/3 code rate. In one of the used ARQ schemes, the user data is encoded with a 1/3 rate convolutional code and interleaved over 3 bursts. Rate-compatible punctured convolutional (RCPC) codes are used [6].

The coding and interleaving are performed in such a way that decoding is possible after one of three bursts has been received. Thus, the effective code rate is 1 after the reception of the first burst. The packets to be transmitted are divided into blocks of 240 bits each, which constitutes the user block including data, cyclic redundancy check (CRC), block number, and encoder tail. If the decoding is not successful, the second burst is sent and decoding is reattempted. After the second burst, the code rate is 1/2. If the decoding is still not successful, the third burst is sent and decoding is performed again, now with the code rate of 1/3. If the decoding is not successful, the burst with the lowest signal-to-noise-and-interference value is retransmitted, and the original burst and the retransmitted burst are combined by maximum ratio combining. This repetition coding is repeated until the decoding is successful. The system parameters for implementing the UDD services are summarized in Table 2.21. Both code pooling and timeslot pooling are considered.

Table 2.21

System Parameters of UDD Services

Service	UDD, 144 Kbps, 384 Kbps, and 2048 Kbps, no delay constraint
User bit rate	Variable
Number of timeslots per frame per user	Variable
Number of codes per timeslot per user	Variable
Burst type	Spread speech/data burst 2
Data modulation	QPSK
Convolutional code rate (inner code)	Variable, 1 1/2, 1/3
Interleaving depth	1,2,3 frames = 1,2,3 bursts
User block size	232 bits for QPSK
Antenna diversity	Yes (2 branches)

In the link-level simulations, an ideal CRC is modeled. The effects of ARQ are included in the system-level simulations. The aim of the link-level simulations is to find the E_b/N_0 values required to achieve certain BERs and BLERs. The following alternatives were investigated in the simulations as extreme cases:

- Allocating one code to a user in the uplink, with one user being active per timeslot.

- Allocating nine codes to a user in the uplink, with one user being active per timeslot.

- Allocating one code to a user in the downlink, with four users being active per timeslot.

- Allocating nine codes to a user in the downlink, with one user being active per timeslot.

- Allocating one code to a user in the uplink, with eight users being active per timeslot.

- Allocating three codes to a user in the uplink, with one user being active per timeslot.

- Allocating three codes to a user in the downlink, with two users being active per timeslot.

- Allocating three codes to a user in the downlink, with three users being active per timeslot.

- Allocating four codes to a user in the uplink, with one user being active per timeslot.

These cases are extreme with respect to code pooling. The performance when pooling other numbers of codes lies between the extreme cases given here. Since it is likely that these applications will be executed on a notebook, antenna diversity in the downlink is also included in the results.

Based on a throughput analysis, the bit rates achievable depending on the average *C/I* are determined. The achievable bit rates are determined by taking into account the necessary retransmissions due to block errors and the related decrease of the effective information bit rate. The ARQ scheme improves the code rate from one transmission to the next from 1 to 1/2 and to 1/3.

The *C/I* values required to achieve the nominal bit rate are summarized in Table 2.22 for UDD 64 Kbps, in Table 2.23 for UDD 144 Kbps, in Table 2.24 for UDD 384 Kbps and in Table 2.25 for UDD 2048 Kbps. *TS* is the number of timeslots per user.

Table 2.22

Required Value of *C/I* for the UDD 64 Kbps Service, Code Rates 1, 1/2, and 1/3

UDD 64	$10 \log_{10}(C/I)$ in dB @ 64 Kbps
	$K_c = 4$, $K = 1$, $TS = 1$ UL and DL
Indoor office A, 3 km/h	-6,2/-5,8
Outdoor to indoor and pedestrian A, 3 km/h	-5,9/-6,5
Vehicular A, 120 km/h	-6,7/-7,1

Table 2.23

Required Value of *C/I* for the UDD 144 Kbps Service, Code Rates 1, 1/2, and 1/3

UDD 144	$10 \log_{10}(C/I)$ in dB @ 144 Kbps
	$K_c = 9$, $K = 1$, $TS = 1$ UL and DL
Vehicular A, 120 km/h	-4,8/-4,8

Table 2.24

Required Value of *C/I* for the UDD 384 Kbps Service, Code Rates 1, 1/2, and 1/3

UDD 384	$10 \log_{10}(C/I)$ in dB @ 384 Kbps
	$K_c = 9$, $K = 1$, $TS = 3$, UL and DL
Outdoor to indoor and pedestrian A, 3 km/h	-8,1/-8,1

Table 2.25

Required Value of *C/I* for the UDD 2048 Kbps Service, Code Rates 1, 1/2, and 1/3

UDD 2048	$10 \log_{10}(C/I)$ in dB @ 2048 Kbps
	$K_c = 9$, $K = 1$, $TS = 13$, UL and DL
Indoor office A, 3 km/h	-5,0/-5,0

2.3.2   Link Budget Templates

Based on the results of the link-level simulations, the maximum distances between the mobile station and base station for the mentioned services are calculated using the link budget templates defined in [4]. All E_b/N_0 values derived in the link-level simulations are associated with the energy per bit needed to achieve the corresponding BER/FER. Thus, the midamble is not included in the E_b/N_0 and, therefore, included in the information rate as explained hereafter. Power control is applied on a frame-by-frame basis. Thus, every 10 ms power is changed if necessary.

QPSK modulation leads to the gross bit rate on the air of 512 Kbps, including midamble and guard period, and 492.8 Kbps including midamble and excluding the guard period. Since no energy is needed to transmit the guard period, 492.8 Kbps is used to determine the information bit rate in the template. The information bit rate is derived by multiplying 492.8 Kbps with the total coding rate used by the service. That means the information rate becomes

- 492.8 Kbps · 0.330 = 162.1 Kbps for speech, data burst 1 (long midamble);

- 492.8 Kbps · 0.580 = 285.8 Kbps for LCD   64, data burst 2 (short midamble);

- 492.8 Kbps · 0.580 = 285.8 Kbps for LCD   144, data burst 2 (short midamble);

- 492.8 Kbps · 0.522 = 257.2 Kbps for LCD   384, data burst 2 (short midamble);

- 492.8 Kbps · 0.638 = 314.4 Kbps for LCD   2048, data burst 2 (short midamble).

As mentioned above, the E_b/N_0 figures are related to one code. Therefore, the total information rate of a particular service is determined by multiplying the information rate for a single code with the number of codes used in the case of the considered service.

The lognormal fade margin and the handover gain in the subsequent templates are obtained by independent quasistatic simulations. The basis of the simulations is a hexagonal cell structure, in which mobiles are randomly distributed over the cell area. Each mobile is subject to fading effects according to the lognormal conditions specified for each test environment (e.g., standard deviation of 10/12 dB for outdoor/indoor environments). Thus, the lognormal fade margin is determined regarding one single cell. In a multiple hexagonal cell layout, one gains from the possibility of maintaining the connection due to several potential serving cells. According to the preset conditions for the link budget template the handoff gain is calculated assuming 50% shadowing correlation. The handover gain depends on the fade margin used in the simulations and shown in Table 2.26.

Table 2.26

Fade Margin and Handoff Gain

Environment	Lognormal fade margin	Hard hand-off gain
Indoor office	15,4	5,9
Outdoor to indoor and pedestrian	11,3	4,7
Vehicular	11,3	4,7

For the following reason, no link budget templates are calculated for UDD services. The proposed hybrid type-II ARQ scheme allows for retransmission in case of unsuccessful data detection. Therefore, no fixed Eb/N_0 or C/I values required to achieve the quality of service (QoS) at the cell border can be defined. However, the range of UDD services is larger than the range of the corresponding LCD service due to ARQ retransmissions. Examples for the link budget calculations in case of the speech service and the LCD service with 144 Kbps are given in the Sections 2.3.2.1 and 2.3.2.2, respectively.

2.3.2.1 Speech Service

		DL	UL	DL	UL	DL	UL
Test environment		Indoor A	Indoor A	Pedestr. A	Pedestr. A	Vehicular A	Vehicular A
Test service		speech	speech	speech	speech	speech	speech
Note	1 TS, 1 Code	burst1	burst1	burst1	burst1	burst1	burst1
Bit rate per traffic channel (incl. midamble)	bps	10133	10133	10133	10133	10133	10133
Average TX per traffic channel	dBm	10	4	20	14	30	24
Max. TX power per traffic channel	dBm	22,0	16,0	32,0	26,0	42,0	30,0
Max. total TX power	dBm	31,1	16,0	41,1	26,0	51,1	30,0
Cable, conn. and combiner losses	dB	2	0	2	0	2	0
TX antenna gain	dBi	2	0	10	0	13	0
TX eirp per traffic channel	dBm	22,0	16,0	40,0	26,0	53,0	30,0
Total TX eirp	dBm	31,1	16,0	49,1	26,0	62,1	30,0
RX antenna gain	dBi	0	2	0	10	0	13
Cable and connector losses	dB	0	2	0	2	0	2
RX noise figure	dB	5	5	5	5	5	5
Thermal noise density	dBm/Hz	-174	-174	-174	-174	-174	-174
RX interference density	dBm/Hz	-1000	-1000	-1000	-1000	-1000	-1000
Total effect. noise plus interf. density	dBm/Hz	-169	-169	-169	-169	-169	-169
Information rate Rb	dbHz	52,1	52,1	52,1	52,1	52,1	52,1
Required $E_b/(N_0+I_0)$	dB	6	3,6	6,1	3,7	8,3	5,8
Receiver sensitivity	dBm	-110,9	-113,3	-110,8	-113,2	-108,6	-111,1
Handoff gain	dB	5,9	5,9	4,7	4,7	4,7	4,7
Explicit diversity gain	dB	0	0	0	0	0	0
Other gain	dB	0	0	0	0	0	0
Lognormal fade margin	dB	15,4	15,4	11,3	11,3	11,3	11,3
Maximum path loss	dB	123,4	119,8	144,2	140,6	155,0	145,5
Maximum range	m	761,1	577,3	805,5	654,7	5206,7	2902,7
Coverage efficiency	km²/site	1,8	1,0	2,0	1,3	17,6	5,5
Concept optimized parameters							
Average TX per traffic channel	dBm	13	10	23	20	30	24
Max. TX power per traffic channel	dBm	25,0	22,0	35,0	30,0	42,0	30,0
TX antenna gain	dBi	2	0	10	0	17	0
RX antenna gain	dBi	0	2	0	10	0	17
Maximum path loss	dB	126,4	125,8	147,2	144,6	159,0	149,5
Maximum range	m	958,1	915,0	957,3	822,3	6651,8	3708,4
Coverage efficiency	km²/site	2,9	2,6	2,9	2,1	28,7	8,9

2.3.2.2   Long Constrained Delay 144 Kbps (LCD 144) Service

		DL	UL
Test environment		Vehicular A	Vehicular A
Test service		LCD 144	LCD 144
Note	1 TS, 9 codes	burst2	Burst2
Bit rate per traffic channel (incl. midamble)	Bps	160776	160776
Average TX per traffic channel	DBm	30	24
Max. TX power per traffic channel	DBm	42,0	30,0
Max. total TX power	dBm	42,0	30,0
Cable, conn. and combiner losses	dB	2	0
TX antenna gain	dBi	13	0
TX eirp per traffic channel	dBm	53,0	30,0
Total TX eirp	dBm	53,0	30,0
RX antenna gain	dBi	0	13
Cable and connector losses	dB	0	2
RX noise figure	dB	5	5
Thermal noise density	dBm/Hz	-174	-174
RX interference density	dBm/Hz	-1000	-1000
Total effect. noise plus interf. density	dBm/Hz	-169	-169
Information rate Rb	dbHz	64,1	64,1
Required Eb/(N0+I0)	dB	4,1	4,1
Receiver sensitivity	dBm	-100,8	-100,8
Handoff gain	dB	4,7	4,7
Explicit diversity gain	dB	0	0
Other gain	dB	0	0
Log-normal fade margin	dB	11,3	11,3
Maximum path loss	dB	147,2	135,2
Maximum range	m	3228,4	1544,3
Coverage efficiency	km²/site	32,7	7,5
Concept optimized parameters			
Average TX per traffic channel	dBm	30	24
Max. TX power per traffic channel	dBm	42,0	30,0
TX antenna gain	dBi	17	0
RX antenna gain	dBi	0	17
Maximum path loss	dB	151,2	139,2
Maximum range	m	4124,5	1973,0
Coverage efficiency	km²/site	53,4	12,2

2.3.3   Results of System-Level Evaluations

This section describes the system performance obtained by system-level simulations of TD-CDMA. The simulations were performed for several services and environments according to ETSI TR 101 112 Annex 2 [4]. The evaluated test cases shown in Table 2.27 represent the most relevant applications of those proposed in Attachment 7 of ITU-R Circular Letter 8 [5].

The UTRA TDD approach uses a 3-dimensional resource space (consisting of frequency, time, and code dimension). When taking only one carrier (5 MHz) into account, the resource space is reduced to the time and code dimension. Since it was found that the downlink is the critical link with respect to limiting system capacity, only the downlink direction was considered. For instance, antenna diversity can be used in the uplink to improve the soft blocking limit significantly. In case of data services LCD and UDD, in the downlink direction antenna diversity is also assumed.

The simulation results presented in this section rely on statistics gathered from about 5,000 simulated calls in the reference cells.

Table 2.27

Simulation Cases

	Vehicular, 120 km/h	Pedestrian, 3 km/h	Indoor, 3 km/h
Speech @ $BER = 10^{-3}$	8 Kbps	8 Kbps	8 Kbps
Circuit-switched data services (LCD) $BER = 10^{-6}$	144 Kbps	384 Kbps	2 048 Kbps
Packet data (UDD) $BER = 10^{-6}$	144 Kbps	384 Kbps	2 048 Kbps

2.3.3.1   Test Environments

All test environment characteristics including the network structure, the cell shape, the antenna patterns, the propagation models (path loss and shadowing, channel model), the mobility models, the traffic models and the QoS criteria for RT and UDD users are chosen according to [4]. In the vehicular environment, which is a macroenvironment, three-sectored sites are used with cell radius 2 km (i.e., site-to-site distance is 6 km). Statistics are collected within the central site, which consists of three reference cells, surrounded by 54 interfering cells. A frequency reuse scheme of 1 is used. In the micro cellular pedestrian deployment environment based on a Manhattan grid given in [4], with low-speed users at 3 km/h, a frequency reuse scheme of 1 is applied. Statistics are gathered from the six cells according to [4]. For the results given in Table 2.28 in the case of the picocellular indoor office environment, which characterizes a three storey office building where users are entering or leaving the office rooms, a frequency reuse scheme of 1 is used. Statistics are gathered from the six cells in the middle floor according to [4]. Traffic and QoS criteria for RT and UDD users are modeled according to [4].

Table 2.28

Summary of the Results of the System Simulations (Downlink)

Environment	Service	Spectrum Efficiency (Kbps/MHz/cell)
Vehicular	Speech	70
	LCD 144	201
	UDD 144	320
Outdoor to indoor pedestrian	Speech	148
	LCD 384	330
	UDD 384	642
Indoor office	Speech	73
	LCD 2048	62
	UDD 2048	400

Note: For LCD and UDD services, antenna diversity is used.

2.3.3.2 Performance Measures

The evaluation criterion for the system-level simulations is in accordance with the satisfied user criterion, which means the following: For speech service with 8 Kbps and 50% voice activity and LCD services, a user is satisfied, if all three constraints described in [4] are fulfilled. The three different UDD packet data services with 144 Kbps, 384 Kbps, and 2,048 Kbps data rates were evaluated according to the satisfied user criterion described in [4].

2.3.3.3 Resource Allocation

An RU in UTRA TDD is a triple consisting of frequency channel, timeslot, and code. For services that require more than one RU a number of codes (multicode), or a number of timeslots (multislot), or a combination of both (mixed allocation) may be allocated.

For RT services, the resources are allocated at session setup and are kept unchanged until session end. On intercell handover the same type of resource is allocated in the new cell. For UDD services the allocation and deallocation is done on the block level driven by the amount of data a user has in the buffer. The resource allocation tries to distribute the allocated codes homogeneously over all frequencies and timeslots (i.e., the timeslot with the minimum number of codes is searched).

2.3.3.4   Power Control

A slow power control with a control interval of 0.5 seconds is used in the high-speed vehicular environment. Enhanced PC is used in the slow speed deployment environments indoor and pedestrian. According to the interface between link-level and system-level simulations (see Section 2.3.3.8), the received power measurements resulting from link-level simulations are used to incorporate the fast-fading conditions. Hence, the transmitting power is controlled by tracking the needed power gain. The average dead time is 5 ms due to open loop power control. The transmitting power is limited to the dynamic range of 30 dB in the downlink and 80 dB in the uplink direction for both types of the control mechanisms. For UDD services no enhanced PC is used.

2.3.3.5   Handover

The handover is based on the power budget, which means the path loss difference between the serving and the neighboring cells, with a handover margin of 3 dB.

2.3.3.6   ARQ

For UDD the hybrid type-II ARQ protocol is used. Here the first transmission of a block is done nearly uncoded. If the first transmission fails, a second transmission contains the coding bits in a way that all blocks of the first and second transmission together result in a lower coding rate (i.e., a better coding). If the second transmission also fails, the burst with the worst raw BER is retransmitted and a maximum ratio combining is done between the original and the retransmitted burst. The code rate is not decreased in this step. For all further retransmissions this maximum ratio combining is used. If the number of retransmissions exceeds a given threshold of 10 to 20, the session is dropped.

2.3.3.7   DTX

For speech services, voice activity detection (VAD) and discontinuous transmission (DTX) are used with an activity factor of 0.5 and mean speech activity periods of 3 seconds, according to [4]. During the silent periods, the transmit power is switched off, but the channels are not released (i.e. DTX is used to reduce interference and, therefore, to relieve the soft blocking limit).

2.3.3.8   Interface Between the System and Link-Level Simulation

Link-level simulations were carried out for fast-fading multipath channels, and the obtained results describe the relationship between carrier-to-interference ratio (CIR) and BER needed in the system-level simulations. Figure 2.16 shows the actual value interface (AVI) between system and link-level simulations, which was taken into account in the system-level simulator to calculate the actual values of

CIR experienced in each burst. In UTRA TDD it is distinguished between intercell CIRinter and intracell CIRintra interference. The CIRinter is the ratio between the power of the desired signal in the reference cell and the sum of the powers of the interfering signals from all other cells in the same frequency and the timeslot. The CIRintra is the ratio between the power of the desired signal in the reference cell and the sum of the powers of the signals from all other users in the same cell in the same frequency and timeslot. Simulations showed that due to joint detection the impact of the intracell interference is negligible. Within the AVI, the burst CIR values are mapped on a raw BER of a burst. The raw BER not only depends on CIRintra and CIRinter, but also on

- The number of codes K_c per user and timeslot used in a certain service (code pooling);

- The number of users K per timeslot.

Depending on the interleaving depth assumed for each service, the average raw BER on a corresponding number of bursts constituting one block is calculated and subsequently mapped on a user BER value of the received block. For ARQ, there is an additional interface function that gives the relationship between the average raw BER and the block error rate (BLER).

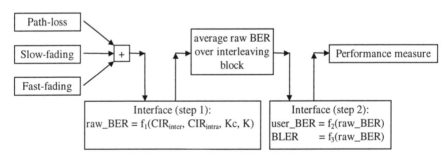

Figure 2.16 Interface between system level and link-level simulations.

2.3.3.9 Simulation Results

The simulation results are summarized in Table 2.28. Two circuit-switched-services, speech and LCD, and one packet service were evaluated for each environment. The performance measure for circuit-switched and packet services is in accordance with Annex D of ETSI TR 101 112 (UMTS 30.03) [4]. The spectrum efficiency or cell capacity, respectively, are derived from the interference limited system load, which fulfills the selected user criterion (i.e., for RT the requirement that a satisfied user must obtain sufficiently good quality more than 95% of the session time).

2.3.3.10   Summary

Simulation results concerning spectrum efficiency of TD-CDMA were presented for real-time and nonreal-time services. The presented values do not include the impact of signaling, which, however, is estimated to be about 10%. On the other hand there are some options that may significantly improve spectrum efficiency:

- Enhanced link adaptation;

- Power control based on quality;

- Turbo coding;

- Smart antennas.

The options mentioned above and the corresponding tradeoff between capacity increase and complexity issues are for further study.

2.4   CONCLUSIONS

In this chapter the physical layer of TD-CDMA, that is, the RTT applied in the unpaired portion of the frequency band of the UTRA concept, was described. The main features of the UTRA TDD mode are:

- Separation of users in the time and code domain;

- Flexibility in downlink/uplink timeslot allocation to meet asymmetric traffic requirements;

- High performance due to intracell interference elimination by joint detection;

- Multiple switching points per frame to allow fast power control;

- Interference dependent resource allocation by dynamic channel allocation;

- Easy implementation of adaptive antennas.

The UTRA FDD mode and the UTRA TDD mode are a part of the European (ETSI) proposal for IMT-2000. Both components of the UTRA proposal were harmonized with respect to basic system parameters such as carrier spacing, chip rate, and frame length. In this way, FDD/TDD dual mode operation is facilitated, which provides a basis for the development of low-cost terminals [1].

References

[1] A. Klein, G.K. Kaleh and P.W. Baier: Zero Forcing and Minimum Mean-Square-Error Equalization for Multiuser Detection in Code-Division Multiple-Access Channels, IEEE Transactions on Vehicular Technology, Vol. 45, No. 2, May 1996, pp. 276-287.

[2] M. Haardt, A. Klein, W. Mohr, and J. Schindler: "Overview of the UTRA TDD Physical Layer Description," FRAMES Workshop, Delft, The Netherlands, Jan. 1999, pp. 14-21.

[3] B. Steiner, P.W. Baier: "Low Cost Channel Estimation in the Uplink Receiver of CDMA Mobile Radio Systems," Frequenz, Vol. 47, Nov./Dec. 1993, pp. 292-298.

[4] ETSI, Technical Report: Universal Mobile Telecommunications System (UMTS); Selection procedures for the choice of radio transmission technologies of the UMTS. TR 101 112, v. 3.2.0, April 1998.

[5] ITU-R, "Circular letter: Request for Submission of Candidate Radio Transmission Technologies (RTTs) for the IMT-2000/FPLMTS Radio Interface," Circular Letter 8/LCCE/47.

[6] P. Frenger, P. Orten, T. Ottosson and A. Svensson: "Rate Matching in Multichannel Systems Using RCPC-Codes," Proceedings of the IEEE Vehicular Technology Conference, Arizona, 1997, pp. 354-357.

[7] A. Klein and P.W. Baier: "Linear Unbiased Data Estimation in Mobile Radio Systems Applying CDMA," IEEE Journal on Selected Areas in Communications, Vol. 11, Sept. 1993, pp. 1058-1066.

Chapter 3

WCDMA

The TD-CDMA concept is presented in detail in Chapter 2. This chapter introduces the WCDMA mode. In May 1999, the Operators Harmonization Group (OHG) achieved global harmonization of WCDMA schemes, resulting in technical framework with three different modes: direct spread (DS), multi-carrier (MC), and TDD [1]. Details are presented in Chapter 12. This harmonization will provide the foundation for accelerated growth of mobile communications toward future wireless information society. The technical parameters of the harmonized 3G CDMA are based on work done in ETSI, ARIB, TTA, TTC, and TIA.

The resulting harmonized 3G CDMA specification should [1]:

- Focus on customer needs for widespread availability of voice and high-speed non-voice services;

- Maximize ability of customers to roam with their services across regions, countries, and systems;

- Minimize 3G costs for the mobile industry;

- Maximize the ability of the information technology, Internet, and personal computer industries to provide mobile applications, solutions, and subscriber devices;

- Provide a smooth and compatible evolution path from the existing infra-structure;

- Be completed in time to meet the commercialization plans of all countries/regions;

- Recognize that there are two well-established core network architectures;

- Minimize the intellectual property right (IPR) impact on the industry;

- Promote the free flow of IPRs to accelerate innovation and create greater customer choice;

- Accommodate regional needs for different spectrum allocations;

- Use technical approaches and parameters that meet customer requirements.

In this chapter, we focus on the DS mode that is based on WCDMA proposal, originally developed as a joint effort between ETSI and ARIB during the second half of 1997 [2]. The ETSI WCDMA scheme has been developed from the FRAMES multiple access scheme 2 (FMA2) in Europe [3-9] and the ARIB WCDMA from the Core-A scheme in Japan [10-15]. The uplink of the WCDMA scheme is based mainly on the FMA2 scheme, and the downlink on the Core-A scheme. In addition, WCDMA received inputs from TTA [16-20]. The main improvements of the WCDMA scheme in the harmonization process are chip rate (3.86 Mchips/s) and downlink pilot structure.

This chapter is structured as follows: the basic concept of WCDMA is explained in Section 3.1. First, the carrier spacing and deployment scenarios are described. The logical and the physical channels are introduced. Then follow subsections on spreading and modulation, multirate scheme, handover, and interoperability between GSM and WCDMA. Some performance results are contained in Section 3.2 and conclusions are drawn in Section 3.3.

Note that currently the standardization of WCDMA is ongoing in the 3GPP resulting in changes in parameters described in this chapter. The reader is advised to refer to the latest 3GPP documents at http:/www.3gpp.org.

3.1 BASIC CONCEPT

This section presents the basic concepts of the WCDMA mode. Table 3.1 lists the main parameters of WCDMA.

3.1.1 Carrier Spacing and Deployment Scenarios

The carrier spacing has a raster of 200 kHz and can vary from 4.2 to 5.4 MHz. The different carrier spacings can be used to obtain suitable adjacent channel protections depending on the interference scenario. Figure 3.1 shows an example for the operator bandwidth of 15 MHz with three cell layers. Larger carrier spacing can be applied between operators than within one operator's band in order to avoid interoperator interference. Interfrequency measurements and handovers are supported by WCDMA to utilize several cell layers and carriers.

Table 3.1

Parameters of WCDMA

Carrier spacing	5 MHz (nominal)
Downlink RF channel structure	Direct spread
Chip rate	3.84 Mcps
Roll-off factor for chip shaping	0.22
Frame length	10 ms
Number of slots/frame	15
Spreading modulation	Balanced QPSK (downlink) Dual channel QPSK (uplink) Complex spreading circuit
Data modulation	QPSK (downlink) BPSK (uplink)
Coherent detection	Pilot symbols/channel
Channel multiplexing in uplink	Control and pilot channel time multiplexed. For the data and control channels I and Q multiplexing.
Multirate	Variable spreading and multicode
Spreading factors	4 – 256
Power control	Open and fast closed loop (1.6 kHz)
Spreading (downlink)	Variable length orthogonal sequences for channel separation. Gold sequences 2^{18} for cell and user separation (truncated cycle 10 ms)
Spreading (uplink)	Variable length orthogonal sequences for channel separation, gold sequence 2^{18} for user separation (different time shifts in I and Q channel, truncated cycle 10 ms)
Handover	Soft handover Interfrequency handover

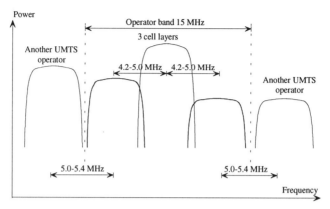

Figure 3.1 Frequency utilization with WCDMA.

3.1.2   Logical Channels

The following logical channels are defined for WCDMA:

- Broadcast channel (BCH) carries system and cell-specific information. The BCH is always transmitted over the entire cell with a low fixed bit rate;

- Paging channel (PCH) for messages to the mobiles in the paging area;

- The forward access channel (FACH) is a downlink transport channel that is used to carry control information from the base station to the mobile station in one cell when the system knows the location cell of the mobile station. The FACH may also carry short user packets. The FACH is transmitted over the entire cell or over only a part of the cell using lobe-forming antennas;

- The random access channel (RACH) is an uplink channel that is used to carry control information from the mobile station. The RACH may also carry short user packets. The RACH is always received from the entire cell;

- The common packet channel (CPCH) is a channel that is used to carry small and medium-sized packets. CPCH is a contention-based, random access channel used for transmission of bursty data traffic. CPCH is associated with a dedicated channel on the downlink, which provides power control for the uplink CPCH;

- The dedicated channel (DCH) is a downlink or uplink channel that is used to carry user or control information between the network and the UE. The DCH thus corresponds to three channels, dedicated traffic channel (DTCH), stand-alone dedicated control channel (SDCCH), and associated control channel (ACCH), defined within ITU-R M.1035. The DCH is transmitted over the entire cell or over only a part of the cell using lobe-forming antennas. The DCH is characterized by the possibility of fast rate change (every 10 ms), fast power control.

3.1.3   Physical Channels

Physical channels typically consist of a three-layer structure of superframes, radio frames, and time slots. Depending on the symbol rate of the physical channel, the configuration of radio frames or time slots varies.

A superframe has a duration of 720 ms and consists of 72 radio frames. A radio frame is a processing unit that consists of 15 time slots. A time slot is a unit that consists of the set of information symbols. The number of symbols per time slot depends on the physical channel.

A physical channel corresponds to a specific carrier frequency, code, and, on the uplink, relative phase (0 or $\pi/2$).

3.1.3.1   Uplink Physical Channels

There are two dedicated channels and one common channel on the uplink. User data is transmitted on the dedicated physical-data channel (DPDCH), and control information is transmitted on the dedicated physical-data channel (DPDCH). The physical random-access channel (PRACH) is a common-access channel.

Figure 3.2 shows the principal frame structure of the uplink DPDCH. Each DPDCH frame on a single code carries 150×2^k bits (15×2^k Kbps), where $k = 0, 1, ..., 6$, corresponding to a spreading factor of $256/2k$ with the 3.84-Mcps chip rate. Multiple parallel variable rate services (= dedicated logical traffic and control channels) can be time multiplexed within each DPDCH frame. The overall DPDCH bit rate is variable on a frame-by-frame basis.

In most cases, only one DPDCH is allocated per connection, and services are jointly interleaved sharing the same DPDCH. However, multiple DPDCHs can also be allocated (e.g., to avoid a too-low spreading factor at high data rates).

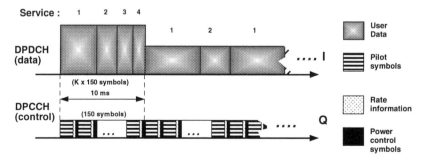

Figure 3.2   WCDMA uplink multirate transmission.

The dedicated physical-control channel (DPCCH) is needed to transmit pilot symbols for coherent reception, power control signaling bits, and rate information for rate detection. Two basic solutions for multiplexing physical control and data channels are time multiplexing and code multiplexing. A combined IQ and code multiplexing solution (dual-channel QPSK) is used in WCDMA uplink to avoid electromagnetic compatibility (EMC) problems with DTX.

The major drawback of the time multiplexed control channel is the EMC problems that arise when DTX is used for user data. One example of a DTX service is speech. During silent periods, no information bits are transmitted, which results in pulsed transmission, as control data must be transmitted in any case. This is illustrated in Figure 3.3. Because the rate of transmission of pilot and power control symbols is on the order of 1 to 2 kHz, they cause severe EMC problems to both external equipment and terminal interiors. This EMC problem is more difficult in the uplink direction, since mobile stations can be close to other electrical equipment, like hearing aids.

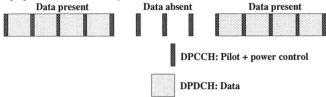

Figure 3.3 Illustration of pulsed transmission with time multiplexed control channel.

The IQ/code multiplexed control channel is shown in Figure 3.4. Now, since pilot and power controls are on a separate channel, no pulselike transmission takes place. Interference to other users and cellular capacity remains the same as in the time-multiplexed solution. In addition, link-level performance is the same in both schemes if the energy allocated to the pilot and the power control bits is the same.

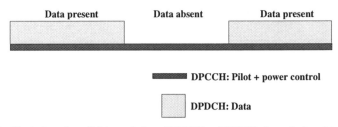

Figure 3.4 Illustration of parallel transmission of DPDCH and DPCCH channel when data is present/absent (DTX).

The WCDMA random access scheme is based on a slotted ALOHA technique with fast acquisition indication. The mobile station can start the transmission at a number of well-defined time-offsets, relative to the frame boundary of every second frame of the received BCH of the current cell. The different time offsets are denoted access slots. There are 15 access slots per two frames and they are spaced 5,120 chips apart. Information on what access slots are available in the current cell is broadcast on the BCH. Before the transmission of a random access request, the mobile terminal should carry out the following tasks:

- Achieve chip, slot, and frame synchronization to the target base station from the SCH and obtain information about the downlink scrambling code, also from the SCH;

- Retrieve information from BCCH about the random access code(s) used in the target cell/sector;

- Estimate the downlink path loss, which is used together with a signal strength target to calculate the required transmit power of the random access request.

It is possible to transmit a short packet together with a random access burst without setting up a scheduled packet channel. No separate access channel is used for packet-traffic-related random access, but all traffic shares the same random access channel. More than one random access channel can be used if the random access capacity requires such an arrangement. The use of the random access burst for packet access is described in Section 3.1.6.

3.1.3.2   Downlink Physical Channels

In the downlink, there are four common physical channels. The common pilot channel (CPICH) is used for coherent detection; the primary and secondary common control physical channels (CCPCH) are used to carry the BCH; the SCH provides timing information and is used for handover measurements by the mobile station.

The primary CCPCH carries the BCH channel. It is of fixed rate and is mapped to the DPDCH in the same way as dedicated traffic channels. The primary CCPCH is allocated the same channelization code in all cells. A mobile terminal can thus always find the BCH, once the base station's unique scrambling code has been detected during the initial cell search.

The secondary physical channel for common control carries the PCH and FACH in time multiplex within the super-frame structure. The rate of the secondary CCPCH may be different for different cells, and is set to provide the required capacity for PCH and FACH in each specific environment. The channelization code of the secondary CCPCH is transmitted on the primary CCPCH.

The SCH consists of two subchannels, the primary and secondary SCHs. Figure 3.5 illustrates the structure of the SCH. The SCH applies short code masking to minimize the acquisition time of the long code. The SCH is masked with two short codes (primary and secondary SCH). The unmodulated primary SCH is used to acquire the timing for the secondary SCH. The modulated secondary SCH code carries information about the long code group to which the long code of the BS belongs. In this way, the search of long codes is limited to a subset of all the codes.

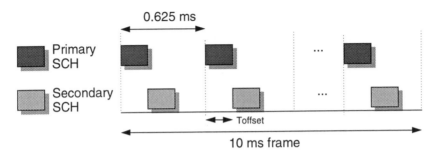

Figure 3.5 Structure of the SCH.

The primary SCH consists of an unmodulated code of length 256 chips, which is transmitted once every slot. The primary synchronization code is the same for every base station in the system and is transmitted time aligned with the slot boundary, as illustrated in Figure 3.5.

The secondary SCH consists of one modulated code of length 256 chips, which is transmitted in parallel with the primary SCH. The secondary synchronization code is chosen from a set of 16 different codes, depending on which of the 32 different code groups the base station downlink scrambling code c_{sc} belongs.

The secondary SCH is modulated with a binary sequence length of 16 bits, which is repeated for each frame. The modulation sequence, which is the same for all base stations, has good cyclic autocorrelation properties.

The multiplexing of the SCH with the other downlink physical channels (DPDCH/DPCCH and CCPCH) is illustrated in Figure 3.6.

The SCH is transmitted only intermittently (one codeword per slot), and it is multiplexed with the DPDCH/DPCCH and CCPCH after long code scrambling is applied on DPDCH/DPCCH and CCPCH. Consequently, the SCH is nonorthogonal to the other downlink physical channels.

There is only one type of downlink dedicated physical channel, the downlink dedicated physical channel (DPCH). Within one downlink DPCH, data is transmitted in time-multiplex with control information generated at layer 1, such as known pilot bits, TPC commands, and an optional transport format combination indicator (TFCI).

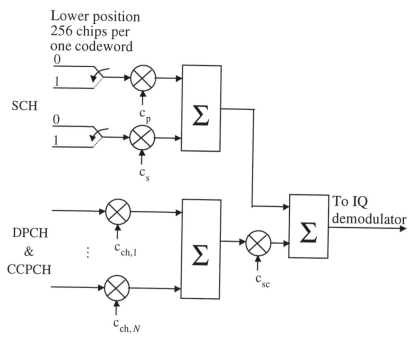

Lower position
256 chips per
one codeword

Figure 3.6 Multiplexing of the SCH (s_p = primary spreading code, s_c = secondary spreading code, c_{ch} = orthogonal code, and c_{sc} = long scrambling code).

3.1.4 Spreading

The WCDMA scheme employs long spreading codes. Different spreading codes are used for cell separation in the downlink and user separation in the uplink. In the downlink, Gold codes of length 218 are used, but they are truncated to form a cycle of a 10-ms frame. The total number of available scrambling codes is 512, divided into 32 code groups with 16 codes in each group to facilitate a fast cell search procedure. In the uplink, either short or long spreading (scrambling codes) are used. The short codes are used to ease the implementation of advanced multiuser receiver techniques; otherwise long spreading codes are used. Short codes are VL-Kasami codes of length 256 and long codes are Gold sequences of length 241, but the latter are truncated to form a cycle of a 10-ms frame.

For channelization, orthogonal codes are used. Orthogonality between the different spreading factors is achieved by the tree-structured orthogonal codes whose construction was described in Chapter 5.

IQ/code multiplexing leads to parallel transmission of two channels, and therefore, attention must be paid to modulated signal constellation and related peak-to-average power ratio (crest factor). By using the complex spreading circuit shown in Figure 3.7, the transmitter power amplifier efficiency remains the same as for QPSK transmission in general.

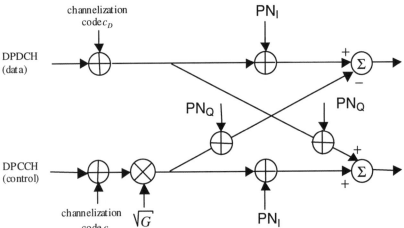

Figure 3.7 IQ/code multiplexing with complex spreading circuit.

Moreover, the efficiency remains constant irrespective of the power difference G between DPDCH and DPCCH. This can be explained with Figure 3.8, which shows the signal constellation for IQ/code multiplexed control channel with complex spreading. In the middle constellation with $G = 0.5$, all eight constellation points are at the same distance from the origin. The same is true for all values of G. Thus, signal envelope variations are very similar to the QPSK transmission for all values of G. The IQ/code multiplexing solution with complex scrambling results in power amplifier output backoff requirements that remain constant as a function of power difference. Furthermore, the achieved output backoff is the same as for one QPSK signal.

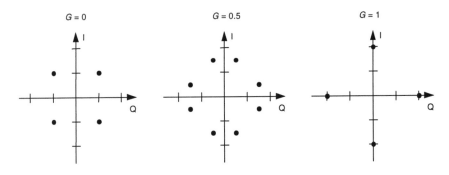

Figure 3.8 Signal constellation for IQ/code multiplexed control channel with complex spreading. G is the power difference between DPCCH and DPDCH.

3.1.5 Multirate

Multiple services of the same connection are multiplexed on one DPDCH. After service multiplexing and channel coding, the multiservice data stream is mapped to one DPDCH. If the total rate exceeds the upper limit for single code transmission, several DPDCHs are allocated.

A second alternative for service multiplexing is to map parallel services to different DPDCHs in a multicode fashion with separate channel coding and interleaving. With this alternative scheme, the power, and consequently, the quality of each service, can be separately and independently controlled.

For BER = 10^{-3} services, convolutional coding of 1/3 is used. For high bit rates, a code rate of 1/2 can be applied. For higher quality service classes requiring quality of service between 10^{-3} and 10^{-6} BER inclusive, parallel concatenated convolutional code is used. Retransmissions can be utilized to guarantee service quality for nonreal-time packet data services.

After channel coding and service multiplexing, the total bit rate can be almost arbitrary. The rate matching adapts this rate to the limited set of possible bit rates of a DPDCH. Repetition or puncturing is used to match the coded bit stream to the channel gross rate. The rate matching for uplink and downlink are introduced below.

For the uplink, rate matching to the closest uplink DPDCH bit rate is always based on unequal repetition (a subset of the bits repeated) or code puncturing. In general, code puncturing is chosen for bit rates less than ≈20% above the closest lower DPDCH bit rate. For all other cases, unequal repetition is performed to the closest higher DPDCH bit rate. The repetition/puncturing patterns follow a regular predefined rule (i.e., only the amount of repetition/puncturing needs to be agreed on). The correct repetition/puncturing pattern is then directly derived by both the transmitter and receiver side.

For the downlink, rate matching to the closest DPDCH bit rate, using either unequal repetition or code puncturing, is only made for the highest rate (after channel coding and service multiplexing) of a variable rate connection and for fixed-rate connections. For lower rates of a variable rate connection, the same repetition/puncturing pattern as for the highest rate is used, and the remaining rate matching is based on DTX where only a part of each slot is used for transmission. This approach is used to simplify the implementation of blind rate detection in the mobile station.

3.1.6 Packet Data

WCDMA has two different types of packet data transmission possibilities. Short data packets can be appended directly to a random access burst. This method, called *common channel packet transmission*, is used for short infrequent packets, where the link maintenance needed for a dedicated channel would lead to an unacceptable overhead.

When using the uplink common channel, a packet is appended directly to a random access burst. Common channel packet transmission is typically used for short, infrequent packets, where the link maintenance needed for a dedicated channel would lead to unacceptable overhead. Also, the delay associated with a transfer to a dedicated channel is avoided. Note that for common channel packet transmission, only open loop power control is in operation. Common channel packet transmission should therefore be limited to short packets that only use a limited capacity. Figure 3.9 illustrates packet transmission on a common channel.

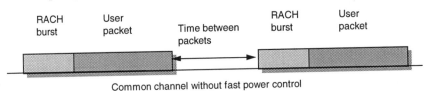

Figure 3.9 Packet transmission on the common channel.

Larger or more frequent packets are transmitted on a dedicated channel. A large single packet is transmitted using a single-packet scheme where the dedicated channel is released immediately after the packet has been transmitted. In a multipacket scheme, the dedicated channel is maintained by transmitting power control and synchronization information between subsequent packets.

3.1.7   Handover

Base stations in WCDMA need not be synchronized, and therefore, no external source of synchronization, like global positioning system (GPS), is needed for the base stations. Asynchronous base stations must be considered when designing soft handover algorithms and when implementing position location services.

3.1.7.1   Soft Handover

Before entering soft handover, the mobile station measures observed timing differences of the downlink SCHs from the two base stations. The structure of SCH is presented in Section 3.1.3. The mobile station reports the timing differences back to the serving base station. The timing of a new downlink soft handover connection is adjusted with a resolution of one symbol (i.e., the dedicated downlink signals from the two base stations are synchronized with an accuracy of one symbol). That enables the mobile RAKE receiver to collect the macrodiversity energy from the two base stations. Timing adjustments of dedicated downlink channels is carried out with a resolution of one symbol without losing orthogonality of downlink codes. Details are discussed in [2].

3.1.7.2   Interfrequency Handovers

Interfrequency handovers are needed for utilization of hierarchical cell structures; macro, micro, and indoor cells. Several carriers and interfrequency handovers may also be used for taking care of high capacity needs in hot spots. Interfrequency handovers are also needed for handovers to second generation systems, like GSM or IS-95. In order to complete interfrequency handovers, an efficient method is needed for making measurements on other frequencies while still having the connection running on the current frequency. Two methods are considered for interfrequency measurements in WCDMA: dual receiver and slotted mode.

The dual receiver approach is considered suitable, especially if the mobile terminal employs antenna diversity. During the interfrequency measurements, one receiver branch is switched to another frequency for measurements, while the other keeps receiving from the current frequency. The loss of diversity gain during measurements must be compensated for with higher downlink transmission power. The advantage of the dual receiver approach is that there is no break in the current frequency connection. Fast closed loop power control is running all the time.

The slotted mode approach depicted in Figure 3.10 is considered attractive for the mobile station without antenna diversity. The information normally transmitted during a 10 ms frame is compressed time either by code puncturing or by changing the FEC rate.

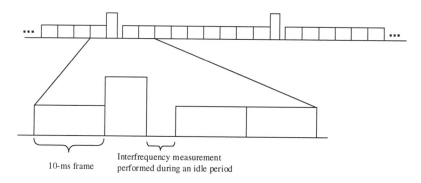

Figure 3.10 Slotted mode structure.

3.1.8   Interoperability Between GSM and WCDMA

The handover between the WCDMA system and the GSM system, already offering worldwide coverage today, has been one of the main design criterion, taken into account in the WCDMA frame timing definition. The GSM compatible multiframe structure, with the superframe being a multiple of 120 ms, allows similar timing for intersystem measurements as in the GSM system itself. Apparently, the needed measurement interval does not need to be as frequent as for GSM terminal operating in a GSM system, as intersystem handover is less critical from an intrasystem interference point of view. Rather, the compatibility in timing is so important, that when operating in WCDMA mode, a multimode terminal is able to catch the desired information from the synchronization bursts in the synchronization frame on a GSM carrier with the aid of frequency correction burst. This way, the relative timing between GSM and WCDMA carriers is maintained similar to the timing between two asynchronous GSM carriers. The timing relation between WCDMA channels and GSM channels is indicated in Figure 3.11, where the GSM traffic channel and WCDMA channels use similar 120 ms multiframe structure. The GSM frequency correction channel (FCCH) and GSM SCH use one slot out of the eight GSM slots in the indicated frames, with the FCCH frame with one time slot for FCCH always preceding the SCH frame with one time slot for SCH, as indicated in Figure 3.11.

A WCDMA terminal can do the measurements either by requesting the measurement intervals in a form of a slotted mode, where there are breaks in the downlink transmission, or then it can perform the measurements independently with a suitable measurement pattern. With independent measurements, the dual receiver approach is used instead of the slotted mode, since then the GSM receiver branch can operate independently of the WCDMA receiver branch.

For smooth interoperation between the systems, information needs to be exchanged between the systems, to allow the WCDMA base station to notify the terminal of the existing GSM frequencies in the area. In addition, more integrated operation is needed for the actual handover where the current service is maintained, taking naturally into account the lower data rate capabilities in GSM, when compared to UMTS maximum data rates reaching all the way to 2 Mbps.

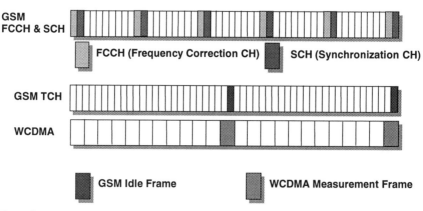

Figure 3.11 Measurements timing relation between WCDMA and GSM frame structures.

The GSM system is likewise expected to also indicate the WCDMA spreading codes in the area to make the cell identification simpler. After that, the existing measurement practices in GSM can be used for measuring the WCDMA when operating in GSM mode.

WCDMA does not rely on any superframe structure as with GSM to find out synchronization, so the terminal operating in GSM mode can obtain the WCDMA frame synchronization once the WCDMA base station scrambling code timing is acquired. The base station scrambling code has a 10 ms period and its frame timing is synchronized to WCDMA common channels.

3.2 PERFORMANCE EVALUATION

Extensive performance evaluation of WCDMA has been carried out in [21]. Here we highlight the main results to demonstrate the advantage of WCDMA. In Table 3.2, spectrum efficiency results of WCDMA are presented for different services. The test cases are the same as used in Chapter 2: speech, LCD and UDD. For more details, refer to [21].

Table 3.2

Spectrum Efficiency Results in Kbps/MHz/cell for UTRA Concepts

Environ- ment	1Service	WCDMA (concept α) Uplink/downlink
	Speech	98 / 78
Vehicular	LCD384	138/85 204/123 (30 dBm MS, 8 Mcps) 138/211 175/211 (30 dBm MS) 204/250 (30 dBm MS, 8 Mcps)
Outdoor-to- indoor pedestrian	UDD384	470 / 565
	Speech	127 / 163 189 / - (C/I-based HO)
Indoor office	Service mix - speech - UDD384	315 / 207 315 / 460 (DL ant div)
	UDD2048	300 / 230 300 / 500 (DL ant div)

3.3 CONCLUSIONS

In this chapter, the physical layer of WCDMA was described. WCDMA has been harmonized with other wideband CDMA modes and provides the full IMT-2000 service set. It also facilitates backward compatibility with major second generation cellular systems such as GSM, IS-95, and PDC.

References

[1] *Harmonized Global 3G (G3G) Technical Framework for ITU IMT-2000 CDMA Proposal*, May, 1999.

[2] T. Ojanperä, and R. Prasad, *Wideband CDMA for Third Generation Mobile Communications*, Norwood, MA: Artech House, 1998.

[3] T. Ojanperä, M. Gudmundson, P. Jung, J. Sköld, R. Pirhonen, G. Kramer, and A. Toskala, "FRAMES – Hybrid Multiple Access Technology," *Proceedings of ISSSTA'96*, Vol. 1, Mainz, Germany, September 1996, pp. 320–324.

[4] T. Ojanperä, P.-O. Anderson, J. Castro, L. Girard, A. Klein and R. Prasad, "A Comparative Study of Hybrid Multiple Access Schemes for UMTS," *Proceedings of ACTS Mobile Summit Conference*, Vol. 1, Granada, Spain, November 1996, pp. 124–130.

[5] Ojanperä, T., J. Sköld, J. Castro, L. Girard and A. Klein, "Comparison of Multiple Access Schemes for UMTS," *Proceedings of VTC'97*, Vol. 2, Phoenix, Arizona, May 1997, pp. 490–494.

[6] T. Ojanperä, A.Klein and P.-O. Anderson, "FRAMES Multiple Access for UMTS," *IEE Colloquium on CDMA Techniques and Applications for Third Generation Mobile Systems*, London, May 1997.

[7] F. Ovesjö, E. Dahlman, T. Ojanperä, A. Toskala, and A. Klein, "FRAMES Multiple Access Mode 2 – Wideband CDMA," *Proceedings of PIMRC97*, Helsinki, Finland, September 1997, pp. 42–46.

[8] E. Nikula, A. Toskala, E. Dahlman, L. Girard, and A. Klein, "FRAMES Multiple Access for UMTS and IMT-2000," *IEEE Personal Communications Magazine*, April 1998, pp. 16–24.

[9] CSEM/Pro Telecom, Ericsson, France Télécom – CNET, Nokia, Siemens, "FMA - FRAMES Multiple Access A Harmonized Concept for UMTS / IMT-2000," *ITU Workshop on Radio Transmission Technologies for IMT-2000*, Toronto, Canada, September 10 – 11, 1997.

[10] ARIB FPLMTS Study Committee, "Report on FPLMTS Radio Transmission Technology SPECIAL GROUP, (Round 2 Activity Report)," Draft v.E1.1, January 1997.

[11] F. Adachi, K. Ohno, M. Sawahashi, and A. Higashi, "Multimedia Mobile Radio Access Based on Coherent DS-CDMA," *Proceedings of 2nd International Workshop on Mobile Multimedia Commun.*, A2.3, Bristol University, UK April 1995.

[12] K. Ohno, M. Sawahashi, and F. Adachi, "Wideband Coherent DS-CDMA," *Proceedings of VTC'95*, Chicago, Illinois, July 1995, pp. 779–783.

[13] T. Dohi, Y. Okumura, A. Higashi, K. Ohno, and F.Adachi, "Experiments on Coherent Multicode DS-CDMA," *Proceedings of VTC'96*, Atlanta, Georgia, April 1996, pp. 889–893.

[14] F. Adachi, M. Sawahashi, and K. Ohno, "Coherent DS-CDMA: Promising Multiple Access for Wireless Multimedia Mobile Communications," *Proceedings of ISSSTA'96*, Mainz, Germany, September 1996, pp. 351–358.

[15] S. Onoe, K. Ohno, K. Yamagata, and T. Nakamura, "Wideband-CDMA Radio Control Techniques for Third Generation Mobile Communication Systems," *Proceedings of VTC97*, Vol. 2, Phoenix, Arizona, May 1997, pp. 835–839.

[16] Y. Han, S. C. Banh, H.-R. Park, and B.-J. Kang, "Performance of Wideband CDMA System for IMT-2000," *2nd CDMA International Conference (CIC)*, Seoul, Korea, October 1997, pp. 583–587.

[17] J.M. Koo, Y. I. Kim, J. H. Ryu, and J. I. Lee, "Implementation of Prototype Wideband CDMA System," *Proceedings of ICUPC'96*, Cambridge, Massachusetts, September/October 1996, pp. 797–800.

[18] Koo, J. M., E. K. Hong, and J. I. Lee, "Wideband CDMA-Technology for FPLMTS," *The 1st CDMA International Conference*, Seoul, Korea, November 1996.

[19] Y.-W. Park, E. K. Hong, T.-Y. Lee, Y.-D. Yang and S.-M. Ryu, "Radio Characteristics of PCS Using CDMA," *Proceedings of IEEE VTC'96*, Atlanta, Georgia, April 1996, pp. 1661–1664.

[20] E.K. Hong, T.-Y. Lee, Y.-D. Yang, B.-C. Ahn and Y.-W. Park, "Radio Interface Design for CDMA-Based PCS," *Proceedings of ICUPC '96*, Cambridge, Massachusetts, September/ October 1996, pp. 365–368.

[21] ETSI UMTS 30.06, "UMTS Terrestrial Radio Access Concept Evaluation," ETSI Technical Report, 1998.

Chapter 4

Advanced Receiver Algorithms

Air-interface specification also depends on the receiver algorithms, since the system performance and capacity depends on the algorithms used. Hence, it is clear that receiver algorithms should be studied in detail before finalizing a system concept [1-33]. In the case of CDMA systems, conventional RAKE receivers are used quite often. A RAKE receiver consists of several branches, each allocated to receive distinct multipath components. Although RAKE receivers require a reasonable amount of hardware (i.e., correlators for despreading the received signals), they are the simplest multipath diversity receivers that can be used in the WCDMA-type systems. Therefore, the WCDMA physical layer has been defined in such a way that the conventional RAKE receivers would give sufficient performance in most cases. It is well known that RAKE receivers suffer from interference caused by other users or data channels. RAKE receivers would require very powerful channel coding to guarantee sufficient performance. By using some other techniques based on interference suppression or cancellation, system capacity can be significantly improved and possibly less channel coding is needed. For those reasons, advanced receiver algorithms capable of suppressing or canceling interference are seen as future enhancements of WCDMA systems.

There are several ways to categorize multiuser detectors (MUDs). For the purposes of this book, three different classifications are used:

- Centralized [11] and decentralized receivers [8];

- Linear [11] and non-linear receivers [1];

- Schemes for short [8] and long spreading sequences [32].

Most of the MUDs are centralized, which means that the signals of all users need to be processed at the receiver. There are, however, some equalizer-type receiver structures, which can be implemented adaptively in decentralized receivers (i.e., only the signal for the user of interest is demodulated).

Linear receivers are multiple-input-multiple-output (MIMO) or single-input-single-output (SISO) equalizers applied to CDMA systems. Both adaptive [9] and nonadaptive [8] techniques exist. Interference cancellation (IC) receivers [1] are the most typical nonlinear receivers. IC receivers can be applied to systems using either short or long spreading codes. Most linear receivers are better suited to short code systems [8-10]. So-called chip equalizers [32] can be applied also in long code systems.

In the FRAMES project, multiuser receivers[1] both for the uplink and downlink have been studied. Most of the known techniques are theoretically interesting but may not be suitable for multipath fading channels. Furthermore, most of them do not provide significant performance improvements in cellular CDMA systems where a large number of users must be served. Based on several published papers [1-5], interference cancellation is the most promising and the most practical technique for uplink receivers. Due to the fact that all user signals are demodulated coherently in WCDMA proposals, the so-called hard-decision parallel interference cancellation (HD-PIC) receiver ([4, 5]) is seen as the most promising IC scheme for the uplink. IC techniques can straightforwardly be applied to multisensor (or adaptive antenna) receivers, which further improves the performance and capacity of CDMA systems.

Most of the MUDs can be applied to uplink receivers ([1, 3]). To assist the application of some advanced receivers also at downlink, FMA2 concept supports both short and long scrambling codes. Short codes are needed in some adaptive decentralized equalizers [7-10] in order to retain cyclostationarity of MAI. Since UTRA FDD does not have optional short scrambling codes, equalizers for chip waveform [32] may be considered to improve the performance at the smallest spreading factors.

The chapter is structured as follows. Section 4.1 presents the limitations of conventional RAKE receivers. Joint detection for TD/CDMA is discussed in Section 4.2. Uplink multiuser detection for CDMA and improved downlink receivers for CDMA are explained in Sections 4.3 and 4.4, respectively. Finally, conclusions are given in Section 4.5.

4.1   LIMITATIONS OF CONVENTIONAL RAKE RECEIVERS

One of the most important requirements for the UMTS is high and variable data rate. In FMA2 and WCDMA systems, it is possible to increase the data rate without bandwidth expansion by reducing the spreading factor. This is referred to as a *variable spreading factor* (VSF) technique. Alternatively, the spreading factor may remain fixed and the data rate is increased by allocating several parallel spreading codes (i.e. data channels) for the same service. This is referred to as a *multicode* (MC) technique. The combination of these two techniques is also supported, which results in hybrid VSF-MC techniques.

In the first phase of the third generation CDMA systems, the receivers will be based on conventional RAKE receivers. From the RAKE receiver perspective, the multirate techniques are quite different. First, separate despreading devices (i.e., correlators) are needed for each data channel. Hence, the MC technique is more hardware-intensive than the VSF technique. Secondly, the mechanism generating interference is quite different.

---

[1] Multiuser receivers are often used to denote both centralized and decentralized receivers.

In the VSF approach, the spreading factor can be as small as 4 in the FMA2 concept. In multipath channels, the small spreading factor will cause so-called *interpath interference* (IPI) due to imperfect spreading sequence autocorrelations. Due to IPI, multipath components are correlated and some diversity is lost, even if the intersymbol interference (ISI) was rather small (e.g., if two multipath components are merging towards one propagation path, there is IPI but almost no ISI and diversity gain is lost). Hence, although ISI may be negligible, IPI can be quite large. The smaller the spreading factor is, the larger the loss due to IPI will be. This phenomenon is illustrated in Figure 4.1. The bit error probability curves were generated by using the characteristic function method described in [7]. Random spreading sequences and one chip delay between the multipath components were used. As we can see, the performance loss is already significant at the signal-to-noise ratio (SNR) of 10 dB. RAKE receivers have traditionally been used in spread-spectrum systems (i.e., in systems with large spreading factors). The future CDMA systems with relatively small spreading factors result in significant performance losses in RAKE receivers due to IPI.

With the MC technique, IPI is rather small due to the higher spreading factor. Instead of IPI, *interchannel interference* (ICI) causes performance degradation in the RAKE receivers in the same way as the multiple-access or multiuser interference in any CDMA system. The bit error probability degradation as a function of the number of parallel data channels is illustrated in Figure 4.2. The results of Figure 4.2 reveal that although the mechanism for generating interference in VSF and MC techniques are quite different, the performance is almost the same when both have the same data rate. This can be seen, for example, by examining the BEPs with different spreading factors at the level of 10^{-2}. When the spreading factor is divided in half, the number of parallel code channels is also halved to obtain the same BEP.

The performance is also similar in a near-far situation. The bit error probabilities for different near-far ratios[2] are presented for a two-user case in Figure 4.3. The near-far resistance is improved approximately by 3 dB when doubling the spreading factor. The performance will therefore be equal with both MC and VSF techniques, assuming the same energy per information bit.

Based on the bit error performance results of both multirate techniques, it can be concluded that the performance of the conventional RAKE receivers is interference limited. With the variable spreading factor approach the interpath interference limits the performance. With the multicode technique, the inter-channel interference is the limiting factor, and both techniques are sensitive to the near-far problem when the RAKE receivers are used. Clearly, the major problem with conventional RAKE receivers is that the performance is degraded as the data rate increases. The performance degradation can be avoided by using some near-far resistant receivers. In the sequel, some near-far resistant receiver techniques both for hybrid TDMA-CDMA systems and WCDMA-type systems will be presented.

---

[2] The near-far ratio is the difference between the power of the desired user and an interfering user.

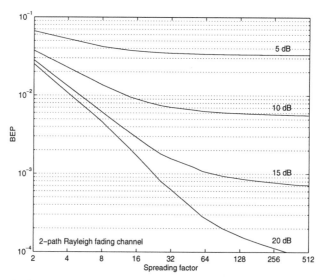

Figure 4.1 BER degradation of the conventional RAKE receiver due to interpath interference in a Rayleigh fading channel, with two paths of the same power as a function of the spreading factor of random sequences, in a single-user system using BPSK modulation with different SNRs.

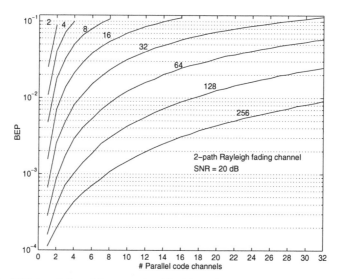

Figure 4.2 BER degradation of the conventional RAKE receiver due to interchannel interference in a Rayleigh fading channel, with two paths of the same power as a function of the number of parallel data channels, in a single-user system using BPSK modulation with different spreading factors (2 - 256) at the SNR of 20 dB.

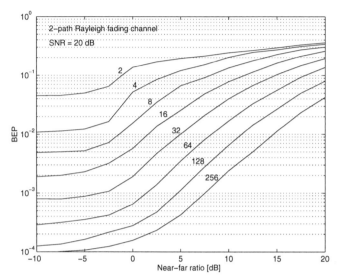

Figure 4.3 BER degradation of the conventional RAKE receiver due to the near-far problem in a Rayleigh fading channel, with two paths of the same power as a function of the near-far ratio for different spreading factors (2 - 256), in a two-user system using BPSK modulation at the SNR of 20 dB for the desired user.

4.2 JOINT DETECTION FOR TD-CDMA

The performance of mobile radio systems applying CDMA is degraded by multiple access interference (MAI), originating in the superposition of user signals, and by ISI caused by multipath propagation. The spectral efficiency of such systems can be significantly increased, if ISI and MAI are mitigated. In systems like TD-CDMA, which, in addition to CDMA utilize TDMA, multiuser detection techniques can be applied that combat these two types of interference.

The important basis for the following considerations is the whitening matched filter (WMF). This filter is introduced in Section 4.2.1. Then two joint detectors, namely the zero-forcing block linear equalizer (ZF-BLE) and the corresponding feedback version; that is, the zero-forcing block decision feedback equalizer (ZF-BDFE), are treated in Sections 4.2.2 and 4.2.3, respectively. The treatment is based on the analytical uplink model of TD-CDMA described in the section *Analytical Model* of Chapter 2. K users are assumed to be simultaneously active, and only single antenna receivers are considered. Two other joint detectors should also be mentioned, namely the minimum mean-square-error block linear equalizer (MMSE-BLE) and the minimum mean-square-error block decision feedback equalizer (MMSE-BDFE), see [12, 13]. For further information on joint detection techniques and their applications, see [12-25].

4.2.1   Whitening Matched Filter

In the section *Analytical Model* of Chapter 2, the system model of TD-CDMA was presented and its discrete-time representation was described. The combined data symbol vector

$$\underline{d} = \left(\underline{d}_1, \underline{d}_2 \ldots \underline{d}_{KN}\right)^{\mathrm{T}} \tag{4.1}$$

that consists of the data symbol blocks of size N of K simultaneously active users, was transmitted. The combined receive vector

$$\underline{e} = \underline{A}\,\underline{d} + \underline{n} \tag{4.2}$$

was derived for the case of multiple receiver antennas, whereas only the single antenna receiver is considered here. The multiuser detection problem

$$\hat{\underline{d}} = \underline{M}\,\underline{e} \tag{4.3}$$

was stated, where $\underline{e} = \underline{A}\,\underline{d} + \underline{n}$ is the received signal, see [12], and $\underline{M}$ is a $(KN) \times (NQ + W - 1)$ matrix that characterizes the data estimation algorithm. The aim of (4.3) is to obtain the estimate

$$\hat{\underline{d}} = \left(\hat{\underline{d}}_1, \hat{\underline{d}}_2 \ldots \hat{\underline{d}}_{KN}\right)^{\mathrm{T}} \tag{4.4}$$

of $\underline{d}$, see (4.1). Based on these derivations, different multiuser data estimators are considered in the following.

The joint detection data estimation techniques presented in Sections 4.2.2 and 4.2.3, which take into account both ISI and MAI, can be interpreted as an extension of the WMF. In the case of the WMF, the continuous valued estimate $\hat{\underline{d}}$ of $\underline{d}$, see (4.1), is given by

$$\hat{\underline{d}}_{WMF} = \underline{M}\,\underline{e} = \underline{A}^{*T} \underline{R}_n^{-1} \underline{e} = \left(\underline{L}^{-1}\underline{A}\right)^{*T} \underline{L}^{-1} \underline{e} \tag{4.5}$$

Here $\underline{L}$ is a lower triangular matrix determined by the Cholesky decomposition, see [13, 26],

$$\underline{R}_n = \underline{L}\,\underline{L}^{*T} \tag{4.6}$$

of the noise covariance matrix $\underline{R}_n$. By the operation $\underline{L}^{-1}\underline{e}$ in (4.5), the received noise $\underline{n}$ is decorrelated or prewhitened, since

$$E\left\{ \underline{L}^{-1}\underline{n}\left(\underline{L}^{-1}\underline{n}\right)^{*T} \right\} = \underline{I} \tag{4.7}$$

The signal vector $\underline{L}^{-1}\underline{e}$ obtained after noise prewhitening is further processed by the operator $\left(\underline{L}^{-1}\underline{A}\right)^{*T}$. This operator corresponds to a filter matched to the response of the prewhitening filter to the sequences $\underline{b}^{(k)}$, $k = 1...K$, see [12]. With $\underline{e} = \underline{A}\,\underline{d} + \underline{n}$, (4.5) can be written in the form

$$
\begin{aligned}
\underline{\hat{d}}_{WMF} &= \underline{A}^{*T} R_n^{-1} \underline{A}\,\underline{d} + \underline{A}^{*T} R_n^{-1} \underline{n} \\
&= \underbrace{diag\left(\underline{A}^{*T} R_n^{-1} \underline{A}\right) \underline{d}}_{\text{desired symbols}} + \underbrace{\overline{diag}\left(\underline{A}^{*T} R_n^{-1} \underline{A}\right) \underline{d}}_{\text{ISI and MAI}} + \underbrace{\underline{A}^{*T} R_n^{-1} \underline{n}}_{\text{noise}}
\end{aligned}
\tag{4.8}
$$

where $\mathbf{diag}(\underline{X})$ is a diagonal matrix containing only the diagonal elements of the matrix $\underline{X}$, and $\overline{\mathbf{diag}}(\underline{X}) = \underline{X} - \mathbf{diag}(\underline{X})$ is a matrix with zero diagonal elements containing all but the diagonal elements of $\underline{X}$. On the right-hand side of (4.8), the first term represents the desired symbols, the second term represents the impact of ISI and MAI, and the third term is due to the received noise $\underline{n}$. The first term is a vector whose jth component is the transmitted symbol $\underline{d}_j$, $\underline{d}_j \in \underline{d}$ (see 4.1), multiplied by a scalar. The term designated as ISI and MAI is a vector whose jth component is a weighted sum of the other transmitted symbols $\underline{d}_{j'}$, $j \neq j'$, $\underline{d}_{j'} \in \underline{d}$. The correlation of the noise term is given by its covariance matrix $\underline{A}^{*T} \underline{R}_n^{-1} \underline{A}$. The signal-to-noise ratio (SNR) per symbol at the output of a detector, including the two disturbing terms, is generally defined as

$$\gamma(k,n) = \frac{\left| E\left\{ \underline{\hat{d}}_n^{(k)} \right\} \right|^2}{E\left\{ \left| \text{ISI} + \text{MAI} + \text{noise} \right|^2 \right\}} \tag{4.9}$$

In the case of the WMF, we obtain

$$\gamma_{\text{WMF}}(k,n)$$

(4.10)

$$= \frac{E\left\{\left|\underline{d}_n^{(k)}\right|^2\right\}\left([\underline{\mathbf{F}}]_{j,j}\right)^2}{[\underline{\mathbf{F}}\,\underline{\mathbf{R}}_d\underline{\mathbf{F}}]_{j,j} - 2\text{Re}\left\{[\underline{\mathbf{F}}\,\underline{\mathbf{R}}_d]_{j,j}\right\}[\underline{\mathbf{F}}]_{j,j} + E\left\{\left|\underline{d}_n^{(k)}\right|^2\right\}\left([\underline{\mathbf{F}}]_{j,j}\right)^2 + [\underline{\mathbf{F}}]_{j,j}}$$

(see [12]). In this expression

$$\underline{F} = \underline{A}^{*T}\,\underline{R}_n^{-1}\,\underline{A}, \ j = n + N(k-1), \ k = 1...K, \ n = 1...N \qquad (4.11)$$

holds, $\underline{R}_d = E\{\underline{d}\,\underline{d}^{*T}\}$ denotes the covariance matrix of $\underline{d}$, and $[\underline{X}]_{j,j}$ represents the element in the jth row and jth column of the matrix $\underline{X}$. Since the matched filter maximizes the SNR for a fixed sampling point for one single transmitted data symbol [13], the maximum SNR per symbol, which cannot be exceeded at the output of any estimator, equals

$$\gamma_{\max}(k,n) = E\left\{\left|\underline{d}_n^{(k)}\right|^2\right\}\left[\underline{A}^{*T}\,\underline{R}_n^{-1}\,\underline{A}\right]_{j,j}$$
$$j = n + N(k-1), \ k = 1...K, \ n = 1...N.$$

(4.12)

Equation (4.12) results from (4.11) by setting the first three terms of the sum in the denominator to zero, which are determined by ISI and MAI. For the special case $\underline{R}_n = \sigma^2\underline{I}$ with $\underline{I}$ the $(NQ+W-1)\times(NQ+W-1)$ identity matrix (i.e., for uncorrelated noise), which can be approximately assumed in practical situations, the continuous valued estimate of (4.5) becomes

$$\hat{\underline{d}}_{\text{WMF}}\Big|_{\underline{R}_n=\sigma^2\underline{I}} = \frac{\underline{A}^{*T}}{\sigma^2}\,\underline{e} \qquad (4.13)$$

In this case, the WMF turns into the conventional matched filter data estimator. If, furthermore, uncorrelated data (i.e., $\underline{R}_d = \underline{I}$) is assumed, the SNR per symbol at the output of the WMF becomes with (4.10)

$$\gamma_{WMF}(k,n)\Big|_{\substack{\underline{R}_n=\sigma^2\underline{I}\\ \underline{R}_d=\underline{I}}}$$

$$= \frac{\left\|\underline{b}^{(k)}\right\|^4}{\left[\underline{A}^{*T}\underline{A}\,\underline{A}^{*T}\underline{A}\right]_{j,j} - \left\|\underline{b}^{(k)}\right\|^4 + \sigma^2\left\|\underline{b}^{(k)}\right\|^2}$$
$$j = n + N(k-1),\ k = 1...K,\ n = 1...N$$

(4.14)

(see [12]). Then the maximum SNR per symbol becomes with (4.12)

$$\gamma_{\max}(k)\Big|_{\substack{\underline{R}_n=\sigma^2\underline{I}\\ \underline{R}_d=\underline{I}}} \stackrel{def}{=} \gamma_{\max}(k,n)\Big|_{\substack{\underline{R}_n=\sigma^2\underline{I}\\ \underline{R}_d=\underline{I}}} = \frac{\left\|\underline{b}^{(k)}\right\|^2}{\sigma^2}$$
$$k = 1...K,\ n = 1...N$$

(4.15)

$\gamma_{MAX}(k)$ is independent of n. Noise whitening followed by matched filtering is only close to the optimum in (4.12), if the ISI and MAI components are negligible, which is the case for large spreading factors (i.e., Q much larger than W), and orthogonal signature sequences $\underline{c}^{(k)}$, $k = 1...K$ (see [12]). Furthermore, the powers of the K received signals should not differ too much. In the following, multiuser detection strategies are presented, which lead to a considerable performance improvement over the WMF and the conventional matched filter data estimator.

4.2.2 Zero-Forcing Block Linear Equalizer

The ZF-BLE investigated in [14] and minimizing the quadratic form

$$\left(\underline{e} - \underline{A}\,\hat{\underline{d}}_{ZF-BLE}\right)^{*T}\underline{R}_n^{-1}\left(\underline{e} - \underline{A}\,\hat{\underline{d}}_{ZF-BLE}\right)$$

(see [13]), leads to a continuous valued unbiased estimate

$$\hat{\underline{d}}_{ZF\text{-}BLE} = \left(\underline{A}^{*T}\underline{R}_n^{-1}\underline{A}\right)^{-1}\underline{A}^{*T}\underline{R}_n^{-1}\underline{e}$$
$$= \underbrace{\underline{d}}_{\text{desired symbol}} + \underbrace{\left(\underline{A}^{*T}\underline{R}_n^{-1}\underline{A}\right)^{-1}\underline{A}^{*T}\underline{R}_n^{-1}\underline{n}}_{\text{noise}}$$

(4.16)

of $\underline{d}$ (see [13]), containing no ISI and MAI components, but only the desired symbols and a noise term. The equalizer leading to the estimate $\hat{\underline{d}}_{ZF\text{-}BLE}$ is termed zero-forcing, since it totally eliminates ISI and MAI irrespective of the noise level. As a presupposition of the ZF-BLE, the matrix $\underline{A}^{*T}\underline{R}_n^{-1}\underline{A}$ has to be nonsingular, which is not necessarily the case in all conceivable situations. However, when considering real-world, time-varying channel impulse responses

$\underline{h}^{(k)}$, $k = 1...K$, and when choosing the signature sequences $\underline{c}^{(k)}$, $k = 1...K$, properly, the probability of a singular matrix $\underline{A}^{*T}\underline{R}_n^{-1}\underline{A}$ is zero. The correlation of the noise term is given by its covariance matrix $(\underline{A}^{*T}\underline{R}_n^{-1}\underline{A})^{-1}$. The variance of the noise is generally larger than the variance of the noise term in (4.8). With the Cholesky decomposition

$$\underline{A}^{*T} \underline{R}_n^{-1} \underline{A} = \left(\Sigma \underline{H} \right)^{*T} \Sigma \underline{H} \qquad (4.17)$$

(see [12, 13]), of the matrix $\underline{A}^{*T}\underline{R}_n^{-1}\underline{A}$ with $\underline{H}$ an upper triangular matrix with ones along the diagonal and Σ a diagonal matrix with real entries only, the estimate $\hat{\underline{d}}_{ZF\text{-}BLE}$ of (4.16) becomes with (4.5) (see [12], [13]),

$$\hat{\underline{d}}_{ZF\text{-}BLE} = \underbrace{\left(\Sigma \underline{H} \right)^{-1}}_{\substack{\text{ISI and MAI} \\ \text{estimator}}} \underbrace{\left(\underline{H}^{*T} \Sigma \right)^{-1}}_{\substack{\text{whitening} \\ \text{filter}}} \underbrace{\underline{A}^{*T} \underline{R}_n^{-1} \underline{e}}_{\substack{\text{whitening} \\ \text{matched filter}}} \qquad (4.18)$$

$$= \left(\Sigma \underline{H} \right)^{-1} \left(\underline{H}^{*T} \Sigma \right)^{-1} \hat{\underline{d}}_{WMF}$$

ZF-BLE can be interpreted as an extension of the WMF described in Section 4.2.1 by a whitening filter and an ISI and MAI eliminator. If no ISI is present, the ZF-BLE turns into the decorrelating detector described in [13]. The SNR per symbol at the output of the ZF-BLE is with (4.9) equal to

$$\gamma_{ZF\text{-}BLE}(k,n) = \frac{E\left\{ \left| \underline{d}_d^{(k)} \right|^2 \right\}}{\sigma^2 [(\underline{A}^{*T} \underline{R}_n^{-1} \underline{A})^{-1}]_{j,j}} \qquad (4.19)$$
$$j = n + N(k-1), \, k = 1...K, \, n = 1...N$$

(see [12, 13]). These SNRs are generally larger than $\gamma_{WMF}(k,n)$ of (4.10). For the special case $\underline{R}_n = \sigma^2 \underline{I}$, the continuous valued unbiased estimate of (4.16) becomes

$$\hat{\underline{d}}_{ZF\text{-}BLE} \bigg|_{R_n = \sigma^2 \underline{I}} = \left(\underline{A}^{*T} \underline{A} \right)^{-1} \underline{A}^{*T} \underline{e} \qquad (4.20)$$

If, additionally, $E\{ | \underline{d}_n^{(k)} |^2 \} = 1$ is assumed, the SNR per symbol of (4.19) becomes (see [12, 13]),

$$\gamma_{ZF\text{-}BLE}(k,n) \bigg|_{\substack{R_n = \sigma^2 \underline{I} \\ E\{|\underline{d}_n^{(k)}|^2\}=1}} = \frac{1}{\sigma^2 [(\underline{A}^{*T} \underline{A})^{-1}]_{j,j}} \qquad (4.21)$$
$$j = n + N(k-1), \, k = 1...K, \, n = 1...N$$

4.2.3   Zero-Forcing Block Decision Feedback Equalizer

For the description of ZF-BDFE, starting from $\underline{e} = \underline{A}\,\underline{d} + \underline{n}$ and (4.17), the modified received sequence

$$
\begin{aligned}
\underline{e}' &= \underline{H}\left(\underline{A}^{*\mathrm{T}}\underline{R}_n^{-1}\underline{A}\right)^{-1}\underline{A}^{*\mathrm{T}}\underline{R}_n^{-1}\underline{e} \\
&= \Sigma^{-1}\,\Sigma^{-1}\left(\underline{H}^{*\mathrm{T}}\right)^{-1}\underline{A}^{*\mathrm{T}}\underline{R}_n^{-1}\underline{e} \\
&= \underline{d} + (\underline{H} - \underline{I})\underline{d} + \Sigma^{-1}\,\Sigma^{-1}\left(\underline{H}^{*\mathrm{T}}\right)^{-1}\underline{A}^{*\mathrm{T}}\underline{R}_n^{-1}\underline{n}
\end{aligned}
\tag{4.22}
$$

is introduced (see [12, 13]). The noise term in (4.22) is uncorrelated and has the covariance matrix $\Sigma^{-1}\Sigma^{-1}$. Taking into account the fact that the matrix $\underline{H}$ is upper triangular, the decision on the symbol d_j is obtained by using past decisions on previous symbols. If decisions are made in reverse order of the index j of the components of $\underline{d}$, the term $(\underline{H} - \underline{I})\,\underline{d}$ in (4.22) depends only on already decided symbols. Decisions are made according to the recursive formula

$$
\hat{\underline{d}}_{\mathrm{ZF\text{-}BDFE},KN} = Q\{\underline{e}'_{KN}\}
$$

$$
\begin{aligned}
&\hat{\underline{d}}_{\mathrm{ZF\text{-}BDFE},KN-j'} \\
&= Q\left\{\underline{e}'_{KN-j'} - \sum_{j''=1}^{j'}[\underline{H} - \underline{I}]_{KN-j',\,KN-j'+j''}\cdot\hat{\underline{d}}_{\mathrm{ZF\text{-}BDFE},KN-j'+j''}\right\} \\
&j' = 1\ldots(KN-1)
\end{aligned}
\tag{4.23}
$$

(see [12, 13]), with $Q\{\cdot\}$ being a quantization operation performed in a threshold detector. In the case of coded transmission, it would be possible to generate soft inputs for the decoder by using the quantized estimates for decision feedback and the continuous valued estimates before quantization for decoding, possibly together with some channel state information. Subtracting the term $(\underline{H} - \underline{I})\hat{\underline{d}}_{\mathrm{ZF\text{-}BDFE}}$, which contains the estimates $\hat{\underline{d}}_{\mathrm{ZF\text{-}BDFE}}$, from (4.22), the vector

$$
\underline{t} = \underline{d} + (\underline{H} - \underline{I})\left(\underline{d} - \hat{\underline{d}}_{ZF-BDFE}\right) + \Sigma^{-1}\,\Sigma^{-1}\left(\underline{H}^{*T}\right)^{-1}\underline{A}^{*T}\underline{R}_n^{-1}\underline{n}
\tag{4.24}
$$

is obtained, which turns into

$$
\underline{t} = \underline{d} + \Sigma^{-1}\,\Sigma^{-1}\left(\underline{H}^{*T}\right)^{-1}\underline{A}^{*T}\underline{R}_n^{-1}\underline{n}
\tag{4.25}
$$

if all past decisions are correct (see [12, 13]). The decision on the data symbol $\underline{d}_j$ is obtained by means of a threshold detector with input $\underline{t}_j$. The operation $(\underline{H} - \underline{I})\hat{\underline{d}}_{ZF-BDFE}$ in (4.24) constitutes the feedback operator. In [27], it is shown that the ZF-BDFE is equivalent to a noise canceling detector derived from $\underline{H}^{-1}\underline{e}'$, cf. (4.22). If continuous valued estimates of $\underline{d}_j$, $j = 2...KN$, are fed back, which are obtained by omitting threshold detection, ZF-BDFE turns into ZF-BLE. The SNR per symbol at the output of ZF-BDFE, is with (4.9), equal to

$$\gamma_{ZF\text{-BDFE}}(k,n) = E\left\{\left|\underline{d}_n^{(k)}\right|^2\right\}\left([\Sigma]_{j,j}\right)^2 \tag{4.26}$$

$$j = n + N(k-1), \; k = 1...K, \; n = 1...N$$

if all past decisions are assumed to be correct. The SNR $\gamma_{ZF\text{-BDFE}}(k,n)$ is in general larger than the $\gamma_{ZF\text{-BLE}}(k,n)$ of (4.19). If a past decision is incorrect, error propagation may occur, whose impairing effect can be reduced by channel sorting (i.e., by reordering the vector $\underline{d}$ in (4.1)), and accordingly the matrix $\underline{A}$ (see [12]), in such a way that decisions are first performed on the more reliable symbols. The derivation of (4.22), (4.24), (4.25), and (4.26) for the special case $\underline{R}_n = \sigma^2 \underline{I}$ and $E\{|\underline{d}_n^{(k)}|^2\}=1$ is straightforward.

4.3   UPLINK MULTIUSER DETECTORS FOR CDMA

In this section, the parallel interference cancellation (PIC) receivers are presented. The following issues are covered: the system model, parallel interference cancellation receiver principles, and residual interference suppression in PIC receivers.

4.3.1   System Model

A standard model for an asynchronous BPSK DS-CDMA system with K users and L propagation paths over M symbol intervals will be considered. The received signal is time-discretized by antialias filtering and sampling at the rate NR_c, where N is the number of samples per chip and R_c is the chip rate:

$$\mathbf{r} = \mathbf{SAb} + \mathbf{n} = \mathbf{SBa} + \mathbf{n} \tag{4.27}$$

where $\mathbf{r} = \left[\mathbf{r}^{T_{(0)}}, \mathbf{r}^{T_{(1)}}, \ldots, \mathbf{r}^{T_{(M)}}\right]^T$ is $(NG(M+1)) \times 1$ sample vector,

$$S = \begin{pmatrix} \mathbf{S}^{(0)}(0) & \mathbf{0} & \cdots & \mathbf{0} \\ \vdots & \mathbf{S}^{(1)}(0) & \ddots & \vdots \\ \mathbf{S}^{(0)}(D) & \vdots & \ddots & \mathbf{0} \\ \mathbf{0} & \mathbf{S}^{(1)}(D) & \ddots & \mathbf{S}^{(M-1)}(0) \\ \vdots & \ddots & \ddots & \vdots \\ \mathbf{0} & \cdots & \mathbf{0} & \mathbf{S}_U^{(M-1)}(D) \end{pmatrix} \qquad (4.28)$$

is $(NG(M+1)) \times (KLM)$ code matrix, $D = \left[\dfrac{T+T_m}{T} \right]$, T_m is the maximum delay spread, $\mathbf{S} = \left[\mathbf{s}_{1,1}^{(m)}, \ldots, \mathbf{s}_{1,L}^{(m)}, \ldots, \mathbf{s}_{K,L}^{(m)} \right]$ and $\mathbf{s}_{k,l}^{(m)}$ is the time-shifted and sampled signature sequence for the kth user lth path,

$$\mathbf{B} = \begin{pmatrix} \mathbf{B}^{(0)} & \cdots & \mathbf{0}_{KL} \\ \vdots & \ddots & \vdots \\ \mathbf{0}_{KL} & \cdots & \mathbf{B}^{(M-1)} \end{pmatrix} \qquad (4.29)$$

is diagonal $(KLM) \times (KLM)$ data matrix with

$$B^{(m)} = \begin{pmatrix} b_1^{(m)} I_L & \cdots & 0_L \\ \vdots & \ddots & \vdots \\ 0_L & \cdots & b_K^{(m)} I_L \end{pmatrix} \qquad (4.30)$$

where $\mathbf{I}_L$ is $L \times L$ identity matrix, $\mathbf{a} = \left[\mathbf{a}^{T_{(0)}}, \mathbf{a}^{T_{(1)}}, \ldots, \mathbf{a}^{T_{(M-1)}} \right]^T$ is $(KLM) \times 1$ channel coefficient vector with $\mathbf{a}^{T_{(m)}} = \left[\alpha_{1,1}^{(m)}, \ldots, \alpha_{1,L}^{(m)}, \ldots, \alpha_{K,L}^{(m)} \right]$, $\alpha_{k,l}(t) = A_{k,l}(t) e^{j\theta_{k,l}(t)}$ is the complex attenuation factor of the kth user's lth path, and $\mathbf{n}$ is $(NG(M+1)) \times 1$ noise vector. The correlation matrix can now be formed as

$$R = S^H S = \begin{pmatrix} R^{(0,0)} & \cdots & R^{(0,D)} & 0_{KL} \cdots & & 0_{KL} \\ \vdots & \ddots & \ddots & \ddots & & \vdots \\ R^{(D,0)} & \ddots & \ddots & \ddots & & 0_{KL} \\ 0_{KL} & \ddots & \ddots & \ddots & & R^{(M-D,M-1)} \\ \vdots & \ddots & \ddots & \ddots & & \vdots \\ 0_{KL} & \cdots & 0_{KL} & & & R^{(M-1,M-1)} \end{pmatrix} \qquad (4.31)$$

where $\mathbf{R}^{(m,m-j)} = \displaystyle\sum_{i=0}^{D-j} \mathbf{S}^{\mathrm{H}(m)}(i)\mathbf{S}^{(m-j)}(i+j), \quad j \in \{0,...,D\}.$

4.3.2   Parallel Interference Cancellation Receiver Principles

There are two possibilities to implement PIC receivers in practice. Interference cancellation can take place either before or after matched filtering. The block diagrams of the two options are given in Figure 4.4 and Figure 4.5. The receivers are mathematically equivalent and have the same performance. However, the receiver structure of Figure 4.5 enables the use of near-far resistant delay estimators without additional complexity if the delay estimator uses the signal after interference cancellation. The algorithm derivations given in the remainder of the chapter are based mainly on the receiver structure of Figure 4.4 to keep notations as simple as possible. Nevertheless, the receiver algorithms can be applied to both receivers regardless of the notations used.

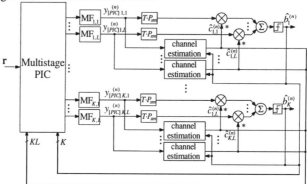

Figure 4.4   Interference cancellation after matched filtering.

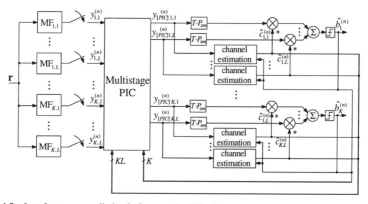

Figure 4.5   Interference cancellation before matched filtering.

There are two possibilities for the MAI estimation in PIC receivers. The first approach is to use the soft-decisions in MAI estimation, which results in the so-called SD-PIC receivers [1].

$$\hat{\mathbf{\Psi}}^{(m)} = \sum_{i=-D}^{D} \left(\hat{\mathbf{R}}^{(m,m+i)} - \delta_{i,0}\mathbf{I}_{KL} \right) \overline{\mathbf{A}\mathbf{b}}^{(m+i)} \qquad (4.32)$$

The SD-PIC receivers use channel coefficient data symbol products in MAI estimation. Hence, the channel coefficients are not estimated in the receiver, which usually are differentially coherent. It was shown in [2] that the SD-PIC receiver with the infinite number of cancellation stages is actually a decorrelating receiver. In order to obtain performance improvement with respect to the optimum SD-PIC receiver with infinite number of stages, which has the same performance as the decorrelator [2], some nonlinearities are required in interference cancellation. The nonlinearities are also needed to reduce the bias of the MAI estimates [3].

The third generation WCDMA systems will also use coherent receivers at the base stations, hence the channel coefficients need to be estimated. Estimated channel coefficients can be used to improve the efficiency of interference cancellation due to more accurate MAI estimates. The MAI estimates in the so-called HD-PIC receivers [4, 5] are expressed as

$$\hat{\mathbf{\Psi}}^{(m)} = \sum_{i=-D}^{D} \left(\hat{\mathbf{R}}^{(m,m+i)} - \delta_{i,0}\mathbf{I}_{KL} \right) \hat{\mathbf{A}}^{(m+i)}\hat{\mathbf{b}}^{(m+i)} \qquad (4.33)$$

The HD-PIC receiver is a natural choice for nonlinear MAI estimation and cancellation (i.e., bias reduction in PIC receivers). The performance of the HD-PIC-based algorithms depends on the quality of the MAI estimates, which can be degraded by inaccurate channel coefficient estimates, data estimates, delay estimates, or incomplete system model. In the sequel, only HD-PIC receivers are considered.

In PIC-based schemes, an estimate of MAI is subtracted from the matched filter outputs ($\mathbf{z}^{(m)} = \mathbf{A}^{(m)}\mathbf{b}^{(m)} + \mathbf{\Psi}^{(m)} + \mathbf{n}^{(m)}$) before detection. The MAI estimates during the mth symbol are written as

$$\hat{\mathbf{\Psi}}^{(m)} = \sum_{i=-D}^{D} \left(\hat{\mathbf{R}}^{(m,m+i)} - \delta_{i,0}\mathbf{I}_{KL} \right) \hat{\mathbf{A}}^{(m+i)}\hat{\mathbf{b}}^{(m+i)} \qquad (4.34)$$

Tentative data decisions are needed to form the MAI estimates. They are obtained from the earlier stages of the multistage detector. The data decisions in the case of BPSK modulation at the $(p+1)$th stage are written as

$$\hat{\mathbf{b}}^{(m)}(p) = \text{sgn}\left[\text{Re}\left\{ \mathbf{C}\hat{\mathbf{A}}^{H\,(m)}(p)\left(\mathbf{z}^{(m)} - \hat{\mathbf{\Psi}}^{(m)}(p) \right) \right\} \right] \qquad (4.35)$$

where $\mathbf{C} = \mathbf{I}_K \otimes \mathbf{1}_L^T$ is the multipath combining matrix, and $\otimes$ denotes the Kronecker product.

4.3.3   Numerical Examples

The performance results of the HD-PIC receivers in the case of mix of services are presented in the sequel. The uplink dedicated physical channel structure as specified in for the UTRA FDD uplink was used with multirate transmission implemented with the VSF principle. A Large Kasami sequence family with processing gain 255 extended with one random chip was used for scrambling. The fading channel was assumed to contain two equal-gain Rayleigh fading taps corresponding to the vehicle speed 80 km/h. Intercell interference was not modeled.

The receiver performance is evaluated both with ideal channel gain information and with linear channel estimators. The bit-error rates in a multirate case with 64 users with spreading factor (SF) 128 and (Figure 4.7) four users with SF 16 (Figure 4.8) were present in the system. Channel gain information is assumed known in the receiver. We see, that with good channel estimation, the HD-PIC is able to provide significant performance improvement from that of the RAKE detector, which is used as the initial stage of the receiver. The same multirate scenario was evaluated with channel estimation applied in the receiver, with results illustrated in Figure 4.9 and Figure 4.10. The channel estimator used was a two-stage estimator presented in Figure 4.6.

The scheme combines the data-aided and decision-directed channel estimators. Pilot symbols are used when they are available, otherwise decision-direction is used. The channel estimator consists of a predictor and a smoother [4]. The predictor provides channel estimates to be used for making the tentative data decisions, which are needed to remove the effect of data modulation when there is no pilot symbol available. The channel estimates produced by the predictor are also fed back to the previous receiver stages for MAI estimation. The channel estimates obtained from the smoother are applied in a maximal ratio combiner, and the final data decisions are made. The estimators lengths used were 10 symbols for the predictor and 21 for the smoother. With higher data rates, the lengths scale accordingly. The lowest curves in Figure 4.9 and Figure 4.10 show the extra gain from the good estimation accuracy of the smoother. Channel information accuracy can be seen to directly affect the receiver performance.

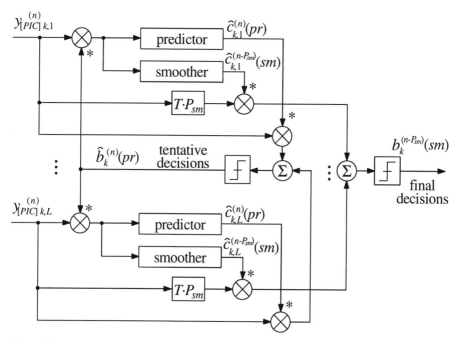

Figure 4.6 Two-stage channel estimator.

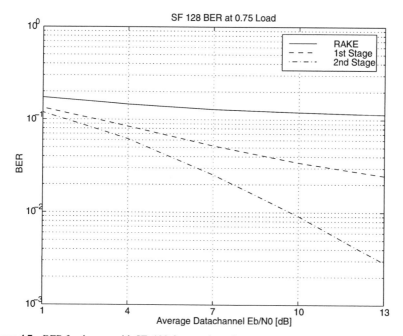

Figure 4.7 BER for the user with SF=128; known channel.

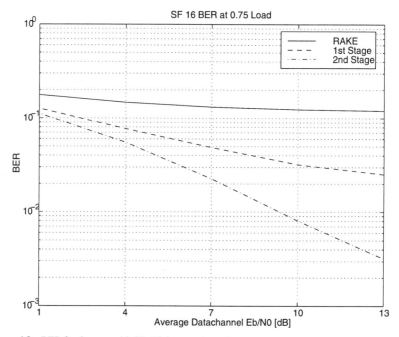

Figure 4.8   BER for the user with SF=16; known channel.

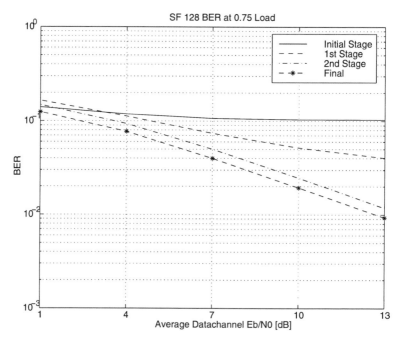

Figure 4.9   BER for the user with SF=128; estimated channel.

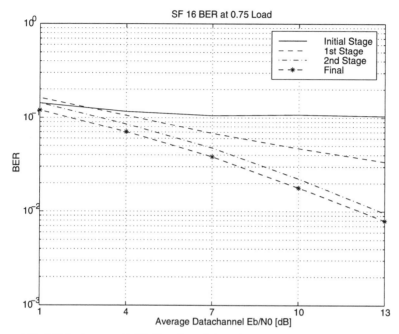

Figure 4.10 BER for the user with SF=16; estimated channel.

4.3.4 Residual Interference Suppression in PIC Receivers

The parallel interference cancellation relies on the knowledge of the number of users and propagation paths. In practice, the exact number of users and paths is not known exactly (e.g., due to intercell interference, unknown propagation paths, or new users trying to connect to the base station). As a result, there will be some residual interference even with perfect cancellation of the known signal components. In many cases, the intercell interference can be large enough to significantly degrade the performance of the PIC receivers. For that reason, residual interference suppression is crucial to guarantee that the PIC receivers can operate reliably. In this section, one possibility for residual interference suppression in PIC receivers is considered. The approach taken is to combine the linear minimum mean-squared error (LMMSE)-RAKE [7] (see Section 4.4.1.1) and PIC receivers. In the so-called LMMSE-PIC receiver, the conventional matched filters are replaced with the LMMSE-RAKE filters.

4.3.4.1 Numerical Examples

The channel model used was a two-path equal energy Rayleigh fading channel with vehicle speeds of 80 km/h and a carrier frequency of 2.0 GHz. The data rate was assumed to be 16 Kbps and the maximum delay spread was one symbol interval. Gold codes of length 31 chips were used, the receiver sampling rate was one sample per chip. There were 16 known users at the SNR of 15 dB and one unknown user. The power of the unknown user was between [-20, 20] dB in comparison to the other users. The blind LS receiver was used in the HD-PIC receiver to suppress residual interference due to unknown signal components. The sample-covariance was estimated recursively by using a forgetting factor value $\gamma=0.999$. Direct inversion of the sample-covariance matrix was performed once per hundred data symbols to speed up simulations.

The BER results for the basic HD-PIC receiver are presented in Figure 4.11 for the average SNRs of 5, 10, and 15 dB as a function of the unknown user power difference with respect to the synchronized users. The results reveal that the BER is degraded significantly with an unknown user with a 10-dB higher power. At SNRs higher than 10 dB, an unknown user with the same power as the synchronized known users is sufficient to cause a significant performance degradation.

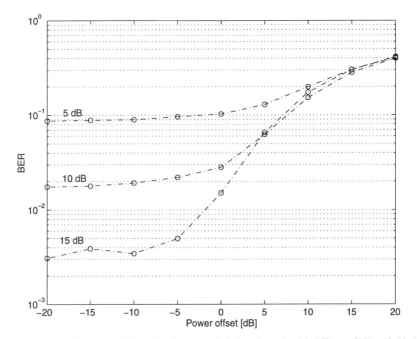

Figure 4.11 BER for the HD-PIC receiver in two-path fading channels with different SNRs (5, 10, 15 dB), as a function of the unknown user power offset with respect to the known users, K = 16 + 1.

The BER results as a function of the number of users at the SNR of 15 dB and unknown user power offset of 10 dB are presented in Figure 4.12. The basic HD-PIC receiver has intolerably high BER in all cases studied. The BERs of the LMMSE-PIC receivers with span one and three are almost the same up to 16 active users. Increasing the number of users up to 32 reveals that the receiver with span three has significantly lower BER than the receiver with span one symbol interval. In fact, the BER of the LMMSE-PIC with span one is almost as high as with the basic HD-PIC with 32 active users. In the case of 28 active users, the BER of the receiver with span one is significantly smaller than with the basic HD-PIC receiver. Hence, the hybrid LMMSE-PIC receiver with blind interference suppression filters of span one symbol interval at every receiver stage can be used to achieve robustness against unsynchronized users. The performance gains with it are significant when the number of active users is less than the processing gain of the system.

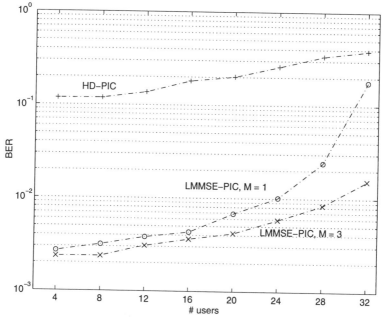

Figure 4.12 BER for the HD-PIC and LMMSE-PIC receivers with span one (M= 1) and three (M= 3) symbol intervals in two-path fading channels at SNR of 15 dB, as a function of the number of users with one unknown user of 10-dB higher power.

4.4 IMPROVED DOWNLINK RECEIVERS FOR CDMA

In a CDMA system with relatively short spreading sequences such that interference is cyclic (cyclostationary), several adaptive equalizers can be applied [7-10]. One approach is to apply LMMSE receiver principles. The third generation CDMA systems do not provide training sequences for adaptation and therefore blind adaptive equalizers must be applied.

If the spreading code period is too long to retain cyclostationarity of interference, adaptive CDMA equalizers suppressing multiple access interference cannot be applied. In such a case, standard equalization techniques used in TDMA systems could be used to equalize the chip waveform [32].

Application of interference cancellation at downlink receivers is not straight-forward. Some form of partial interference cancellation taking into account only the dominant interferers would be needed. Without the assistance of the network, it might be quite difficult and complicated to identify dominant interferers. By applying SD-PIC (parallel interference cancellation without channel estimation and hard tentative data decisions in interference estimation) a receiver [30], for four virtual users for example, with partial interference cancellation, can be used.

4.4.1   LMMSE Receivers in Multipath Fading Channels

When considering downlink receivers, only the desired user signal should be demodulated while suppressing the interference due to other users. The linear minimum mean-squared-error-single-user receivers [8-10] are one option for the downlink receivers. The adaptive versions of the LMMSE receivers are usually defined in such a way that only one user is demodulated, as desired in the downlink. The standard LMMSE receiver [8-10] minimizes the mean-squared-error between the receiver output and the true transmitted data sequence. The LMMSE receivers are capable of handling both interpath and interchannel interference under severe near-far situations. The coefficients of the standard LMMSE receiver [8] depend on the channel coefficients of all users, and hence it must be adapted as the channel changes. If the fade rate of the channel is fast enough, the standard adaptive LMMSE receivers need to be updated continuously. Thus, the standard LMMSE receivers will have severe convergence problems in relatively fast-fading channels.

The optimization criterion can be modified to overcome the convergence problems of the LMMSE receiver. The modified optimization criterion, which leads to the LMMSE-RAKE receiver [6], minimizes the MSE between the receiver output and the channel coefficient data symbol product for each path. Hence, it assumes that the channel parameters of the desired user are known or estimated, as is the case in the conventional coherent RAKE receiver. LMMSE-RAKE receiver depends only on the normalized signature sequence crosscorrelations and the average channel profiles of the users. Since the delays and the average channel profiles change rather slowly, the adaptation requirements of the LMMSE-RAKE receiver are significantly less stringent than those of the adaptive LMMSE receivers [10]. What is more, the complexity of the conventional coherent RAKE receiver is increased only moderately. The adaptive implementations of the LMMSE-RAKE receiver do not necessarily require training sequences, since the decisions made by the conventional RAKE receiver can often be used to train the adaptive receiver. Thus, the LMMSE-RAKE receiver can be viewed as an add-on feature in the conventional coherent RAKE receivers.

4.4.1.1   LMMSE-RAKE Receiver

The received signal is time-discretized by antialias filtering and sampling continuous-time signal at the rate, where S is the number of samples per chip. The received discrete-time signal over a data block of N_b symbols is

$$\mathbf{r} = \mathbf{SCAb} + \mathbf{n} \in C^{SGN_b} \tag{4.36}$$

where $\mathbf{S} \in R^{SGN_b \times KLN_b}$ is the sampled spreading sequence matrix, $\mathbf{C} \in C^{KLN_b \times KN_b}$ is the channel coefficient matrix, $\mathbf{A} \in R^{KN_b \times KN_b}$ is the matrix of total received energies, $\mathbf{b} \in \Xi^{KN_b}$ is the data vector, and $\mathbf{n} \in C^{SGN_b}$ is the channel noise vector.

The LMMSE-RAKE receiver is obtained by minimizing each element of $E(|\mathbf{h} - \hat{\mathbf{h}}|^2)$, where $\mathbf{h} = \mathbf{CAb}$ and $\hat{\mathbf{h}} = \mathbf{M}^T \mathbf{r}$, which results in a linear receiver

$$\mathbf{M} = \mathbf{S}\left(\mathbf{R} + \sigma^2 \, \Sigma_{\mathbf{h}}^{-1}\right)^{-1} \in R^{SGN_b \times KLN_b} \tag{4.37}$$

where $\Sigma_{\mathbf{h}} = \mathrm{diag}\left[A_1^2 \, \Sigma_c, \ldots, A_K^2 \, \Sigma_c\right] \in R^{KLN_b \times KLN_b}$ is a diagonal matrix and it consists of user energies and the average channel tap powers, with $\Sigma_c = \mathrm{diag}\left[E(|c_1|^2), \ldots, E(|c_L|^2)\right] \in R^{L \times L}$, where $E(|c_l|^2)$ is the average energy of the lth propagation path. The output of the modified LMMSE filter is

$$\mathbf{y}_{[M]} = \left(\mathbf{R} + \sigma^2 \, \Sigma_{\mathbf{h}}^{-1}\right)^{-1} \mathbf{S}^T \mathbf{r} \in C^{KL} \tag{4.38}$$

where $\mathbf{S}^T \mathbf{r}$ is the matched filter bank output vector without multipath combining. As we see from [4], the modified receiver does not depend on the instantaneous values of channel complex coefficients but on the average power profiles of the channels. The receiver is exactly of the same form as the standard LMMSE receiver in a nonfading AWGN channel. The adaptation requirements are now much milder and the receiver can be made adaptive even in fading channels. The adaptive implementations result in RAKE receivers with additional interference suppression filters in each receiver branch. Hence it is called *LMMSE-RAKE receiver*.

4.4.1.2   Bit Error Probability Analysis of WCDMA Downlink With the Conventional RAKE and LMMSE-RAKE Receivers

Assumptions and Parameters Used

Perfect channel estimation and ideal receivers are used to obtain the bit error probability lower bounds. The channel model used is a two-path fading channel (equal energies) with maximum delay spread of 2 μs and velocity of 5 km/h for the spreading factors $G = 2$ and 4; delay spread of 7 μs and velocity of 50 km/h for $G = 8, 16, 32$. The actual delay values were randomly selected. Root raised cosine filtering with roll-off factor 0.22 was used and the number of samples per chip has been 4.

The control channels were not included in the analysis, which means that the data modulation for a single data channel case has been BPSK. The bit error probability for QPSK modulation is the same as for BPSK, and hence, the results for BPSK can be used also for the QPSK case. The channel bit rates studied were 2.048 Mbps - 128 Kbps ($G = 2 - 32$) for a single code channel. Neither coding nor power control was included in the analysis.

The conclusions to be drawn from the analysis are based on the raw BEP target value of 0.01 and a two-path Rayleigh channel with equal energies and maximum delay spreads of 2 or 7 (μs). In practical systems, the target raw BEP can be higher (up to 0.1), depending on the channel code being used. Note that the channel multipath profile as well as the fast power control loop used to compensate for the fast fading have significant impact on the BEP results. Therefore, the analysis results presented in this section can be considered only indicative when considering the practical WCDMA system. Nevertheless, the comparisons between the conventional RAKE receiver and the LMMSE-RAKE receiver are still justified.

Numerical Results

The bit error probabilities as a function of the number of users for different spreading factors in the single data channel case are given in Figure 4.13 for the conventional RAKE receiver, and in Figure 4.14 for the LMMSE-RAKE receiver, respectively. If the target raw bit error rate is 0.01 at 20 dB, the conventional RAKE can support roughly 50% fewer users than the LMMSE-RAKE receiver.

The bit error probabilities as a function of the average SNR for different numbers of active users in single data channel case are presented in Figure 4.15 and Figure 4.16 for the spreading factors 4 and 32. In the single-user case, the BEP of the conventional RAKE saturates with small spreading factors due to the bad average autocorrelation properties of the combined Walsh code and the scrambling code. If the number of users is half of the spreading factor, the BEP of the conventional RAKE is the same regardless of the spreading factor. If the target raw BEP at 20 dB is 0.01, the capacity with the conventional RAKE is half of the spreading factor in the case with no near-far problem. The capacity with the LMMSE-RAKE detector is approximately 100% of the spreading factor. Alternatively if the target BEP is 0.01 with a half load ($K=G/2$), the conventional RAKE receivers would require 20 dB SNR, whereas the LMMSE-RAKE receiver needs only 10 dB. This is valid for all spreading factors studied.

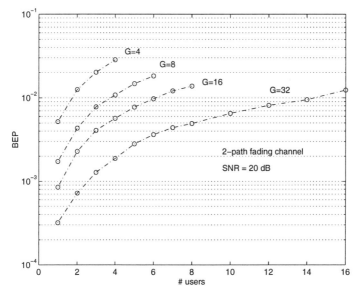

Figure 4.13 BEPs with different spreading factors (G) for RAKE at SNR 20 dB, data rates 128 Kbps - 1.024 Mbps.

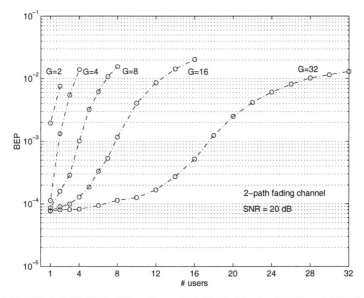

Figure 4.14 BEPs for LMMSE-RAKE receiver at SNR 20 dB, data rates 128 Kbps - 2.048 Mbps.

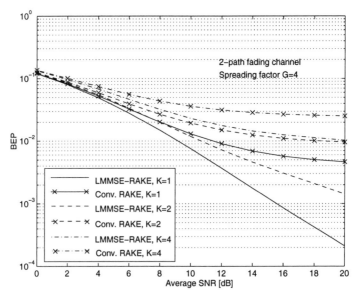

Figure 4.15 BEPs for conventional RAKE and LMMSE-RAKE with different numbers of users (K = 1, 2, 4), data rate 1.024 Mbps.

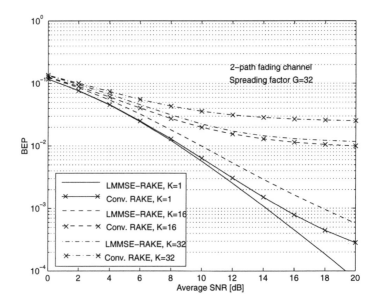

Figure 4.16 BEPs for conventional RAKE and LMMSE-RAKE with different numbers of users (K = 1, 16, 32), data rate 128 Kbps.

In near-far situations, the BEP of the conventional RAKE receiver collapses completely. Under rather mild near-far conditions, where 50% of users have 6 dB higher power (Figure 4.17), only the spreading factor 32 results in acceptable performance. The absolute minimum spreading factor that may be used in this case is 16 for the conventional RAKE to achieve the BEP of 0.01. The bit error performance for the LMMSE-RAKE detector is the same with the near-far cases at high SNRs, and hence, the results for the LMMSE-RAKE receiver can be read from the Figure 4.14.

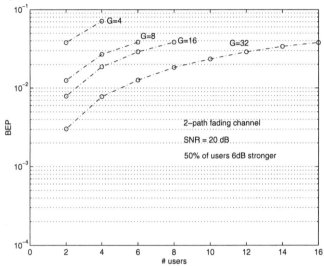

Figure 4.17 BEPs with different spreading factors (G) for RAKE at SNR 20 dB, data rates 128 Kbps - 1.024 Mbps; 50 % of users have 6-dB higher energy.

The BEPs as a function of the near-far ratio with different SNRs are given in Figure 4.18. The LMMSE-RAKE receiver has similar performance regardless of the near-far ratio at large SNRs. The conventional RAKE receiver is sensitive to the near-far problem and has relatively high BEP with the lowest spreading factors even with small near-far ratios. With mix of services with different QoS requirements, the high data rate channels require higher power, which causes a near-far problem for the low data rate users, e.g., a data channel with a spreading factor of $G=8$ will be seen at a 15 dB higher power for the data channels with the spreading factor of $G=256$. Based on this observation, the maximum dynamic range for the downlink power control should be reduced in a mixed data rate system since the average near-far ratio is a combination of the power control and the data rate caused power difference between the users.

Large near-far ratios can be avoided by reducing the power control dynamic range and by allocating different data rate groups to different cells or different carrier frequencies. By extrapolating the numerical result of Figure 4.18 to the largest spreading factors (64-256), can be estimated that the maximum power control dynamic range is 20 dB for the spreading factor of 256 in a two-user case.

Assuming that there are the data rate groups A (spreading factors 8-32, 4 is not used) and B (spreading factors 64-256), the maximum allowable power control dynamic range would be approximately 5 dB in the group A and 15 dB in the group B. Although this result is deduced starting from a simple two-user two-path case, the message is clear: service mix per carrier should be reduced when using conventional RAKE receivers.

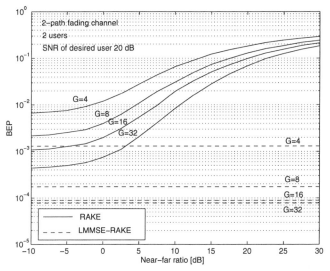

Figure 4.18 BEPs as a function of the near-far ratio for RAKE and LMMSE-RAKE receiver in 2-user case with different spreading factors (G) at SNR 20 dB.

4.4.2   Chip Waveform Equalization

The performance results of the previous section clearly show the interpath interference problem with small spreading factors. One possibility to reduce the problem is to decorrelate the propagation paths of the desired user. In other words, IPI is forced to zero and diversity gain is increased. The BEP improvements at high SNRs (LMMSE and decorrelating receivers are asymptotically equal) can be estimated in the single-user case by comparing Figure 4.13 and Figure 4.14. Implementation of the decorrelating detector for the desired user's propagation paths would require computation of the inverse of the cross correlation matrix of size $ML \times ML$, where L is the number of propagation paths and M is the length of the detector in symbol intervals. In [11] it was shown that decorrelators can be realized even in long code systems, and hence the decorrelation of the propagation paths is one possibility to enhance the performance and the capacity of the WCDMA downlink. However, the problems related to the robustness of the decorrelator in the presence of timing errors [28] would demand high-resolution timing estimators for such receivers. In [29], a timing estimator suitable to iterative decorrelating receivers has been studied.

Another possible solution to suppress IPI would be to use adaptive channel impulse response matched filters prior to signal despreading. After such a channel matched filter, the channel that caused distortion is compensated and only one correlator is needed for despreading. Effectively, the adaptive channel matched filter equalizes the chip waveform. In synchronous downlink with orthogonal channel separation, MAI is zero with the optimum sampling instant from the channel matched filter. Since the channel matched filter is the same for all users at a mobile terminal, the filter can be adapted by using the pilot channel bits as a training sequence. The signals of other users can also be used to increase the training signal energy. Data detection for those users would be required unless only the pilot symbols are utilized. This approach fails at base station receivers and also in the case when downlink beamforming is used. The adaptive channel matched filter can be applied also in a long code system since the filter depends only on the channel impulse response. The drawback of this approach is that the filter weights depend on the channel phases, which causes convergence problems in fading.

4.4.2.1 LMMSE Chip Equalizers

The system model must be slightly modified for the definition of chip equalizer. The model for the received signal can be written as

$$\mathbf{r} = \sum_{k=1}^{K} \mathbf{DCA}_k \mathbf{S}_k \mathbf{b}_k + \mathbf{n} \in C^{SGN_b} \tag{4.39}$$

where $\mathbf{D} \in R^{SGN_b \times GN_b L}$ is the matrix of sampled and delay chip waveforms for each propagation path over N_b symbols, C is the channel coefficient matrix, A_k contains user powers and b_k data symbols. Note that the chip waveform and channel coefficient matrices are the same for all users.

It is easy to show [32] that the LMMSE chip can be approximated as

$$\mathbf{F} = \left(s^2 \sum_{k}^{K} A_k^2 \mathbf{DCC}^H \mathbf{D}^H + \sigma_n^2 \mathbf{I} \right)^{-1} \cdot \mathbf{DC} \in {}^{SGN_b \times GN_b} \tag{4.40}$$

where s^2 is the squared value of a chip. The LMMSE equalizer (A) is an approximation of the exact version (B), which takes advantage of the correlation between successive chips. Note that the LMMSE equalizer version A is exactly the same as the standard LMMSE receiver [8] for IPI alone.

Numerical Examples

BEPs were evaluated in Rayleigh fading frequency-selective channels by applying a semianalytical method, where for a random bit pattern, spreading sequences and channel gain realization, a noise-free decision variable was calculated. BEP was evaluated based on the decision variable realization and variance of additive Gaussian noise in the input of the decision device. To obtain final results, BEPs were averaged over a 10,000 bit pattern, spreading sequences and channel gain realizations. The same realizations were used for all receivers.

All link parameters from UTRA FDD were used. The power of all users was the same in the numerical evaluations, and no channel coding was assumed. A spreading factor of 16 was used. In the figures, BEPs are presented for the conventional RAKE receiver, as well as for the chip equalizers. The chip equalizers used are based on two LMMSE criteria and a ZF criterion.

In Figure 4.19, BEPs are presented for 12 simultaneous users in a two-path Rayleigh fading channel with equal energies and a tap delay difference of 300 ns.

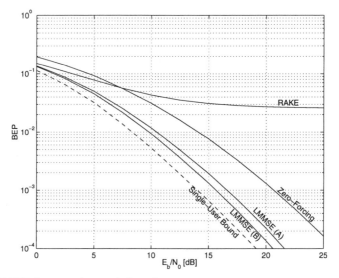

Figure 4.19 BEPs in a general two-path channel.

In Figure 4.20, ITU indoor A channel is used. As seen from the presented results, the LMMSE receivers offer significant BEP improvement when compared to the conventional RAKE receiver. Also, the number of users considerably affects the performance of the RAKE receiver, but has only a small effect on the performance of the LMMSE receivers.

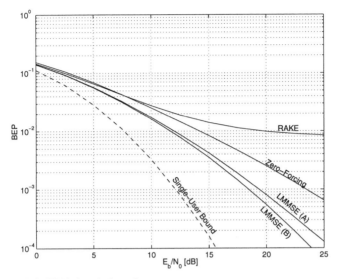

Figure 4.20 BEPs in ITU indoor A channel.

4.4.3   Comparing Selected UTRA Downlink Receiver Concepts

Many advanced receiver concepts, such as, the LMMSE receiver in [6], rely on the cyclostationarity of the interference statistics and, consequently, neglect the aperiodic scrambling codes used in the UMTS UTRA downlink spreading signals. Here, we compare two conventional receiver (CR) schemes with a new interference cancellation (IC) approach in terms of the BER for the UTRA downlink, where both schemes take into account the aperiodic scrambling codes. The main objective is to reduce the error floor of the CR in the high SNR region where MAI and intercell interference ultimately limit the overall system performance. The IC scheme exploits the hierarchical code structure of the orthogonal variable spreading factor (OVSF) codes in the following way. In an attempt to find the intracell interfering signals being correlated with the signal of interest (SOI), the 4-chip segments of the basic code tree branches are used in combination with the estimated transmission channel parameters being common to all users to reconstruct and subtract the interference. Note that the information about the number of active users as well as their spreading sequences need not be communicated to the mobile station. Since the IC requires no more knowledge than the CR, it belongs to the class of schemes being *blind* with regard to the structure of the interfering signals. The use of long scrambling codes suggests carrying out the IC in front of the correlators, as has been done in the case of uplink channel parameter estimation [33].

4.4.3.1   Conventional Receiver With Multiple Antennas

The CR with several antennas is depicted in Figure 4.21. During pilot transmission the channel parameters are estimated adopting a maximum likelihood approach within each slot. The resulting channel coefficients are assumed to be constant within one slot. Based on these channel estimates a coherent temporal maximum ratio combining (MRC) is carried out at each sensor. Identical front ends (i.e., identical noise variances) are assumed for all antennas. Therefore, the spatial MRC reduces to a simple sum of the outputs of the signals at each antenna branch. A threshold detector is used for symbol estimation.

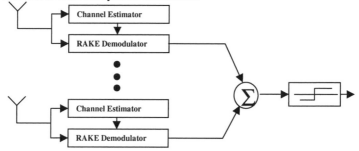

Figure 4.21 Conventional receiver with multiple antennas.

4.4.3.2   Downlink Receiver With Interference Cancellation

The OVSF codes used for downlink transmission within one cell are shown in Figure 4.22.

While two codes of different lengths taken from the tree are orthogonal as long as the shorter does not constitute a signed segment of the longer one, two codes with identical lengths are always orthogonal. Consequently, in a frequency-nonselective channel, the conventional detector with perfect channel parameter estimates gives the optimal BER performance. With multipath propagation, however, the performance is degraded substantially due to ISI and MAI. Clearly, multiuser estimation and detection schemes with IC (e.g., [33]) can be readily applied as soon as the number of active users as well as their channelization codes are known at the receiver. The overhead required for transmitting the corresponding information in the downlink is not reasonable in view of the highly time-variant symbol rates and the resulting code assignment. The minimum spreading factor defined in UTRA FDD is $\chi_{\min} = 4$. The fact that all sequences in the OVSF code contain signed code segments with $\chi = 4$ suggests a strategy to cancel potential MAI terms: The codes $c_{4,\kappa}$, $\kappa \in \{1,...,4\}$ in Figure 4.22 with $\chi = 4$, which do not constitute segments of the first user's code, are assumed to be assigned to three other users currently active in the system. The underlying idea relies on the fact that a longer code originating from one of these codes in the tree and being indeed assigned to an active user can be reconstructed and subtracted using the corresponding short segment. In the following, the approach is employed for both channel parameter estimation during the preamble, as well as for symbol detection during information transmission.

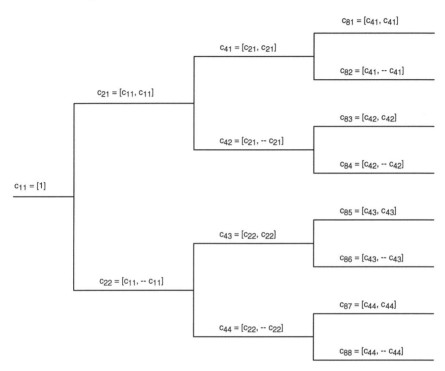

Figure 4.22 OVSF channelization codes.

For channel parameter estimation, the short segments, which can be parts of different longer codes, are multiplied at the transmitter by user-specific gains and the individual quarternary phase-shift keying (QPSK) symbol values. Thus, the resulting weighting factor can take a large variety of complex values and is therefore modeled as a complex number. The latter has to be estimated for each segment and each symbol index separately. Since the path delay values are identical for all users, they can be estimated initially as in the conventional receiver. A detailed description of the algorithm can be found in [30].

The procedure for detecting the information symbols is analogous to the aforementioned estimation scheme. An *iteration cycle* consists of two steps. First, the channel parameter estimates calculated from the preamble are used to detect the symbols of the users $k = 2, 3, 4$. Then, after detection of a certain symbol, the corresponding signal parts in the received signal are reconstructed and subtracted. To improve the accuracy of the MAI estimates, two iteration cycles are carried out. The scheme for estimating the complex amplitudes of the interfering signals is identical to the corresponding one for preamble transmission. After the two iteration cycles, the symbols of the SOI are detected. In opposite to the two iteration cycles of the IC for channel parameter estimation, the symbols of the first user are detected only once using a conventional maximum-ratio combining after IC. Note that the cancellation of the ISI terms of the first user can be included in the detection scheme. In the following simulation results, however, mainly MAI limits the BER performance and, thus, ISI is not taken into account in the IC.

4.4.3.3   Numerical Examples

In this section, the BER performances of the IC receiver with one sensor and two CR schemes are compared for different situations. The two CRs are equipped with $M=1$ and $M=3$ antenna sensors, respectively. First, we consider a system with $K=4$ users and spreading factor $\chi_k=16$ for $k=1,...,4$. The time-variant CIR are created using the stochastic radio channel model in [31] with $L=4$ transmission paths. We have chosen a delay spread of 3 µs where each transmission path contains a total of 10 waves impinging at the receiver from different directions. In Figure 4.23, the absolute value of the CIR used in the simulations is shown.

The total distance covered by the first user is approximately 6m. Note that due to the movement of the user, the spatial dependence of the CIR is transformed into a temporal one. In Figure 4.24, the BER of the first user obtained with the different receivers are shown as a function of γ_1 which represents the first user's SNR at the first sensor averaged over time.

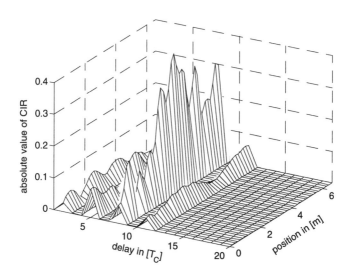

Figure 4.23 Absolute value of the CIR used in the simulations.

The dashed line indicates a loose BER lower-bound for the case $M = 1$, which results from averaging the BER for the AWGN case over the received instantaneous SNR. It can be observed that for low values γ_l, the single-sensor CR (SSCR) and the IC scheme give approximately the same BER, while the multiple-sensor CR (MSCR) provides substantially lower BER values. The BER of the SSCR saturates approximately at BER $= 1.5*10^{-3}$ for $\gamma_1 = 35$ dB, while the corresponding values for the IC scheme are given by BER $= 1.5*10^{-4}$ for $\gamma_1 = 40$ dB. Moreover, the MSCR is outperformed for $\gamma_1 \geq 35$ dB. Consequently, in this case, the performance of a CR with three sensors can be obtained by the IC scheme with only one sensor. The BER for a similar situation are shown in Figure 4.25, where only the value $\chi_k = 4$ for $k = 1,...,4$ has been changed with regard to Figure 4.24. Since this situation corresponds to a fully loaded system with short spreading sequences, the BER saturation level is drastically increased.

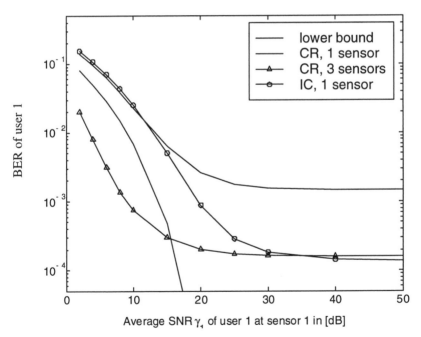

Figure 4.24 BER for K = 4, $\chi_k=16$.

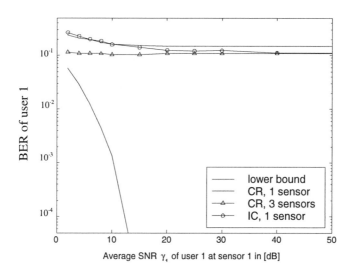

Figure 4.25 BER for K = 4, $\chi_k=4$.

If the system load and thus the total amount of MAI is reduced, the IC scheme gives approximately the same BER as the SSCR, as seen for $K = 2$ and $\chi_k = 16$ in Figure 4.26.

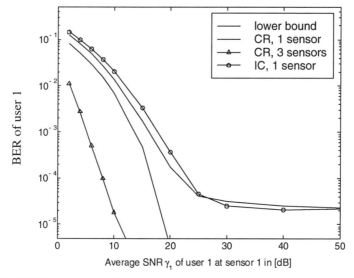

Figure 4.26 BER for K = 2, $\chi_k=16$.

In this case, the assumption of three other users with $\chi_k = 4$ being in the system is not satisfied. Consequently, the BER performance of the SSCR is not improved by the IC scheme. Note, however, that the amount of interference *added* in the IC by erroneously canceling MAI terms being absent can be neglected as compared to the SSCR. At the same time, the MSCR clearly outperforms the other receivers in such a situation.

Another important issue present in a practical system is power control (PC). To investigate the influence of PC on the BER, we consider Figure 4.27, where ideal PC is realized, and Figure 4.28, where PC is absent, for $K = 8$ and $\chi_k = 16$. For low SNR values, the power level variations lead to a performance loss as compared to the ideal PC case. However, for sufficiently high SNR values, the total amount of MAI relative to the SOI is independent of the power level and thus, the BER are the same in Figure 4.28 and Figure 4.27.

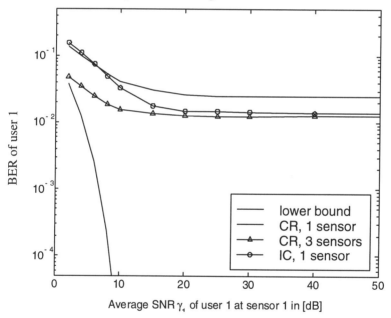

Figure 4.27 BER with ideal power control for K = 8, χ_k=16.

At cell boundaries, the intercell interference represents a major limiting factor. Therefore, a scenario with two cells is considered in Figure 4.29, where $K = 8$ as above and the signals from the neighboring cell base station (BS) are received with the same power as the intracell BS signal. Obviously, the performance gain of the IC scheme is offset by the interference being essentially uncorrelated with the intracell signals.

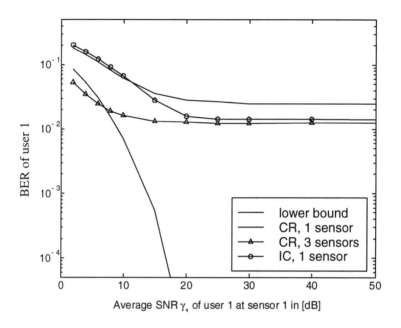

Figure 4.28 BER for K = 8, χ_k =16.

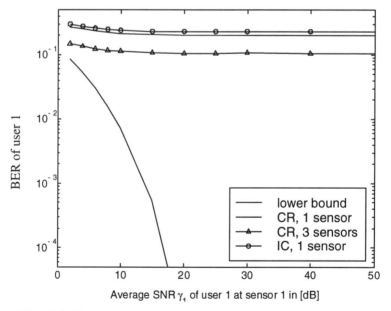

Figure 4.29 BER for K = 8, χ_k =16, high intercell interference.

4.5   CONCLUSIONS

Performance of conventional RAKE receivers is interference limited. By applying interference suppression/cancellation techniques, significant performance and capacity improvements can be obtained. Several joint detection schemes based on block processing were presented in Section 4.2 suitable to TD-CDMA system. Applicable techniques for FRAMES WCDMA uplink reception were presented in Section 4.3 and for the downlink in Section 4.4.

Although the schemes presented in this chapter provide enhanced performance, their performance should be validated in a complete radio network. Implementation losses need to be explored as well, and hence some hardware trials should be carried out to obtain final figures of their performance.

References

[1]   P. Patel and J. Holtzman, "Analysis of Simple Successive Interference Cancellation Scheme in a DS/CDMA System," *IEEE Journal on Selected Areas in Communications*, Vol. 12, No. 10, Oct 1994, pp.796-807.

[2]   R.M. Buehrer and B.D.Woerner, "The Asymptotic Multiuser Efficiency of M-stage Interference Cancellation Receivers," In: *Proc. IEEE International Symposium on Personal, Indoor, and Mobile Radio Communications (PIMRC)*, Helsinki, Finland, Sept. 1-4, 1997, Vol. 2, pp. 570-574.

[3]   D. Divsalar, M.K. Simon and D. Raphaeli, "Improved Parallel Interference Cancellation for CDMA," *IEEE Transactions on Communications*, Vol. 46, No. 2, Feb. 1998, pp. 258-268.

[4]   M. Juntti, "Multiuser Demodulation for DS-CDMA Systems in Fading Channels," Vol. C106 of Acta Universitatis Ouluensis, Doctoral Thesis, University of Oulu Press, Oulu, Finland, 1997.

[5]   M. Latva-aho and J. Lilleberg, "Parallel Interference Cancellation in Multiuser CDMA Channel Estimation," *Wireless Personal Communications*, Kluwer Academic Publishers, Vol.7, No.2/3, Aug. 1998, pp.171-195.

[6]   M. Latva-aho, "Advanced Receivers for Wideband CDMA Systems," Vol. C125 of *Acta Universitatis Ouluensis*, Doctoral thesis, University of Oulu Press, Oulu, Finland, 1998.

[7]   M. Latva-aho, "Bit error Probability Analysis for FRAMES WCDMA Downlink Receivers," *IEEE Transaction on Vehicular Technology*, Vol. 47, No. 4, Nov. 1998, pp.1119-1133.

[8]   U. Madhow and M.L. Honig, "MMSE Interference Suppression for Direct-Sequence Spread-Spectrum CDMA," *IEEE Transactions on Communications*, Vol. 42, No. 12, Dec. 1994, pp. 3178-3188.

[9]   S.L. Miller, "An Adaptive Direct-Sequence Code-Division Multiple-Access Receiver for Multiuser Interference Rejection," *IEEE Transactions on Communications*, Vol. 43, No. 2/3/4, Feb./Mar./Apr. 1995, pp.1746-1755.

[10]   P.B. Rapajic and B.S. Vucetic, "Adaptive Receiver Structures for Asynchronous CDMA Systems," *IEEE Journal on Selected Areas in Communications*, Vol. 12, No. 4, May 1994, pp.685-697.

[11]   M. Juntti, B. Aazhang and J. Lilleberg, "Iterative Implementation of Linear Multiuser Detection for Asynchronous CDMA Systems," *IEEE Transactions on Communications*, Vol. 46, No. 4, Apr. 1998, pp. 503-508.

[12]   A. Klein, G.W. Kaleh, P.W. Baier, "Zero Forcing and Minimum Mean-Square-Error Equalization for Multiuser Detection in Code-Division Multiple-Access Channels," *IEEE Transactions on Vehicular Technology*, Vol. 45, No. 2, May 1996, pp. 276-287.

[13] A. Klein, "Multi-User Detection Algorithms of CDMA-Signals and Their Application to Cellular Mobile Radio," *Fortschrittsberichte VDI*, Vol. 10, No. 423, Düsseldorf, VDI-Verlag, 1996.

[14] A. Klein, P. W. Baier, "Linear Unbiased Data Estimation in Mobile Radio Systems Applying CDMA," *IEEE Journal on Selected Areas in Communications*, Vol. 11, Sept. 1993, pp. 1058-1066.

[15] P. W. Baier, J. J. Blanz, M. M. Naßhan, A. Steil, "Realistic Simulations of a CDMA Mobile Radio System Using Joint Detection and Coherent Receiver Antenna Diversity," *Proceedings of the IEE Colloquium on Spread Spectrum Techniques for Radio Communication Systems*, London, 1994, pp. 1/1-1/5.

[16] P. W. Baier, J. J. Blanz, A. Papathanassiou, "Joint Detection CDMA and Antenna Diversity Techniques," *Proceedings of the IEE Colloquium on CDMA Techniques and Applications for Third Generation Mobile Systems*, London, 1997, pp. 1/1-1/7.

[17] J. J. Blanz, A. Klein, M.M. Naßhan, A. Steil, "Performance of a Cellular Hybrid C/TDMA Mobile Radio System Applying Joint Detection and Coherent Receiver Antenna Diversity," *IEEE Journal on Selected Areas in Communications*, Vol. 12, 1994, pp. 568-579.

[18] J. J. Blanz, A. Klein, M. M. Naßhan, A. Steil, "Cellular Spectrum Efficiency of a Joint Detection CDMA Mobile Radio System," *Proceedings of the International Zurich Seminar on Digital Communications*, Zürich, 1994, Vol. 783, pp. 184-195.

[19] P. W. Baier, M. M. Naßhan, "Recent Results Concerning the Benefit of Joint Detection in CDMA Systems," *Proceedings of the IEE Colloquium on Spread Spectrum Techniques for Radio Communication Systems*, London, 1993, pp. 5/1-5/4.

[20] A. Klein, P. W. Baier, "Simultaneous Cancellation of Cross Interference and ISI in CDMA Mobile Radio Communications," *Proceedings of the IEEE International Symposium on Personal, Indoor and Mobile Radio Communications (PIMRC'92)*, Boston, 1992, pp. 118-122.

[21] A. Klein, G. K. Kaleh, P. W. Baier, "Equalizers for Multi-User Detection in Code Division Multiple Access Mobile Radio Systems," *Proceedings of the IEEE Vehicular Technology Conference (VTC'94)*, Stockholm, 1994, pp. 762-766.

[22] A. Klein, B. Steiner, A. Steil, "Known and Novel Diversity Approaches in a JD-CDMA System Concept Developed Within COST 231," *Proceedings of the IEEE International Symposium on Personal, Indoor and Mobile Radio Communications (PIMRC'95)*, Toronto, 1995, pp. 512-516.

[23] M. M. Naßhan, "On the Effects of Adjacent Channel Interference on the Link Level Performance of a JD-CDMA Mobile Radio System," *Proceedings of the IEEE Vehicular Technology Conference (VTC'97)*, Phoenix, 1997, pp. 218-222.

[24] A. Steil, J. J. Blanz, "Spectral Efficiency of JD-CDMA Mobile Radio Systems Applying Coherent Receiver Antenna Diversity with Directional Antennas," *Proceedings of the International Symposium on Spread Spectrum Techniques & Applications (ISSSTA'96)*, Mainz, 1996, pp. 313-319.

[25] P. Höher, "On Channel Coding and Multiuser Detection for DS-CDMA," *ITG-Fachbericht*, Vol. 124, Sept. 1993, pp. 545-566.

[26] S. W. Marple, "Digital Spectral Analysis with Applications," Englewood Cliffs; NJ, Prentice-Hall, 1987.

[27] A. Duel-Hallen, "Decorrelating Decision-Feedback Multiuser Detector for Synchronous Code-Division Multiple-Access Channel," *IEEE Transactions on Communications*, Vol. 41, Feb. 1993, pp. 285-290.

[28] S. Parkvall, E. Ström and B. Ottersten, "The impact of timing errors on the performance of linear DS-CDMA receivers," *IEEE Journal on Selected Areas in Communications*, Vol. 14, No. 8, Oct. 1996, pp.1660-1668.

[29] J. Lilleberg, E. Nieminen and M. Latva-aho, "Blind iterative multiuser delay estimator for CDMA," *Proc. IEEE International Symposium on Personal, Indoor, and Mobile Radio Communications (PIMRC)*, Taipei, Taiwan, Oct. 15-18, 1996, Vol.2, pp. 565-568.

[30]   D. Dahlhaus and A. Jarosch, "Comparison of Conventional and Adaptive Receiver Concepts for the UTRA Downlink," *Proc. UMTS Workshop*, Schloss Reisensburg, Germany, Nov. 1998, pp. 233-242.

[31]   R. Heddergott, U. P. Bernhard, and B.H. Fleury, "Stochastic Radio Channel Model for Advanced Indoor Mobile Communication Systems," In *Proc. of the 8th IEEE Int. Symp. on Personal, Indoor and Mobile Radio Communications (PIMRC'97)*, Sept. 1997, Vol. 1, Helsinki, Finland, pp. 140-144.

[32]   K. Hooli, M. Latva-aho, M. Juntti, "Comparison of LMMSE Receivers in WCDMA Downlink," *ACTS Mobile Summit 1999*, Sorrento, Italy, June 8-11, 1999.

[33]   D. Dahlhaus, A. Jarosch, B.H. Fleury and R. Heddergott, "Joint Demodulation in DS/CDMA Systems Exploiting the Space and Time Diversity of the Mobile Radio Channel," In *Proc. of the 8th IEEE Int. Symp. on Personal, Indoor and Mobile Radio Communications (PIMRC'97)*, Vol. 1, Sept. 1997, Helsinki, Finland, pp. 47-52.

Chapter 5

Modulation and Coding

Two major goals of wireless communication systems are high capacity and high quality. Both these goals are dominated by the choice of modulation and channel coding. Traditionally, high capacity was obtained when the bandwidth of the transmitted signal was kept narrow, but with the introduction of wideband techniques like CDMA, this is not necessarily true anymore. A narrow spectrum gives preference to high coding rates and high signaling constellations, which does not necessarily give high quality on a wireless channel. In CDMA, a low coding rate is not a problem, since the signal is spread in bandwidth anyway. However, it is of utmost importance that many users can exist simultaneously in the given system bandwidth. Large signaling constellations often give rise to problems with large amplitude variations in amplifiers and high sensitivity to noise and interference, which can be avoided by using smaller constellations. These are just some examples to show the importance of modulation and coding.

In FRAMES, a lot of effort has been directed towards improving modulation and coding, and some of the results will be reported in this chapter. In Section 5.1, a new multilevel modulation method is presented that gives a good possibility of trading spectral efficiency and bit error probability. Multicode CDMA, which is proposed in UTRA/FDD to handle large data rates, is described in Section 5.2. It has the disadvantage of large amplitude variations, but here a precoding scheme is given that at least partly overcomes this problem.

Turbo coding is a relatively new channel coding technique, which has been shown to have a performance close to the theoretical limits. This technique has been evaluated in great detail for UTRA/TDD, and some of the findings are summarized in Section 5.3. Various aspects of traditional convolutional codes are studied in Sections 5.4 to 5.7. It has been found that the convolutional codes in most textbooks can be slightly improved by changing the design criterion. Some of these new codes are discussed in Section 5.4. New multimedia types of services require that the source data rate is matched to an often limited set of channel data rates, and this may be done efficiently by convolutional codes as described in Section 5.5. In order to obtain very low error probabilities with convolutional codes, the constraint length needs to be large and the decoding complexity becomes prohibitive with Viterbi decoding. Then sequential decoding is an alternative technique, which is studied for wireless channels in Section 5.6. Section 5.7 is devoted to low rate codes that can be used for combined spreading and error correction in CDMA systems. It has been found that large capacity improvements are possible with this technique. Both convolutional codes and a special class of block codes named TCH codes are studied. Finally, in Section 5.8, different types of ARQ schemes for packet transmission are described and evaluated. Packet data transmission is believed to be the major switching technique in future tele-communication systems, and here we show that large throughput improvements are possible by adapting the code rate to the channel.

The reader is assumed to be familiar with the modulation and coding techniques that are discussed throughout this chapter. There exist many good textbooks on these topics. We recommend [1] as an excellent text in the basics of digital communications. For more details on channel coding, we recommend [2] as a general text, and [3] for the details on turbo coding. Since some parts of this chapter are specifically devoted to the CDMA system, we can also recommend [4] and Chapters 2 to 3 of this book.

5.1 MALGMSK MODULATION SCHEMES

Continuous phase modulations (CPMs) [5] combine the characteristics of being constant envelope and bandwidth efficient modulation schemes. Among CPM, Gaussian minimum shifting keying (GMSK) is an important scheme for its spectral properties, being used for the GSM and DECT systems. A drawback of GMSK is that for low values of $B_b T^1$, which guarantee high spectral efficiency, a strong ISI heavily degrades the error performance. Therefore, $B_b T$ is set to 0.5 for DECT and to 0.3 for GSM. Linearized GMSK (LGMSK) is a linear modulation scheme that uses the pulse shaping function $C_0(t)$, derived by a linear approximation of GMSK [6]. The main advantage of LGMSK with respect to conventional GMSK is the possibility to implement a quadrature receiver which successfully eliminates the ISI for any value of BbT, therefore allowing an increase of the spectral efficiency at the expense of amplitude variation [7].

A further increase of the spectral efficiency can be achieved by multi-amplitude linearized gaussian minimum shifting keying (MALGMSK) which is the multilevel linear modulation derived directly from binary LGMSK. The complex envelope of the MALGMSK signal is:

$$s_{\mathrm{MALGMSK}}(t) = \sum_i A_{2i} C_0(t - 2iT) + j \sum_i A_{2i+1} C_0(t - (2i+1)T) \qquad (5.1)$$

where $A_i = \{-3, -1, 1, 3\}$ are the transmitted symbols.

As for any other multiamplitude modulation, the increased efficiency is attained at the price of an increase in the *peak-to-average ratio* (PAR), which is defined as the ratio of the peak signal power to the average signal power. The PAR is related to the backoff required for operation with nonlinear power amplifiers, in order to prevent extra spectrum spreading. For nonconstant-envelope modulation

---

[1] $B_b T$ is the normalized 3-dB-down bandwidth of the premodulation Gaussian low-pass filter used.

schemes, a nonlinear power amplifier causes degradation of the performances: the signal undergoes distortions so its demodulation is less efficient and the power spectrum is widened, thus the adjacent channel interference is increased.

To evaluate the power spectral efficiency of a modulating scheme, we used the *out-of-band power* (OBP), which indicates the fraction of power outside a certain bandwidth. In the case of MALGMSK, an adequate choice of the receiver filter, which satisfies the first Nyquist criterion, allows the complete cancellation of the ISI. This receiver filter $H(f)$ is not matched to the transmitted pulse $C_0(f)$. Therefore, it is necessary to introduce a correction factor δ, which indicates the performance degradation with respect to the ideal performance of a matched filter receiver. In Table 5.1, some results are summarized for MALGMSK for different values of $B_b T$ and for quaternary offset QAM (Q-O-QAM) with raised cosine pulse shaping.

Table 5.1

Comparison for PAR, OBP, and δ for Different Instances of MALGMSK and Q-O-QAM

	MALGMSK (BbT=0.1)	MALGMSK (BbT=0.15)	MALGMSK (BbT=0.25)	Q-O-QAM
PAR [dB]	5.01	4.16	3.29	5.71
OBP-10dB	0.10	0.11	0.14	0.12
OBP-20dB	0.15	0.18	0.21	0.15
δ [dB]	1.59	0.49	0.09	0

Figure 5.1 shows the OBP for MALGMSK with $B_b T$=0.15 and for Q-O-QAM with rolloff factor 0.35. The results are obtained considering a solid state nonlinear power amplifier as modeled in [8], to evaluate the impact of the different values of PAR on the two modulation techniques for different values of output backoff. While the performances of both modulations deteriorate when compared to the linear amplifier case, MALGMSK performs better because of a smaller value of the PAR.

MALGMSK has flexible performance with respect to the choice of the value of the parameter $B_b T$. In comparison with Q-O-QAM: $B_b T$ smaller than 0.15 allows better spectral performance at the expense of an increase in terms of BER; $B_b T$ larger than 0.15 gives better BER performance (the factor δ decreases and also the PAR) but also lower spectral efficiency. The choice of a certain value of $B_b T$ is then the ultimate question about MALGMSK, which depends on the overall system requirements.

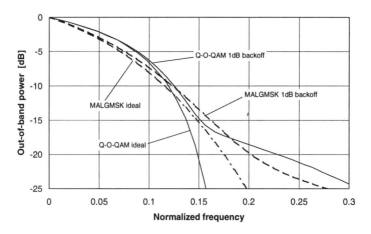

Figure 5.1 Out-of-band power for Q-O-QAM and MALGMSK (BbT=0.15). The results are plotted for the ideal case (linear amplifier) and with nonlinear, solid-state amplifier with 1dB backoff.

5.2   MULTICODE CDMA WITH PRECODING SCHEMES

In the recent WCDMA standard, variable spreading factors and multicode transmission are used to obtain multiple data rates. Variable spreading factors are used for the low and medium-high data rates, and are combined with multicode transmission for the highest data rates. The reason for this split is that for high data rates, the spreading factor in the variable spreading factor scheme will be very small and thus the performance will significantly degrade due to intersymbol interference [9]. On the other hand, using many parallel codes in the multicode scheme will result in a large envelope variation. This envelope variation results because the transmitted signal is a sum of many independently spread signals [9-11].

In a cellular phone, most of the power is consumed by the RF power amplifier, and thus the power efficiency of this amplifier is very important. A power amplifier that has high power efficiency operates near the saturation point. However, in this region, the amplitude input-output characteristics (AM-AM) of the amplifier is very nonlinear, and this nonlinearity has negative effects such as increased out-of-band radiation (spectral spreading) and decreased performance (increased bit error rate) due to bad modulation accuracy. The degree of the out-of-band radiation depends on the envelope variation of the input signal. A large variation results in a large out-of-band radiation and thus decreased spectral efficiency of the overall system. Alternatively, a large variation will decrease the power efficiency, since it forces the amplifier into the less power-efficient linear (or active) region. These arguments are the main reasons for the prevailing use of constant (or near constant) amplitude modulation schemes in cellular communication systems. It is thus also obvious that the envelope variations introduced by the multicode transmission may prohibit effective use in wireless communications, at least in the uplink, where it is of utmost importance that the handsets have a high power efficiency. It is especially true when many parallel codes are used. In this section, a precoding scheme that reduces the envelope variation is reviewed.

It was shown in [9] and [11] that the envelope variations can be drastically reduced by the introduction of a nonlinear block code called a precoder. This precoder of course depends on the set of used spreading codes, and thus the precoder must be designed specifically for that user and the spreading codes used. However, by the introduction of concatenated spreading codes, where a user-specific spreading code is element-wise multiplied by a set of Hadamard-sequences of the same length as the number of parallel codes, the precoder becomes independent of the user-specific codes [9,11]. The transmitter and receiver pair for an (n,k)-precoder are shown in Figure 5.2. Here $u(t)$ is the user-specific spreading code, $\mathbf{h}_n(t)$ the vector spreading code (a column in the Hadamard matrix of size n) of size n (the number of parallel codes), and $\mathbf{b}$ is the precoded codeword.

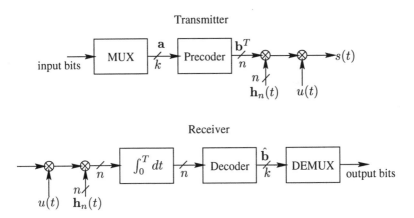

Figure 5.2 Transmitter and receiver pair for a precoded multicode DS-CDMA system using BPSK modulation.

Procedures for designing precoders are given in [11]. It is shown that the $(4,3)$-precoder simply is the parity-check code $\mathbf{b} = (b_1, b_2, b_3, b_4)^{\mathrm{T}}=(a_1, a_2, a_3, -a_1 a_2 a_3)^{\mathrm{T}}$, where $\mathbf{a} = (a_1, a_2, a_3)^{\mathrm{T}}$ is the binary information word.

As a measure of the degree of envelope variations, we use the crest factor (CF), defined as the ratio between the maximum absolute value and the RMS (root mean square) value of the transmitted signal $s(t)$, respectively. Usually the CF is given in decibels. The lowest value of the crest factor is 3 dB and obtained by an unmodulated carrier. For example, a 3-code multicode scheme has a CF of 7.8 dB, while a 1-code scheme has a crest factor of 3 dB. The $(4,3)$-precoder introduced above has a crest factor of 3 dB, which implies that we can transmit 3 bits in parallel using four codes with the same low CF as for one-code transmission. In addition to the decrease in envelope variation, a coding gain of about 1.5 dB compared to one-code transmission is introduced by the precoder.

Examples of other existing precoder parameters are given in Table 5.2. Given are also the CF in dB, the uncoded CF using k-code transmission, and the reduction in crest factor due to the precoder. As seen, the (10,4) and (16,9) precoders also results in the optimum CF of 3 dB. See [11] for more details regarding construction and performance of these precoders.

Table 5.2

Some Possible Precoder Parameters

(n,k)	CF [dB]	$CF_{uncoded}$ [dB]	$CF-CF_{uncoded}$ [dB]
(4,3)	3	7.8	4.8
(6,5)	5.6	10	4.4
(8,7)	9.5	11.5	2
(8,6)	6	10.8	4.8
(10,9)	7.4	12.6	5.2
(10,4)	3	9	6
(12,11)	7.8	13.4	5.6
(12,8)	4.3	12	7.7
(16,9)	3	12.6	9.6

5.3 EVALUATION OF TURBO CODES WITH TD-CDMA

In the following sections, the evaluation of turbo codes in the TD-CDMA concept to be used as the UTRA TDD mode is presented. First, in Section 5.3.1 the performance of turbo codes is compared to the performance of concatenated Reed-Solomon (RS) and convolutional codes by means of the results of link-level simulation. In the subsequent sections, the influence of different turbo code parameters on the system performance is considered. In detail Section 5.3.2 deals with the required minimum interleaver block size of turbo codes, Section 5.3.3 considers turbo codes with optimized interleaving, and in Section 5.3.4 the performance of turbo codes using constituent recursive-systematic-codes (RSCs) with constraint lengths 3 and 4 is compared. Finally, conclusions are drawn in Section 5.3.5.

Based on the link-level simulation programs that have been developed by the Research Group for RF Communications at the University of Kaiserslautern during the past several years, a new simulation program has been written, that is capable of performing link level simulations for TD-CDMA. This program is implemented in FORTRAN90 and was used to obtain the simulation results presented in the following sections.

5.3.1 Performance of Turbo Codes Versus Concatenated Reed-Solomon and Convolutional Codes

In the current section, the performance of concatenations of outer RS codes and inner convolutional codes and mainly the promising turbo codes is evaluated and presented. The basic principles of turbo codes are assumed to be known and not considered here, see [12, 13]. LCD services with 300 ms delay and LDD services with 50-ms delay using data rates of 64, 144, 384, 512 Kbps, and 2 Mbps are considered.

All previously enumerated service implementations have been simulated for three different mobile environments. These are characterized by the mobile radio channel models *indoor office A* (abbr. indoor*), outdoor to indoor and pedestrian A* (abbr. pedestrian) and *vehicular high antenna A* (abbr. vehicular), and were specified by the ITU, [14]. In the case of the indoor and the pedestrian channels, a mobile speed of 3 km/h and a closed loop frame-by-frame power control were used in the simulations. In the case of the vehicular channel, a mobile speed of 120 km/h was chosen and no power control was applied. The uplink simulations were carried out using two omnidirectional receiver antennas. In the downlink simulations, no antenna diversity was used. Except for the 64 Kbps service simulated with two simultaneously active users, there is no difference between the uplink and the downlink, because multiple access interference is only caused by pooled codes of one and the same user.

As an example, the performances of a turbo code and a concatenated RS and convolutional code for an LCD service with 300-ms delay and 144 Kbps data rate, evaluated for different environments, will be discussed. A turbo code with constraint length 3 and a total code rate of 0.58 is implemented, using optimum maximum-a-posteriori (MAP) symbol estimation in the decoder and a nonoptimized interleaving of the user data block. The user data block for one channel interleaving frame consists of 43,200 bits and is split into five single user data blocks of 8,640 bits which are coded and decoded separately. This means that the turbo code interleaver block size is in this case 8,640 bits. After five iterations the service implementation using the turbo code needs about 4.3 dB E_b/N_0 at the receiver input to reach the target bit error probability of 10^{-6}.

The service protected by the code concatenation is implemented with a (127,120) RS code using 8-bit symbols, and a convolutional code with constraint length 9 and mother code rate 0.5. The total code rate of the code concatenation is 0.58 and the target bit error probability of 10^{-6} is reached at 5.9 dB. In this case, the turbo code implementation of the considered service is about 1.5 dB better than the simulated code concatenation. The computational complexity of the two considered schemes is approximately the same.

For almost all simulated services and environments, the turbo code implementation of the considered service shows a gain of approximately 1 to 3 dB compared to the concatenated codes. In some cases for LDD services and the vehicular channel simulated without power control at a mobile speed of 120 km/h, the bit error curve according to the code concatenation even runs into an error floor above the target bit error probability of 10^{-6}. The turbo code reaches the target bit error probability even with these simulation parameters.

Since the concatenation of the RS code and the convolutional code is not optimized, the performance of the code concatenation could be improved by harmonizing the codes. Nevertheless, the service implementation with turbo codes is still expected to require an E_b/N_0 at the receiver input, to reach the target bit error probability of 10^{-6}, which is significantly below that of an optimized concatenated RS and convolutional code.

5.3.2 Minimum Required Turbo Code Interleaver Block Size

In order to evaluate the influence of the interleaver block size of the turbo code interleaver on the error correction capability of the turbo code, see also [15], several simulations were carried out, and the obtained results are presented in this section. One LCD service with 300 ms delay and one LDD service with 50 ms delay were investigated, both providing a data rate of 144 Kbps and using one time slot per frame with pooling of nine CDMA codes. In the case of the LCD service, the channel interleaving frame consists of 30 frames and the total user data block size for one interleaving block is 43,200 bits. For the LDD service, the user block size of one channel interleaving frame is 7,200 bits, and the block is interleaved over five consecutive frames.

For all simulations the indoor channel at a mobile speed of 3 km/h was chosen, using frame by frame power control implemented in a closed loop fashion. No antenna diversity is applied, and since there is only one active user, the simulations are valid for the downlink as well as for the uplink. A turbo code with constraint length 3 and optimum MAP symbol estimators in the decoder is considered, and five decoding iterations are carried out. For fixed parameter settings, bit error probability curves for different sizes of the turbo code interleaver were simulated using the indoor channel at 3 km/h. In the case of the 50 ms delay service, block lengths of 200, 400, 600, 1,200, 2,400, 3,600 and 7,200 bits were considered and for the LCD service with 300 ms delay block lengths of 200, 400, 600, 1,200, 2,400, 4,800, 14,400 and 43,200 bits were evaluated. In the case of the maximum possible interleaver size of 7,200 bits for the LDD service and 43,200 bits for the LCD service, the user data block of one channel interleaving frame was coded as a whole. For smaller interleaver sizes, the channel interleaving user data block was split into smaller partitions and each partition is coded and decoded separately. The pseudorandom turbo code interleavers are not optimized and were implemented according to the scheme described in Section 5.3.3.

Both for the service with 300 ms delay and for the service with 50 ms delay, the bit error curves get steeper with increasing block length and reach the target bit error probability of 10^{-6} at decreasing E_b/N_0. For the LDD service, an E_b/N_0 of 9.5, 9 dB, approximately 8.6 and 8.2 dB is needed to achieve the target bit error rate of 10^{-6} in the case of the block lengths of 200 bits, 400 bits, 600 bits, and 1,200 bits, respectively. Using a block length of 1,200 bits instead of a block length of 200 bits we achieve a gain of 1.3 dB. Larger block lengths as 1,200 bits lead only to little additional performance improvement. Therefore, in the case of the LDD services with 50 ms delay, a block length of 1,200 bits is sufficient, if we assume non-optimized turbo code interleavers. As will be shown in Section 5.3.3, it is possible to achieve the same performance for block lengths shorter than 1,200 bits, if the interleaver is optimized.

In case of the LCD services with 300 ms delay and the block lengths of 200 bits, 400 bits and 600 bits the target bit error probability of 10^{-6} is reached at 7.9 dB, approximately 7 dB and approximately 6.5 dB E_b/N_0, respectively. An E_b/N_0 of 5.8 and 5.2 dB is required for the block lengths of 1,200 bits and 2,400 bits, respectively. Using block lengths larger than 2,400 bits does not significantly enhance performance. For the LCD services a block length of 2,400 bits is sufficient to achieve acceptable performance. If optimized turbo interleavers are applied, even shorter block lengths can be used.

5.3.3 Optimization of Turbo Code Interleaving

In this section, the performance enhancement achievable by an optimization of the turbo code interleaver will be presented. In all simulated cases an LCD service with 300 ms delay and a data rate of 144 Kbps is considered. The simulations were carried out using the indoor channel model with a mobile speed of 3 km/h. A frame by frame closed loop power control was deployed, and only one receiver antenna is used. Since code pooling in one time slot is used to reach the envisaged data rate and only one user is active within one time slot, the presented results are valid for both the uplink and the downlink, respectively.

Three different turbo code interleaver block sizes were evaluated, namely a 600-, 1,200-, and a 4,800-bit interleaver. The nonoptimized interleavers were chosen as follows: Five different interleavers were obtained by a random generator. The interleavers were superficially tested using identical simulation parameters, and the best one was chosen and used for this evaluation. Therefore, the nonoptimized pseudo random interleavers used here are considered as interleavers with average performance. The optimization process of the optimized interleavers used in the current evaluation is described in [16] and based on the weight distribution of the overall turbo code and bit error probability simulations carried out for an AWGN channel. The particular turbo code interleaver with the most favorable performance at a bit error probability of 10^{-6} for each block size was chosen.

The simulation results will be discussed in the following. For an interleaver size of 600 bits at the target bit error probability of 10^{-6}, the turbo code using the optimized interleaver structure shows a gain of about 1 dB compared to the non-optimized turbo code. For the larger interleaver sizes of 1,200 bits and 4,800 bits, which were simulated equivalently to the size of 600 bits, the gain decreases. For the block size of 1,200 bits, the gain at the same target bit error probability is in the order of 0.5 dB, whereas no gain can be obtained in the case of an interleaver block size of 4,800 bits, as the simulations have shown. Therefore, interleaver optimization is only necessary and feasible for small block sizes and, therefore, for low delay or low data rate services.

5.3.4   Constraint Length of Constituent Recursive-Systematic-Codes

In order to evaluate the performance improvement of turbo codes using constituent RSCs with constraint length 3 as compared to constraint length 4, an LCD service with 300 ms delay and a data rate of 144 kpbs was simulated for both constraint lengths. The indoor channel at a mobile speed of 3 km/h was chosen and frame by frame power control was applied. Antenna diversity was not taken into account. The simulated bit error probability, with five decoding iterations and constraint length 3, reaches a bit error rate of 10^{-6} at about 5 dB. The performance with constraint length 4 shows an improvement of about 0.1 dB for all bit error rates.

The computational complexity of the turbo decoder directly depends on the number of states of the trellis diagram of the constituent RSCs, whereas the number of states can be determined from the constraint length of the codes. The constraint lengths 3 and 4, which are related to a 4 and 8-state trellis diagram, respectively, are most interesting because implementation is feasible and performance is satisfying. As shown in Section 5.3.1, a turbo code using RSCs with constraint length 3 performs significantly better than other coding schemes with comparable complexity. Considering current simulation results, the performance of the 8-state turbo code is slightly better than the performance of the 4-state turbo code, with a gain about 0.1 dB for all bit error probabilities. This performance gain cannot be justified by complexity.

5.3.5   Concluding Remarks

The performance comparison presented in Section 5.3.1 showed that the turbo code outperforms the concatenated RS and convolutional code in almost all considered environments and parameter sets by 1 to 3 dB. Optimizing the code concatenation will reduce the gain of the turbo code by a small amount. For the presented services, a turbo code interleaver size of 2,400 bits is sufficient, if the interleaving is not optimized. Using optimized interleaving, the sufficient block size can be reduced to 600 bits. The complexity of the channel coding scheme is of major importance for the later implementation. In Section 5.3.4 it was shown that a turbo code using RSCs with constraint length 3 performs almost as well as one using RSCs with constraint length 4. In additional simulations the performance of a suboptimum turbo code implementation was evaluated, yielding the result that the performance does not significantly degrade and implementation is therefore feasible thanks to the reduced complexity. The presented results clearly show the superiority of the turbo codes in terms of performance and complexity compared to the concatenation of an RS code with a convolutional code. For more detailed results on the presented topics, please refer to [17].

5.4   OPTIMUM DISTANCE SPECTRUM CONVOLUTIONAL CODES

Since there is no known approach to find good convolutional encoders analytically, good convolutional encoders are normally found by performing a computer search. Such a search can be quite tedious since a large number of combinations must be tested. For a rate $R=1/n$ encoder with constraint length K, there are 2^{Kn} possible combinations. Due to this, previous work has often been restricted to finding maximum free distance (d_f) codes. The search must then go on until a code with maximum free distance is found (fulfilling the Heller bound) or until all combinations are tested.   Since there usually are several codes with maximum free distance this approach is significantly less complex than an exhaustive search. The maximum free distance is an important parameter for Gaussian channels, but there are also other parameters that have impact on the error rate of convolutional codes.   This has motivated a search for improved convolutional codes that have good performance on both AWGN and Rayleigh fading channels [18, 19].

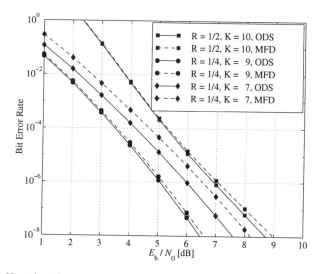

Figure 5.3   Upper bound on the bit error rate on an uncorrelated flat Rayleigh fading channel for ODS and maximum free distance (MFD) codes.

The codes described in this chapter are found using a criterion denoted the optimum distance spectrum (ODS) criterion. When performing a search using the ODS criterion, we compare the error weight coefficients of all possible encoders with maximum free distance d_f and select the one(s) with the smallest error weight coefficient value(s). If more than one encoder remains after this selection, we continue comparing the coefficients for d_f+1, d_f+2, ... until one encoder, or a set of equivalent encoders, remain(s). For more details on this search and the code tables, the reader is referred to [18, 19] and the reference therein. ODS codes are used as the basis for finding different types of good rate-compatible convolutional codes for CDMA systems in later sections. Figure 5.3 shows a comparison of the union upper bound on the bit error probability for maximum free distance codes found in [20] and [21], compared with ODS codes [18, 19] on a perfectly interleaved Rayleigh fading channel. We see that the improvement we get at no complexity cost is up to 0.6 dB.

5.5   CONVOLUTIONAL CODES FOR RATE MATCHING

Wireless communication systems are usually designed so that a channel has a fixed data rate. For example, in a TDMA system, a time slot has a fixed number of bits over a given time period and thus fixed data rate. Similarly, a carrier frequency in a FDMA system, and a spreading code in a CDMA system with a fixed spreading factor correspond to a constant data rate, respectively. Multiple data rates can thus be achieved by allocating several of these fixed data rate channels, resulting in a multichannel system. A typical value of the data rate of a channel is in the order of 10 to 50 Kbps. However, the variability in, for example, speech is in steps in the order of a few Kbps, requiring a rate matching system as an interface to the allocated channels. In this section we propose using *rate-compatible convolutional* (RCC) codes for rate matching and error control in wireless multichannel systems. For further details, see [18, 22-24].

RCC codes are constructed such that lower code rates make use of the same code symbols as the higher code rate plus some extra redundancy symbols. This can easily be obtained by repeating symbols. However, repetition usually results in worse performance than nesting or puncturing. Thus, we have chosen to use *rate-compatible punctured convolutional codes* (RCPC) for the higher code rates and *nested convolutional codes* for the lower code rates.

RCPC codes [25] are constructed by puncturing a convolutional code of rate $R=1/n$ and constraint length K, called the parent code. This code is completely specified by its generator polynomials. The puncturing is done according to a rate compatibility criterion, which requires that lower rate codes use the same coded bits as the higher rate codes plus one or more additional bit(s). The bits to be punctured are described by puncturing matrix consisting of zeros and ones, where zero means that the corresponding coded bit is punctured (deleted). The number of columns, or the puncturing period p, determines the number of code rates and the rate resolution that can be obtained. Generally, from a parent code of rate $1/n$, we obtain a family of $(n-1)p$ different codes with rates from $1/n$ up to $p/(p+1)$. Due to the rate compatibility criterion, the code rate of RCPC codes can be changed during transmission and thus unequal error protection can be obtained [25, 26]. Another application for these codes is in hybrid type-II ARQ systems, as discussed in Section 5.8.2 and [27].

Nested convolutional codes [28, 29] are obtained by extending a code of rate $1/n$ to a rate $1/(n+1)$ code by searching for the best additional generator polynomial. It is obvious that this type of code family is rate-compatible, and the big advantage is the modular code design that reduces the complexity of the search for low-rate codes.

In [18, 23] we present RCPC codes with constraint lengths $K=7$-11 using rate $R=1/4$ ODS parent encoders. Furthermore, these code sets are extended using nested codes down to rate $R=1/512$. It is also interesting to note that the nested codes are maximum free distance codes. It is furthermore shown that a puncturing period of $p=8$ and parent code rate of $R=1/4$ is a good compromise between a flexible coding scheme with many rates and good performance. Given these parameters, the RCPC codes were shown to have a performance close to that of the unpunctured ODS codes at rates 1/2, 1/3, and 1/4. We can therefore conclude that the performance loss due to puncturing is very small.

As an example of rate matching, consider a DS-CDMA system where multiple data rates are obtained by a *multicode* scheme [9], letting each user use multiple spreading codes and transmit data on these in parallel. With fixed spreading there will be a number of fixed-rate channels available. If the system is to support any source rate, there is a need for matching the source rate to a multiple of the channel rate. By using the proposed RCC codes, we have many different code rates with different levels of error protection, and a flexible means of matching the source data rate to the rate of the parallel channels [22]. However, when lower rate coding is applied, more subchannels are needed to transmit the channel symbols. This results in more interference to the other users of the system. It is therefore of interest to investigate the performance of the multicode DS-CDMA system as the code rate is decreased. The efficiencies obtained for a bit error rate of 10^{-6} on a Gaussian and Rayleigh fading channel, respectively, are shown in Figure 5.4 for spreading factor $N=128$, and $E_b/N_0=10$ dB. As we see, the efficiency is increased with decreasing code rate. Thus, the extra coding gain obtained by reducing the code rate is larger than the reduction in the effective signal-to-noise ratio. We also see that the performance difference between Gaussian and Rayleigh fading channels decreases with decreased code rate. This is due to the increased diversity gain provided by the channel code as the code rate is reduced. With high diversity order, the performance on the Rayleigh fading channel approaches that of the Gaussian channel.

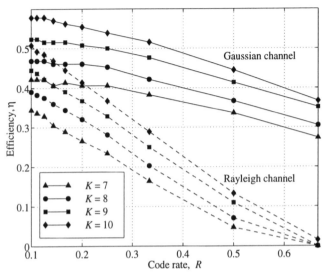

Figure 5.4   Achievable bandwidth efficiency for a multicode DS-CDMA system using rate-compatible codes on Gaussian (solid lines) and Rayleigh fading channels (dashed line).

5.6   SEQUENTIAL DECODING OF CONVOLUTIONAL CODES ON FADING CHANNELS

Sequential decoding is a decoding technique for convolutional codes that received a lot of attention before the Viterbi algorithm was known. To obtain extremely low error probability with convolutional codes, very long constraint lengths must be applied. The Viterbi algorithm is then too computationally intensive. For such systems, sequential decoding is still an attractive decoding method, making decoding of constraint length 30 to 50 convolutional codes possible. Overviews of sequential decoding techniques can be found, for instance in, [2]. With sequential decoding, only a fraction of the code tree is actually searched, reducing the average computational load considerably. The great disadvantage is that the number of computations needed to decode a bit is a random variable with a Pareto distribution [2]. Therefore, we can never avoid input buffer overflows. However, the overflow rate is reduced by increasing the buffer size, or increasing the speed of the decoder. It is possible to evaluate for which E_b/N_0, corresponding to $\rho=1$ and $\rho=2$ respectively, the mean and variance of the number of computations are limited. Table 5.3 shows those values for a number of code rates using BPSK modulation on a Rayleigh fading channel. For more details on how these results are obtained, the interested reader is referred to the papers by Orten and Svensson [30, 31] where also other modulation methods and hard decision decoding are reported.

Table 5.3

Theoretical limits for sequential decoding of convolutional codes on Rayleigh fading channels for different code rates. Energy per information bit E_b/N_0 values are given in dB.

R	1/4	1/3	1/2	2/3	3/4	4/5	5/6	6/7	7/8	8/9
$\rho=1$	3.8	4.4	5.7	7.5	8.8	9.8	10.6	11.3	11.9	12.4
$\rho=2$	5.9	6.5	8.0	10.0	11.3	12.3	13.2	13.8	14.5	15.0

Figure 5.5 shows the overflow rate as a function of E_b/N_0 for hard and soft decisions on a Rayleigh fading channel using BPSK modulation obtained by simulations. We see that simulation results and theoretical limits match quite well. All theoretical results assume a memoryless channel that can be obtained only by perfect (infinite) interleaving. In Figure 5.6, we have therefore plotted the overflow rate for different interleaving depths. As we see from the figure, the loss is quite small if we can afford to use a 50×50 interleaver. Naturally, the loss increases as the interleaver size is reduced.

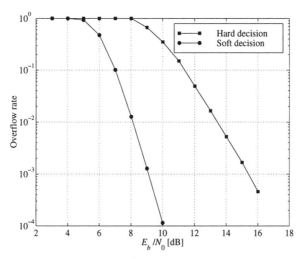

Figure 5.5   The overflow rate as a function of average E_b/N_0 for coherent BPSK with perfect interleaving, obtained by simulations. Results are shown for a hard decision and 8-level soft decisions. An overflow occurs when the total number of forward searches exceeds two times the data block length of 500 information bits.

For a Rayleigh fading channel, the calculation of the Fano metric used with sequential decoding is quite cumbersome. A natural way of avoiding this complexity is to quantize the signal and do a table lookup for each soft decision to find the metric. In [31] there is an evaluation on how the quantization should be done and the loss induced by quantization.

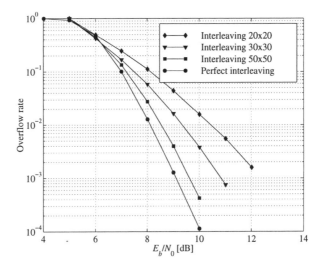

Figure 5.6 Overflow rate as function of E_b/N_0 for BPSK modulation with different interleaving depths and 8-level soft decisions, normalized Doppler frequency fdTs = 0.01.

5.7 CODE SPREAD CDMA

5.7.1 Low-Rate-Coded CDMA Based on Convolutional Codes

In DS-CDMA systems, the bandwidth is expanded by using spreading codes. Bandwidth expansion can also be obtained by the redundancy added by error correcting codes. In a conventional narrowband communication system, this bandwidth expansion is generally an undesired feature. However, for spread-spectrum systems, it has been shown that high efficiency is achievable by employing low-rate channel codes alone for bandwidth expansion [32, 33]. The conventional direct-sequence spreading operation may be viewed as repetition coding followed by randomization. It is of course possible to replace the repetition encoder by a designed channel encoder of the same rate. We will refer to spreading by channel codes only as *combined coding and spreading* or *code-spreading*. A limiting factor has been the lack of good low-rate codes. The work of [32] proposes the use of *orthogonal* convolutional codes [34], and in [35] these codes were modified into the class of *superorthogonal* convolutional codes.

Figure 5.7 shows a schematic of a code-spread CDMA system, where all bandwidth expansion is achieved by a low-rate (rate $1/n$) convolutional code, producing n-coded symbols per information bit. These symbols are then interleaved, and randomized by a nonspreading, user-specific, long pseudo-random scrambling sequence. The kth user's signal, $s_k(t)$, is transmitted over the mobile radio channel, and at the receiver side, similar signals from other users (MAI) are added, producing the resulting received signal.

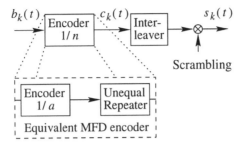

Figure 5.7 The code-spread CDMA system. The very low-rate MFD codes used for combined coding and spreading can be viewed as a higher rate $1/a$ code followed by a device repeating different encoded bits.

Frenger et al. presented low-rate convolutional codes for combined coding and spreading in [36, 37]. These codes are nested codes, obtained by starting with a rate $R = 1/4$ parent convolutional encoder. An advantage with these new codes is that they, in addition to having MFD, have an unequal repetition structure. Thus, the codes can be constructed as a rate $1/a$ code, where $a<<n$, followed by an unequal repeater. As an example, we show in Table 5.4 the generators of the low-rate code family with $K = 8$. We see that for a total spreading of 200, one generator is used 57 times, another 57 times, and other generators are used 28, 27, and 27 times, respectively. This structure greatly reduces the encoder and decoder complexities and may also used to facilitate synchronization.

Table 5.4

Example of encoders for rate $R = 1/n$ codes with constraint length $K = 8$, illustrating the unequal repetition structure of the code family. The table shows the generator polynomials in octal format (top row) and their frequencies.

n	231	247	273	275	327	337	345	353	373	375
4	1	0	1	0	1	0	0	0	0	1
17	1	3	1	3	3	0	3	2	1	1
18	1	3	1	3	3	1	3	2	1	1
100	1	28	1	14	25	12	13	4	1	1
200	1	57	1	28	53	27	27	4	1	1
500	1	143	1	71	139	69	70	4	1	1

In Figure 5.8, we show the calculated efficiency (the integer number of users that the system can support divided by the total spreading) on a Rayleigh fading channel at BER $= 10^{-6}$ and $E_b/N_0 = 10$ dB, assuming perfect interleaving. The low-rate nested code used for the code-spread system has constraint length $K = 10$. Also shown are the results obtained with symbol and chip-interleaved conventional systems with a rate $R = 1/4$ ODS encoder of the same constraint length, and direct-sequence spreading with a random spreading sequence to the same total bandwidth. The efficiency of code-spread systems employing orthogonal codes and superorthogonal codes is also plotted. We assume random spreading sequences for the conventional systems, and that all systems have an outer scrambling sequence that is pseudorandom and much longer than the bit duration. We further assume BPSK modulation and that the users transmit asynchronously. We see that the code-spread system using the presented low-rate encoders and no conventional spreading outperforms all the other schemes. Due to the dependence between code rate and constraint length, the superorthogonal codes can compete only when the spreading is 256. The conventional symbol-interleaved DS-CDMA system with rate 1/4 coding achieves less than 70 % of the efficiency obtained by the proposed code-spread system. As expected, the orthogonal codes have even worse performance, and achieve at best the same efficiency as the conventional system at a spreading factor equal to 256. For more information see [36, 37].

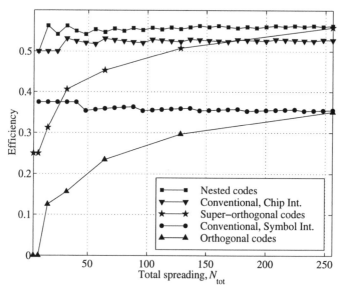

Figure 5.8 The efficiency of a code-spread system with constraint length $K = 10$ nested encoders for different spreading factors and bit error rate 10^{-6}. Also shown are the efficiency results of a conventional system using a $R = 1/4$, $K = 10$ convolutional code and code-spread systems using orthogonal and superorthogonal convolutional codes.

5.7.2   Interference Cancellation for Low-Rate Coded CDMA

The BER performance of a code-spread CDMA system employing PIC [38] is shown in Figure 5.9. The outputs of the channel decoders of each user are used to regenerate the signals to be canceled in the next IC stage. Results from simulations are shown using solid lines, and white and black markers are used for one stage and two stages of PIC, respectively. Using dashed lines, we also show analytical results obtained by approximating the interference from the other users as

Gaussian distributed and employing the union upper bound of the convolutional channel encoder. We define the load as the ratio between the total data rate of all users and the chip rate. We see that when reducing the number of users in the system, the BER decreases rapidly. Eventually a plateau is reached where the BER stays constant. At this plateau, effectively all interference is removed by the interference cancellation scheme, and thus we have reached the single user bound for the low-rate channel code used. We see that the performance of code-spread CDMA systems is greatly enhanced when combined with this kind of interference-reduction technique. Further details are given in [39].

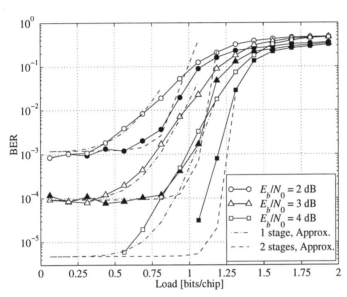

Figure 5.9 Bit error rate for the code-spread system with PIC versus the load for $E_b/N_0 = 2$, 3, and 4 dB. Simulated results for one stage (white markers) and two stages (black markers) of PIC, as well as analytical results (dash-dotted and dashed lines) are shown.

5.7.3 TCH Codes in CDMA

The new family of TCH (Tomlinson, Cercas, Hughes) codes [40, 41] was studied for the radio interface of a mobile receiver. TCH codes are a class of binary, nonlinear, nonsystematic, and cyclic block codes. The code word length is $n=2^m$, where m is any positive integer. The first polynomial in a TCH code, designated as *basic TCH polynomial* or B-TCH, is generated analytically and is then extended to increase the code set using known methods. It was shown in [41] that B-TCH polynomials can only be generated for specific values of the code length n, which are exactly the Fermat numbers minus one, i.e., 2, 4, 16, 256, and 65,536. A very important property of these polynomials is that its autocorrelation function is always three-valued with values 0, –4, and *n*, regardless of the sequence length *n*, which makes its autocorrelation function nearly ideal for large *n*.

TCH codes were initially applied for error correction, either alone or concatenated with a Reed Solomon code. A tight upper bound for its probability of information bit error on an AWGN is available [41]. The performance has also been determined for other channel models, such as a Rayleigh fading channel [42]. The performance of TCH codes is similar to the performance of other error correction codes of the same length and rate, such as BCH codes.

The great advantage of TCH codes is therefore not in terms of coding gain but the simplicity and efficiency of its receiver. In fact, the structure of the receiver depicted in Figure 5.10 performs maximum likelihood soft-decision decoding in the frequency domain with a very small number of correlations. The TCH receiver shown needs only a cycle of one complex FFT, a complex multiplication, and a complex IFFT to perform 4*n* correlations.

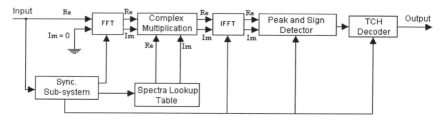

Figure 5.10 TCH receiver or TCH maximum likelihood decoder.

The TCH receiver shown needs only a cycle of one complex FFT, a complex multiplication, and a complex IFFT to perform 4*n* correlations. For instance, if we take a TCH(256, 15) code that has 32,768 code words, the TCH receiver needs only a maximum of 32 correlations in the frequency domain to identify the most likely code word sent, that is, a complexity gain in the order of 1,000.

The fact that TCH codes possess a good autocorrelation and, simultaneously, a well-behaved crosscorrelation function suggested its use in CDMA. In this case, the code words are taken as possible sequences. Moreover, it was found that several sets of sequences derived from B-TCH polynomials are orthogonal, although its use should be more appropriate for a synchronous CDMA system. A new approach was then considered for testing TCH codes in a normal asynchronous CDMA environment. Since TCH codes are inherently low-rate codes, which can be difficult to find in some applications [36], we took advantage of that fact and tested them in a so-called code spreading application, (i.e. with all the spreading done by the FEC code) (compare Section 5.7.1).

This situation was simulated for an asynchronous AWGN DS-CDMA channel, with an E_b/N_0 of 10 dB and an overall spreading factor of $N_{tot} = 32$. The results are shown in Figure 5.11, where we can observe that TCH codes allow a larger or equal system load than the other considered codes.

Similar results were obtained for an uncorrelated flat Rayleigh fading channel with perfect channel estimates [43]. Concluding, we say that TCH codes may be applied in both FEC and in a DS-CDMA system exhibiting a reasonable performance. Their major advantage relies on the structure of the TCH receiver, both in terms of simplicity of implementation and correlation speed, as desired for the mobile receivers of future generations.

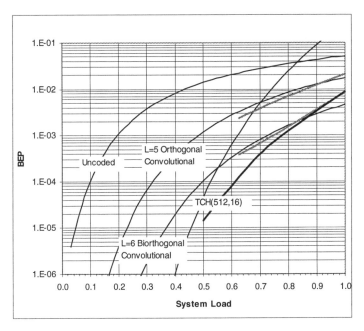

Figure 5.11 Upper bounds and simulation results on performance of various codes in an asynchronous AWGN DS-CDMA channel, $E_b/N_0 = 10$ dB, $N_{tot} = 32$.

5.8 CODING FOR PACKET DATA TRANSMISSION

In some data applications, very low bit error probabilities are required. When a feedback channel is available, this may be obtained by using ARQ protocols. In these protocols, the transmitted data is encoded for error detection rather than error correction and detected errors at the receiver results in a generation of a retransmission request. This protocol is normally used with packet data switching, where a data packet of fixed length is transmitted independent of all other data over the channel. A good overview of this coding technique is given in [2].

The simplest form of ARQ, where the encoded information packet is retransmitted until it is received as a code word, is referred to as simple ARQ. By choosing an error detection code with a large minimum distance or free distance, the undetected error probability (the transmitted information is changed into another code word) can be made as small as required. The main drawback with

ARQ is that the throughput, defined as the inverse of the average number of coded bits needed to transmit one single information bit, becomes quite low unless the channel error probability is low.

Therefore, it is quite common to use error correction coding in combination with error detection coding in ARQ. These kinds of schemes are normally referred to as hybrid type-I ARQ schemes. Since some errors are corrected by the error correction code, the channel as seen by the error detection code improves and the number of retransmissions decrease. The main drawback is that the throughput is reduced by the code rate of the error correction code.

In general, the maximum throughput is given as the product of the code rates of the error correction code and the error detection code. In order to maximize throughput, this means that as little channel coding as possible should be used. This may be quite easy to obtain on static channels, where the channel coding can be optimized for the particular signal-to-noise ratio to be used. On time varying channels, like, for example, a mobile radio channel, the signal-to-noise ratio will instantaneously change such that some packets will be transmitted over a good channel, while others will be transmitted over a bad channel. Then there is no single coding scheme that optimizes the throughput, but an adaptive scheme is needed.

This can be obtained by using hybrid type-II ARQ schemes, which is a kind of code combining scheme. Code combining refers to a scheme, where for each information packet, previously received packets found to be in error are combined with the last received packet before decoding. This is equivalent to using an error correction repetition code and adjusting the code rate to the highest rate that gives no detected errors in the receiver. It is however, well known that repetition codes are inefficient for the given code rate, and therefore optimum error correction codes should be used in place of the repetition code. These optimum codes have to be designed such that they can be used with code combining and therefore need to be rate compatible. Below we consider such schemes based on both turbo codes (Section 5.8.1) and convolutional codes (Section 5.8.2). With convolutional codes, we also look at sequential decoding for long constraints lengths (Section 5.8.3), which makes the error detection code obsolete. In this scheme, the coding rate for error correction can be adapted to the channel and no error detection coding is needed. Therefore, this scheme has the prospect of optimizing the throughput.

5.8.1 Turbo Codes Applied to Type-II Hybrid ARQ Protocols

In order to evaluate the performance of turbo codes for packet data transmission with adaptive coding, two equivalent type-II hybrid ARQ schemes were defined, the first one using convolutional codes and the second one using turbo codes. These ARQ schemes will be described in Section 5.8.1.1. Identical link level simulations with TD-CDMA were carried out for both schemes, yielding comparable results. The obtained simulation results are described and discussed in Section 5.8.1.2, followed by some conclusions in Section 5.8.1.3.

5.8.1.1   Considered ARQ Schemes

A type-II hybrid ARQ protocol is considered using a CRC code for error detection and either a turbo code or a convolutional code for error correction. In both cases, the user information data block with a length of about 200 bits is first encoded for error detection. A CRC code with characteristics (232, 216) and the code polynomial $g_{ANSI}(x)=x^{16}+x^{15}+x^2+1$ generates 16- parity check bits appended to the original user data block, compare with [44]. In the case of the convolutional code, after CRC encoding, 8 tail bits are attached, which are used for trellis termination in the Viterbi decoder inside the receiver.

Then, the CRC encoded data block is encoded for error correction either by a convolutional encoder or a turbo encoder. The convolutional encoder has code rate $1/3$, constraint length 9, and generator polynomials (557, 663, 711) in octal. The turbo encoder has also code rate 1/3 and consists of two identical RSCs with generators (5, 7) in octal and constraint length 3. The outputs of the convolutional or turbo encoders are then multiplexed and mapped to three code words of the same length. The code words are transmitted using the selective repeat ARQ protocol. The number of buffer elements in the transmitter and receiver are assumed infinite. Additionally, the feedback channel is assumed to be error-free. Three different code rates 1, $1/2$, and $1/3$ can be achieved depending on the actual channel conditions.

It should be mentioned that all parameter settings and protocol-specific attributes were pragmatically chosen with the aim of not achieving an absolute performance measure but obtaining comparable results for the application of turbo codes and convolutional codes for ARQ protocols. For more detailed information on type-II hybrid ARQ protocols with rate-compatible, punctured convolutional codes, refer to [25, 27], and Section 5.8.2.

5.8.1.2   Simulation Results and Discussion

For the indoor and the pedestrian channels (as defined in Section 5.3.1) with a mobile speed of 3 km/h, the scheme using turbo codes needs an E_b/N_0 of 8.1 dB to obtain 144 Kbps, whereas the scheme using the convolutional code requires 8.6 dB. For the vehicular channel at 120 km/h, an E_b/N_0 of 10.0 dB and 10.5 dB is needed in the case of the turbo code and the convolutional code, respectively. These results are with a single receiver antenna. By introducing antenna diversity with two omnidirectional receiver antennas, a diversity gain of about 4 dB can be achieved, but the 2 dB loss for the vehicular channel still remains. The general conclusion based on the obtained results is that the gain of the turbo code is 0.5 dB. The comparison is justified by the fact that the computational effort of the two considered schemes are almost equal when five iterations are used with the turbo code as discussed in Section 5.3.1.

The performance of turbo codes strongly depends on the choice of the turbo interleaver and the block size. It was shown that the performance of turbo codes in the case of circuit switched services and in terms of the measured bit error probability can be significantly improved by using optimized interleavers, especially in the case of very short block sizes around and below 600 bits [17]. In the current performance comparison, optimized interleavers were not utilized. Therefore, we assume that the performance of the turbo code using type-II hybrid ARQ protocol can be significantly improved if optimized turbo interleavers are applied. The use of longer data packets should also lead to a performance improvement in the case of the turbo code. Throughout the current evaluation, an ordinary turbo decoder was used, but the iterative turbo decoding process also has the potential of further performance improvement [45, 46].

5.8.1.3   Conclusions

The superiority of turbo codes over comparable coding schemes like concatenated Reed-Solomon and convolutional codes was already shown for the case of circuit switched services for various data rates and environments [17]. The current evaluation has proven that turbo codes are also well suited for the protection of packet data in the case of type-II hybrid ARQ protocols. It was shown that with turbo codes, a slightly higher throughput can be achieved than with convolutional codes, and it was stated that the performance of turbo codes can improve significantly. Although turbo codes are more beneficial in the case of larger block lengths, they are also superior in the case of packet data with relatively short blocks.

5.8.2   Convolutional Coding With ARQ

We present a comparison between four different hybrid type-II ARQ schemes, which are all based on RCPC codes. We will refer to them as schemes 2-5 [27]. The schemes are shown graphically in Figure 5.12.

Scheme 1 is a simple ARQ scheme with code combining and is used as reference. Data bits plus parity bits for error detection, form a packet of length L_C. Schemes 2-5 are hybrid type-II ARQ schemes that use RCPC codes. Scheme 2 was proposed in [47] and is used here as a comparison. The parent code rate is 1/3, the generator polynomials are (133, 165, 171) in octal form, the constraint length is 7 and a puncturing period of 2 is employed. Information bits, encoder tail bits, and CRC bits form a packet of length L_C or $2L_C$, respectively. This packet is encoded in the rate 1/3 parent code, resulting in a coded sequence of length $3L_C$ or $6L_C$, respectively. The coded sequence is now punctured such that the highest code rate is obtained (half of the bits are punctured for schemes 2-3 and one third of the bits are punctured for schemes 4-5). The nonpunctured bits are interleaved and transmitted over the channel. To transmit the coded bits, three channel blocks are used for scheme 3, two channel blocks are used for scheme 5, while for the other schemes only one channel block is used. The fading is assumed independent between channel blocks, but correlated within the block. The received samples are decoded in a Viterbi decoder and we assume that perfect channel state information (CSI) is available to the decoder. When the code rate is 1, the only available redundant information is the encoder tail. If the CRC fails, a retransmission is requested, otherwise the packet is assumed to be received correctly.

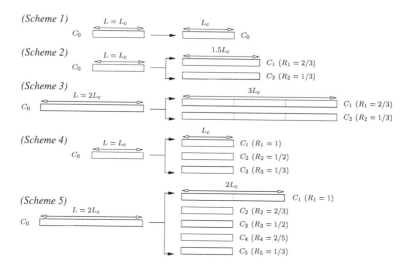

Figure 5.12 Schematic of the considered schemes.

When a retransmission is requested, the second highest code rate is used in the second decoding attempt. The code sequence is punctured according to the second highest code rate, but only those coded bits in the remaining code sequence that were punctured at the highest code rate are interleaved and transmitted. Now only one channel block is used to transmit the coded bits, except for scheme 3 where three blocks are used. In the receiver, these new bits are combined with the previously received bits and decoded in the Viterbi decoder. If the CRC fails, a retransmission is requested and the schemes repeat in a similar way (as shown in Figure 5.12). When all the rate 1/3 coded bits have been transmitted, the procedure is repeated over again. In the receiver, the new received bits are combined with all previously received bits before the Viterbi decoding is done.

In Figure 5.13, the throughput performance of schemes 1-5 on a Rayleigh fading channel with normalized (with the period of the coded symbol) Doppler frequency 0.001 are shown. The number of CRC bits is 16 in all cases and L_C=64. We see that scheme 5 performs better than all the other schemes for all SNRs. By starting at code rate 1, it maximizes the throughput at high SNRs when no error correcting coding is needed, which is not the case with schemes 2-3. At smaller SNRs, it tries to maximize the throughput by adapting the code rate to the SNR and only the increment between the code rates restricts this maximization. Moreover, the transmission delay of these new schemes is comparable to the delay of simple ARQ.

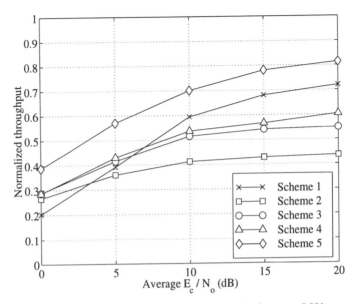

Figure 5.13 Simulated throughput for L_C=64 and normalized Doppler frequency 0.001.

These types of schemes have been thoroughly examined in [27, 48, 49]. The throughput dependency on Doppler frequency, packet length, and constraint length have been evaluated. It was found that a packet length on the order of 150 to 200 bits in most cases gives the best throughput. Furthermore, the constraint length of the code should not necessarily be chosen as large as possible, since the loss due to the extra tail bits in many cases is larger than the gain in throughput due to the increased error correction capability of the code. Scheme 5 was shown to be superior among the chosen schemes also for larger Doppler frequencies, and the throughput gain compared to simple ARQ, for example, becomes larger, since slow fading is advantageous for simple ARQ. Tailbiting Viterbi decoders [50] that do not require tail bits to be transmitted were also tested, but in almost all cases the throughput was reduced due to the increased bit error probability of these decoders. Optimum RCRC codes [51] in combination with RCPC codes were also evaluated but proved to have inferior throughput.

5.8.3 Hybrid ARQ Based on Sequential Decoding of Long Constraint-Length, Tailbiting Convolutional Codes

The conventional ARQ schemes described above require CRC bits in order to determine whether the packet contained errors or not. These check bits carry no information and they therefore reduce the system throughput. By using sequential decoding of long constraint length convolutional codes, the CRC bits are not needed. This is because the computational load, or decoding time, depends on the channel quality. Decoding time can thus be used to indicate erroneous packets (timeout condition), see [52, 53] and the references therein.

Due to the long constraint lengths applied with sequential decoding, a large number of tailbits must be applied in conventional schemes to terminate the data packets and make them independent of each other. Since the number of tailbits is large, the reduction in throughput caused by the many tailbits is significant. By using long constraint length *tailbiting* convolutional codes and sequential decoding, it should therefore be possible to eliminate the loss caused by both CRC bits and tailbits. *Tailbiting* convolutional codes are obtained when using the last $K-1$ information bits in the data packet as the initial state of the convolutional encoder. The encoder is then forced into its initial state as it encodes the last part of the data packet. Thus, flush bits are not needed, and the throughput is significantly increased for long constraint length codes. The disadvantage with tailbiting codes is that the initial state must be estimated by the decoder. In the initial state estimation algorithm proposed by Orten in [52] and [53], a systematic encoder is applied, and $K-1$ systematic bits are used as an estimate on the initial state. If decoding can get started and proceed into the code tree, the initial state is assumed to be correct. Otherwise, a new estimate is obtained by shifting one bit position in the input data stream, and a new decoding attempt is made.

Figure 5.14 shows the throughput as a function of the packet length, with signal to noise ratio E_c/N_0=5dB, which gives a Pareto exponent ρ=2. As we see, analytical and simulated results match rather well. Tailbiting gives higher throughput than using a known tail for packet lengths above 300. The rapid degradation for packet lengths below 300 is due to initial state estimation failure. The algorithm requires that there must be $K-1$ consecutive error-free channel symbols in order to start decoding. By reducing the constraint length slightly, the probability of correct initial state estimation increases considerably, but this will increase the undetected error rate. However, the loss for the shorter packets is only critical for the lower SNRs. By increasing the E_c/N_0 to that required for rate 2/3 and 3/4, the initial state estimation has a low failure rate also for the shorter packets. For plots of the throughput in this case, see [52, 53].

Figure 5.15 shows the throughput as function of E_c/N_0 using an incremental redundancy ARQ scheme. Here we used packet length L_p = 200, L_p = 300, and L_p = 500 and a constraint length K=36 code. The results are obtained analytically. As we see, the gain in throughput by using tailbiting is increasingly significant as the packet length is reduced. However, for the lower signal-to-noise ratios the tailbiting scheme experiences some loss due to initial state estimation failure. For packet length L_p=200 and E_c/N_0 < 6 dB, the tailbiting scheme gives lower throughput than the conventional scheme with known tail. However in most cases, though, it will be more interesting to operate at a signal-to-noise-ratio yielding a higher throughput. Figure 5.15 shows that for a throughput equal to η=0.75 and packet length L_p=200, the gain achieved with tailbiting is close to 5 dB. More details can be found in [52] and [53].

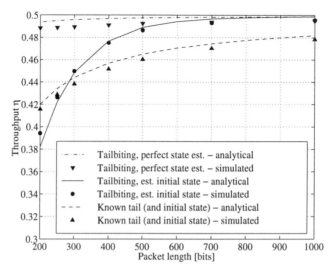

Figure 5.14 Throughput as function of the packet length in information bits. Signal-to-noise ratio is E_c/N_0=5dB, giving channel bit error rate approximately 0.064, and the convolutional encoder used for simulations is a constraint length K=36 systematic encoder with generator 714461625313 (octal notation).

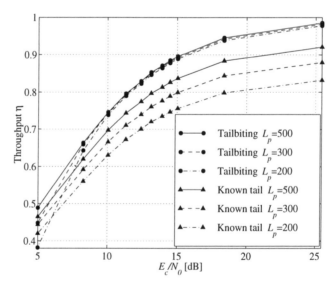

Figure 5.15 Analytical evaluation of the throughput as a function of the signal-to-noise ratio. Results are shown for packet lengths equal to $L_p = 200$, $L_p = 300$, and $L_p = 500$, and constraint length K=36, with and without known initial state (tail) and uncorrelated fading assuming an incremental redundancy ARQ scheme.

References

[1]   J. G. Proakis, *Digital Communications*, McGraw-Hill: New York, 1995.

[2]   S. Wicker, *Error Control Systems for Digital Communications and Storage*, Englewood Cliffs, New Jersey, Prentice Hall, 1995.

[3]   C. Heegard and S. B. Wicker, *Turbo Coding*, Kluwer Academic Publishers, Norwell, 1998.

[4]   T. Ojanperä and R. Prasad, *Wideband CDMA for Third Generation Mobile Communications*, Norwell MA: Artech House, 1998.

[5]   J. B. Anderson, T. Aulin, and C.-E. Sundberg, *Digital Phase Modulation*, Plenum Press, 1986.

[6]   P. A. Laurent, "Exact and Approximate Construction of Digital Phase Modulations by Superimposition of Amplitude Modulate Pulses (AMP)," *IEEE Transactions on Communication*, Vol. COM-34, No. 2, Feb. 1986, pp.150-160.

[7]   M. Moretti, G.J.M. Janssen, and R. Prasad, "Binary and Multilevel Linearised GMSK: Spectrum Efficient Modulation Schemes for Personal Communications," *Proceedings FRAMES Workshop*, Göteborg, Sweden, 1998, pp. 47-52.

[8]   M. Honkanen and S.G. Haggman, "New Aspects on Nonlinear Power Amplifier Modeling in Radio Communication System Simulations," *Proceedings IEEE Personal, Indoor and Mobile Radio Communications Conference*, Helsinki, Finland, 1997, pp. 844-848.

[9]   T. Ottosson and A. Svensson, "On Schemes for Multirate Support in DS-CDMA Systems," *Wireless Personal Communications*, Kluwer Academic Publishers, Vol. 6, No. 3, March 1998, pp. 265-287.

[10]   T. Ottosson, "Precoding in Multicode DS-CDMA Systems," *Proceedings IEEE International Symposium on Information Theory*, Ulm, Germany, 1997, p. 351.

[11]   T. Ottosson, "Precoding for Minimization of Envelope Variations in Multicode DS-CDMA Systems," *Wireless Personal Communications*, Kluwer Academic Publishers, in press.

[12]   C. Berrou, A. Glavieux, and P. Thitimajshima, "Near Shannon Limit Error-Correcting Coding and Decoding: Turbo Codes (1)," *Proceedings of IEEE International Conference on Communications*, Geneva, Switzerland, 1993, pp. 1064-1070.

[13]   C. Berrou and A. Glavieux, "Turbo Codes: General Principles and Applications," *Proceedings of the 6th International Workshop on Digital Communications*, Tirrenia, Italy, 1993, pp. 215-226.

[14]   ETSI Technical Report, "Universal Mobile Telecommunications System (UMTS); Selection Procedures for the Choice of Radio Transmission Technologies of the UMTS," TR 101 112, v. 3.2.0, April 1998.

[15]   H. Koorapaty, Y. P. E. Wang, and K. Balachandran, "Performance of Turbo Codes with Short Frame Sizes," *Proceedings of IEEE Vehicular Technology Conference*, Phoenix, Arizona, May 1997, pp. 329-333.

[16]   P. Jung, J. Plechinger, M. Doetsch and F. M. Berens, "A Pragmatic Approach to Rate Compatible Punctured Turbo Codes for Mobile Radio Applications," *Proceedings of the 6th International Conference on Advances in Communications and Control: Telecommunications/Signal Processing*, Grecotel Imperial, Corfu, Greece, 1997.

[17]   T. Bing and F. Berens, "Parameter Evaluation for Turbo-Codes in the UTRA-TDD-Mode," *FRAMES Workshop*, Delft, The Netherlands, 1999, pp. 260-266.

[18]   P. Frenger, P. Orten, T. Ottosson and A. Svensson, "Multirate Convolutional Codes," Tech. Rep. 21, ISSN-02083, Communication Systems Group, Department of Signals and Systems, Chalmers University of Technology, Sweden, April 1998.

[19]   P. Frenger, P. Orten and T. Ottosson, "Comments and Additions to Recent Papers on New Convolutional Codes," submitted to *IEEE Transactions on Information Theory*, March 1999.

[20]   K. J. Larsen, "Short Convolutional Codes with Maximal Free Distance for Rates 1/2, 1/3 and 1/4," *IEEE Transactions on Information Theory*, Vol. IT-19, May 1973, pp. 371-372.

[21] J. P. Odenwalder, *Optimal Decoding of Convolutional Codes*, Ph.D. Dissertation, School of Engineering and Applied Sciences, University of Califormia, Los Angeles, 1970.

[22] P. Frenger, P. Orten, T. Ottosson and A. Svensson, "Rate Matching in Multichannel Systems Using RCPC-Codes," *Proceedings IEEE Vehicular Technology Conference*, Phoenix, Arizona, 1997, pp. 354-357.

[23] P. Frenger, P. Orten, T. Ottosson, and A. Svensson, "Rate-Compatible Convolutional Codes for Multirate DS-CDMA Systems," *IEEE Transactions on Communications*, in press.

[24] P. Frenger, P. Orten, and T. Ottosson, "Convolutional Codes with Optimum Distance Spectrum," submitted to *IEEE Communications Letters*, July 1997.

[25] J. Hagenauer, "Rate-Compatible Punctured Convolutional Codes (RCPC Codes) and Their Applications," *IEEE Transactions on Communications*, Vol. 36, No. 4, May 1973, pp. 389-400.

[26] J. Hagenauer, N. Seshadri, and C.-E. Sundberg, "The Performance of Rate-Compatible Punctured Convolutional Codes for Digital Mobile Radio," *IEEE Transactions on Communications*, Vol. 38, No. 7, July 1990, pp. 966-980.

[27] S. Falahati and A. Svensson, "Hybrid Type-II ARQ Schemes for Rayleigh Fading Channels," *Proceedings International Conference on Telecommunications*, Vol. I, Chalkidiki, Greece, 1998, pp. 39-44.

[28] P. J. Lee, "New Short Constraint Length Rate 1/N Convolutional Codes Which Minimize the Required SNR for Given Desired Bit Error Rates," *IEEE Transactions on Communications*, Vol. COM-33, No. 2, February 1985, pp. 171-177.

[29] S. Lefrancois and D. Haccoun, "Search Procedures for Very Low Rate Quasi-Optimal Convolutional Codes," *Proceedings IEEE International Symposium on Information Theory*, Trondheim, Norway, 1994, p. 278.

[30] P. Orten and A. Svensson, "Sequential Decoding in Future Mobile Communications," *Proceedings IEEE International Symposium on Personal, Indoor and Mobile Radio Communications*, Helsinki, Finland, 1997, pp. 1186-1190.

[31] P. Orten and A. Svensson, "Sequential Decoding of Convolutional Codes for Rayleigh Fading Channels," *submitted to IEEE Transactions on Communications,* March 1999.

[32] A. J. Viterbi, "Very Low Rate Convolutional Codes for Maximum Theoretical Performance of Spread-Spectrum Multiple-Access Channels," *IEEE Journal on Selected Areas in Communications*, Vol. 8, No. 4, May 1990, pp. 641-649.

[33] J. Y. N. Hui, "Throughput Analysis for Code Division Multiple Accessing of the Spread Spectrum Channel," *IEEE Journal on Selected Areas in Communications*, Vol. SAC-2, No. 4, July 1984, pp. 482-886.

[34] A. J. Viterbi, "Orthogonal Tree Codes for Communication in the Presence of White Gaussian Noise," *IEEE Transactions on Communications*, Vol. COM-15, No. 2, April 1967, pp. 238-242.

[35] A. J. Viterbi, *CDMA Principles of Spread Spectrum Communication*, Reading, MA: Addison Wesley, 1995.

[36] P. Frenger, P. Orten and T. Ottosson, "Code-Spread CDMA Using Maximum Free Distance Low-Rate Convolutional Codes," submitted to *IEEE Transactions on Communications*, February 1998, revised February 1999.

[37] P. Frenger, P. Orten and T. Ottosson, "Code-Spread CDMA Using Low-Rate Convolutional Codes," *Proceedings IEEE International Symposium on Spread Spectrum Techniques and Applications*, Sun City, South Africa, 1998, pp. 374-378.

[38] S. Moshavi, "Multi-User Detection for DS-CDMA Communications," *IEEE Communications Magazine*, Vol. 34, No. 10, October 1996, pp. 124-136,

[39] P. Frenger, P. Orten and T. Ottosson, "Code-Spread CDMA with Interference Cancellation," accepted for publication in *IEEE Journal on Selected Areas in Communications*, 1999.

[40] F. Cercas, M. Tomlinson and A. Albuquerque, "TCH: A New Family of Cyclic Codes Length 2^m," *Proceedings IEEE International Symposium on Information Theory*, San Antonio, Texas, January 1993, p.198.

[41] F. A. B. Cercas, *A New Family of Codes for Simple Receiver Implementation*, Ph.D. Thesis, Technical University of Lisbon, Instituto Superior Técnico, Lisbon, March 1996.

[42] P. Sebastião, *Efficient Simulation of the Performance of TCH Codes Using Stochastic Models* (in Portuguese), M.Sc. Thesis, Instituto Superior Técnico, Lisbon, October 1998.

[43] L. Antunes, *The Application of TCH Codes in CDMA Systems for Mobile Communications* (in Portuguese), M.Sc. Thesis, Instituto Superior Técnico, Lisbon, 1999

[44] S. Lin and D. J. Costello Jr., *Error Control Coding: Fundamentals and Applications*, Englewood Cliffs, New Jersey: Prentice Hall, 1983.

[45] K. R. Narayanan and G. L. Stüber, "A Novel ARQ Technique Using the Turbo Coding Principle," *IEEE Communications Letters*, Vol. 1, No. 2, March 1997.

[46] K. R. Narayanan and G. L. Stüber, "Turbo Decoding for Packet Data Systems," *Proceedings. of the Communication Theory Mini Conference*, Phoenix, Arizona, 1997, pp. 44-48.

[47] H. Lou and A. S. Cheung, "Performance of punctured channel codes with ARQ for multimedia transmission in Rayleigh fading channels," *Proceedings of IEEE Vehicular Technology Conference*, Atlanta, Georgia, 1996, pp. 282-286.

[48] S. Falahati, T. Ottosson, A. Svensson, and L. Zihuai, "Hybrid Type-II ARQ Schemes Based on Convolutional Codes in Wireless Channels," *Proceedings FRAMES workshop*, Delft, The Netherlands, 1999, pp. 225-232.

[49] S. Falahati, T. Ottosson, A. Svensson, and L. Zihuai, "Convolutional Coding and Decoding in Hybrid Type-II ARQ Schemes on Wireless Channels," *Proceedings IEEE Vehicular Technology Conference*, Houston, Texas, 1999.

[50] R.V. Cox and C.-E. Sundberg, "An Efficient Adaptive Circular Viterbi Algorithm for Decoding Generalized Tailbiting Convolutional Codes," *IEEE Transactions on Vehicular Technology*, vol. 43, no. 1, February1994, pp. 57-68.

[51] L. Zihuai and A. Svensson, "New Rate Compatible Repetition Convolutional Codes," submitted to *IEEE Transactions on Information Theory*, February 1999.

[52] P. Orten, "Sequential Decoding of Tailbiting Convolutional Codes for Hybrid ARQ on Rayleigh Fading Channels," submitted to *IEEE Journal on Selected Areas in Communications*, March, 1999.

[53] P. Orten, "Sequential Decoding of Tailbiting Convolutional Codes for Hybrid ARQ on Wireless Channels," *Proceedings IEEE Vehicular Technology Conference*, Houston, Texas, 1999.

Chapter 6

UTRA Transport Control Function

This chapter describes the UTRA transport control function and the principles applied in the research work that has been carried out to define the model. Compared with prior systems, transmission solutions for data originating from both circuit-switched and packet-switched domains are already provided by second generation systems such as GSM/GPRS. What makes the third generation transport control unique in its properties and capabilities is the ability to seamlessly combine an arbitrary number of different variable-rate data sources with a flexible set of transport characteristics. The span is from traditional circuit-switched to traditional packet-switched into radio-efficient transmission respecting the quality requirements of each service. This approach aims to provide service combinations far beyond the reach of second generation systems.

While second generation standards are very precise in describing bit-exact coding for a given radio bearer service, the approach for the third generation has been more to define concepts that can support ranges of parameter values. This results in many alternative ways to map a set of traffic and QoS parameters for radio transmission. Definitions of service capabilities will restrict these limits for given types of terminals, but still maintain the flexibility to provide different kinds of network implementations and operator parameterizations without violating the compatibility of equipment conforming to the standard.

One of the central building blocks of the UTRA transport control is the multirate support provided by the physical layer. The physical layer is able to execute a change of data rate combinations with the maximum frequency of once in a radio frame of 10 ms. Thus, the natural task for medium access control (MAC) is to select the combination to be applied based on offered load from the set of logical channel inputs defined above it. Radio link control (RLC) provides segmentation and retransmission services for both user and control data. Radio resource control (RRC) handles all configuration operations with peer-to-peer control signaling between the network and the terminal, and by acting as a management entity and configuring the operation of all lower radio layers.

The following sections provide further insight into this revolutionary and unique new system created to provide a flexible platform supporting new definitions of end-user services and service combinations far into the next millenium, [1-6]. These references are available on the 3GPP web page (http://www.3GPP.org).

6.1   RADIO INTERFACE PROTOCOL ARCHITECTURE

A simplified protocol architecture of UTRA is illustrated in Figure 6.1. Blocks represent instances of the respective protocols, and service access points (SAPs) for peer-to-peer communication are marked with circles at the interfaces between sublayers. Due to limitations of space, all material throughout Chapter 6 concentrates on core functionalities of the radio protocols striving for a complete and coherent view on the architecture presented. In both research and standardization work, a number of new channels and functional elements have been proposed and added to extend the structure, but for understanding the essential interactions of the different sublayers, the scope used herein is considered optimal.

The radio interface is functionally split into three layers:

- The physical layer (L1);

- The data link layer (L2);

- The network layer (L3).

Layer 2 is further divided into two sublayers, RLC and MAC. The sublayer of layer 3 that belongs to the radio interface protocols is known as RRC [1]. The separation between control plane and user plane is visible down to the level of MAC. RRC together with the RLC entities serving transportation between peer RRC entities belong to the control plane. In MAC, data from both planes can be multiplexed to the same transport resource, resulting in the unification of user and control plane.

Figure 6.1 shows the transport channels provided by physical layer and the logical channels provided by MAC. Transport channels are defined according to how the information is transferred [2], whereas logical channels are defined according to the type of information that is transferred. On the transport channel level, control and user data do not need to be separated (i.e., one and the same transport channel may very well carry RRC signaling messages as well as user data). What distinguishes different types of transport channels are attributes related to physical layer processing, such as whether the bit rate is fixed or variable, or whether power control and/or beamforming is possible. Logical channels, on the other hand, distinguish between control and user data, independent of how the data is to be transmitted.

All transport channels are defined as unidirectional. The broadcast, paging, forward access, and dedicated channels exist in downlink, and random access and dedicated channels in the uplink. Logical channels are considered bidirectional where applicable (broadcast control and paging control only exist in downlink).

The functions defined for RLC are specific to one logical channel, which is why the behavior of RLC is described through one entity (shown as a shadowed block in Figure 6.1) as connected to one logical channel. The functions of MAC address either one common channel or one terminal including the operation on dedicated channels. Therefore, no functional entities specific to one stream of data are shown on MAC.

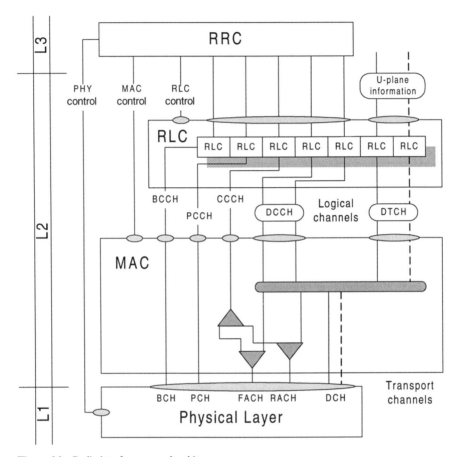

Figure 6.1 Radio interface protocol architecture.

RRC is responsible for configuring all layers of the radio interface protocols. Specific SAPs denoted as PHY control, MAC control, and RLC control are defined to describe the control interfaces. The overall description of the attributes configured by RRC for each lower layer is explained in the associated sections.

6.2 UTRAN ARCHITECTURE

To properly describe the behavior and termination of the protocols, a few words on the assumed radio access network architecture and the related protocol terminations are necessary. The architecture is illustrated in Figure 6.2. A BS controls a number of cells. Each BS is connected to at least one RNC. The connection of the RNC to the core network is outside the scope of this chapter. Where "network" (NW) is used alone, it refers to the UTRAN as presented in Figure 6.2.

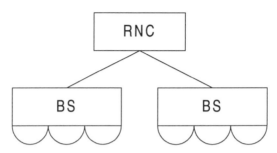

Figure 6.2 UTRAN architecture.

The termination of protocols for common channels and dedicated channels is somewhat different. On common channels the physical layer is terminated in the BS. On dedicated channels, the functions of the physical layer are assumed to include macrodiversity combining in the case of soft handover, effectively placing a part of the physical layer above the macrodiversity combining point in the RNC. All the protocols described in this chapter reside in the RNC.

Furthermore, the role and resulting functions of the RNC can be described as either related to functions specific to a terminal or to a given BS and the cells that exist under it. This explains the split of MAC functions into dedicated and common, as described in Section 6.5. In the scope of the model presented here, RRC, RLC, and the terminal-specific MAC-d always exist in the terminal-specific part of the RNC above the macrodiversity combining point. The other functional blocks of MAC exist in the cell-specific part and have no relation to macrodiversity combining.

6.3 RRC CONNECTION AND MOBILITY

Before any radio bearers can be established, the terminal has to exit from idle mode by establishing an RRC connection. The procedure is triggered by a request from higher layers in the terminal to establish the first signaling connection for the terminal. The establishment of an RRC connection includes an optional cell re-selection, an admission control, and a radio bearer establishment for signaling purposes. The release of an RRC connection can be initiated by a request from higher layers or by the RRC layer itself in the case of an RRC connection failure. After the RRC connection has been established, RRC becomes responsible for handling the mobility of the terminal [1, 5].

The radio interface control functions have been designed to effectively support a large number of terminals using packet data services by providing flexible means to utilize statistical multiplexing. The appeal of packet data services for increasing the number of supported terminals in a network is based on the intermittent nature of data transfer in many interactive applications.

Based on the traffic and QoS characteristics of radio bearers, the UTRA transport control function is designed to extract maximum capacity benefit from the periods of low activity. To accomplish this, a number of states within the RRC connection have been defined.

The RRC state [5] of a given terminal is impacted by the level of activity associated with that terminal. The delay requirement of the most demanding radio bearer will dictate the states applicable to the terminal (e.g., interactive speech service normally means that a dedicated physical channel connection must exist at all times).

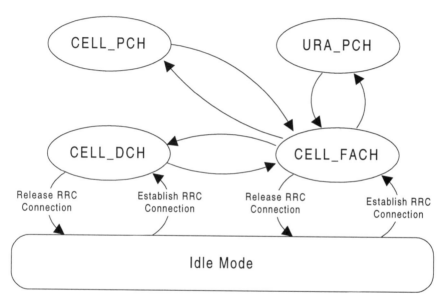

Figure 6.3 RRC states during RRC connection and the relation to idle mode.

The RRC states are shown in Figure 6.3. A brief explanation of the different states and their relation to mobility is given:

- Cell_DCH: The location of the terminal is known on the cell level, and handover procedures are used to track the movements of the terminal. Dedicated physical channels are used for data transmission. Transmission can start immediately without additional allocation delay. If enough delay is allowed, transmission on user-plane radio bearers can be temporarily suspended by the network to assist in radio resource management. The RRC state can be changed to Cell_FACH due to low activity, if all allocated radio bearers can cope with the increased delay in reestablishing dedicated physical resources.

- Cell_FACH: The location of the terminal is known on the cell level and a cell update procedure is applied when a new cell is entered. No dedicated physical connection is maintained, but small amounts of user data can be exchanged on common channels. If the offered load exceeds a threshold parameter, a transition to Cell_DCH is performed. If there is no activity, the RRC transits to Cell_PCH.

- Cell_PCH: The location of the terminal is known on the cell level, and cell updates are performed. No uplink activity is possible within this state. To perform a cell update or to initiate data transmission, the terminal autonomously transits to Cell_FACH. If the transition was due to a cell update, Cell_PCH state is resumed immediately when the procedure is completed. If a cell update counter exceeds a threshold, a transition to URA_PCH via Cell_FACH is performed.

- URA_PCH: The location of the terminal is known on URA level (UTRAN registration area). URA can consist of several cells, reducing the need for mobility-initiated signaling to the network during periods of low activity. For any uplink activity, a transition to Cell_FACH is necessary, as in the Cell_PCH state.

During the lifetime of an RRC connection the identification of the terminal on RACH and FACH is also handled by the RRC. For this purpose, the RRC allocates a radio network temporary identity (RNTI). RNTI is referred to in the section for MAC functions related to common channels (6.5.3) [1, 5].

6.4 PHYSICAL LAYER INTERFACE

Our objective in the coming sections is to explain the interface toward the physical layer, as seen from the layers above, namely the MAC and RRC layers.

6.4.1 Transport Channels

The physical layer offers data transfer services to the MAC layer. The transport channels have been introduced to model these services, representing pipes for information transfer between peer MAC entities. Several types of transport channels have been defined, each representing certain characteristics of how the physical layer transfers the information over the radio interface. A certain transport channel type is only characterized by attributes related to the layer that is offering the transport channel (i.e., the physical layer) [2].

In the following, the different types of transport channels are listed. They are classified into two groups:

- Common channels;

- Dedicated channels (where the UEs can be unambiguously identified by the physical channel; that is code and frequency).

Common transport channels are:

- BCH, characterized by existence in downlink only, low fixed bit rate, and the requirement to be broadcast in the entire coverage area of the cell;

- PCH, characterized by existence in downlink only, association with a physical layer signal, the page indicator, to support efficient sleep mode procedures, and the requirement to be broadcast in the entire coverage area of the cell;

- FACH, characterized by existence in downlink only, the beamforming possibility, the possibility to use slow power control, the possibility to change rate quickly (each 10 ms), and lack of fast power control;

- RACH, characterized by existence in uplink only, limited data field, collision risk, and use of open-loop power control.

The only dedicated transport channel is:

- DCH, characterized by existence in uplink or downlink, the possibility to use beamforming, the possibility to change rate quickly (each 10 ms), and fast power control.

6.4.2   MAC and RRC Interfaces to the Physical Layer

Each transport channel is associated with one or more transport formats, where a transport format is defined by a set of parameters, representing physical layer service attributes such as bit rate, transfer delay, and quality[1] on the transport channel. Before describing the transport format more exactly, we need a few definitions [2].

- A *transport block* is the basic unit exchanged between MAC and the physical layer on a transport channel. The physical layer provides for error detection per transport block.

- A *transport block set* is a set of transport blocks that are exchanged between MAC and the physical layer at the same time instance on the same transport channel.

- *Transmission time interval* is defined as the interarrival time of transport block sets transferred by the physical layer over the radio interface. It can take the values 10, 20, 40, or 80 ms and corresponds to the interleaving period.

---

[1]   The quality (e.g., BER), is not determined by the transport format alone. It is also affected by the target value used in the power control loop.

Now we can define the transport format more precisely. The transport format parameters are grouped into two parts, a dynamic part and a semistatic part.

- Dynamic part:

  - Transport block size, which is defined as the number of bits in a transport block;

  - Transport block set size, which is defined as the number of transport blocks in a transport block set.

- Semistatic part:

  - Transmission time interval;

  - Type of channel coding and coding rate (turbo or convolutional coding, rate 1/2 or 1/3);

  - Static rate matching, specifying the amount of rate matching to apply relatively to other parallel transport channels. The static rate matching is used to balance the quality between transport channels that are mapped onto the same physical channel, and therefore cannot be individually power controlled.

In short, the transport format defines the coding and bit rate mapping in each transmission time interval. The dynamic parameters transport block size and transport block set size, and the semistatic parameter transmission time interval together correspond to the transport channel bit rate. Variable transport channel bit rate is thus achieved by changing either the transport block size, the transport block set size or both (i.e., the number of bits per transmission time interval). A variable bit rate transport channel is therefore assigned a set of transport formats, denoted *transport format set*. Within a transport format set the dynamic parameters differ between the transport formats, while the semistatic parts are the same. A terminal may use several parallel transport channels simultaneously (e.g., one for control signaling, another for a speech service, and yet another for a video service). These transport channels are typically multiplexed onto the same physical channel within the physical layer. The combination of transport formats used at a given point of time on the parallel transport channels is denoted *transport format combination*. It consists of one transport format per transport channel (i.e., one transport format out of each transport format set). The set of allowed transport format combinations is called a *transport format combination set*. Figure 6.4 illustrates the relations of these different ways to group transport formats.

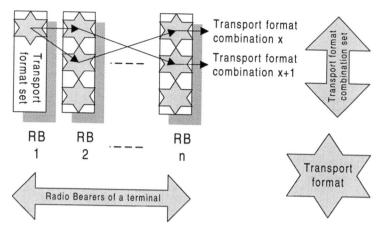

Figure 6.4 Transport format concepts.

Assignment and configuration of the transport format combination set is a task for RRC. Each time a transport channel is added, released, or needs to be reconfigured, the transport format combination set is changed. Both the physical layer as well as the MAC layer are configured by RRC accordingly. Given the transport format combination set, the MAC can select between transport format combinations, depending on source rates and priorities of the services mapped onto the corresponding transport channels. As the complete transport format combination set is already known at both the transmitting as well as at the receiving end, through the configuration by RRC, no extensive configuration signaling is needed when changing between different transport format combinations. Instead, there is a physical layer field in each 10-ms frame, the *transport format combination indicator (TFCI)*, pointing to the transport format combination used in this particular frame within the transport format combination set. The mapping between TFCI values and transport format combinations is defined by RRC in the configuration of the transport format combination set. Through the detection of the TFCI, the physical layer on the receiving side can decode and transfer the information to MAC on the appropriate transport channels.

6.4.3 Examples of L1 Data Transmission

An example of how data is delivered between MAC and the physical layer on three parallel transport channels is illustrated in Figure 6.5. In the example, the transmission time intervals are different for the three transport channels. At each transmission time interval, a transport block set consisting of one or more transport blocks is delivered on each transport channel. On the first transport channel, only one transport block is delivered at a time, and the transport block size varies. The example may correspond to a transport channel carrying a real time service, with relatively low bit rate, requiring data delivery on a constant time basis (e.g., a speech service). The second transport channel on which several transport blocks (i.e., a transport block set) are delivered at a time corresponds to a transport channel carrying a real-time service with higher bit rate. The data has been split into several transport blocks per transmission time interval to avoid too large blocks for physical layer processing and error detection. Finally, a transport channel with constant transport block size is illustrated. In this case, the number of transport blocks, and thus also the transmission block size, is changed when changing the bit rate on the transport channel. This is typical for a transport channel carrying a nonreal-time service employing ARQ, where it has ensured that at least one transport block fits into the lowest bit rate configured for the transport channel. Thereby, resegmentation of transport blocks to be retransmitted is avoided, even if the allowed transport channel bit rate is decreased.

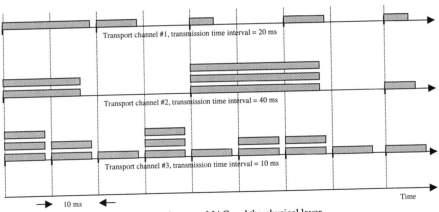

Figure 6.5   Example of data exchange between MAC and the physical layer.

6.5   MEDIUM ACCESS CONTROL (MAC)

The primary service of the UTRA MAC sublayer [3] is unacknowledged data transfer between peer MAC entities. The most important functions to provide the service are:

- Mapping between logical and transport channels, including dynamic switching between different transport channels;

- Selection of appropriate transport format for each transport channel depending on instantaneous source rate.

During normal operation, MAC monitors the instantaneous source rates on each of the dedicated logical channels and provides measurement reports to RRC. Based on these traffic volume reports, RRC can switch the connection of logical channels to different types of transport channels, which can be either common or dedicated. When configured to multiplex several logical channels into one transport channel, MAC provides the header to enable demultiplexing in the receiving MAC. On common transport channels, MAC also needs to provide identification of the terminal.

Once set to operate on dedicated channels, MAC selects the transport format combination (TFC) for each transmission time interval (TTI) from the transport format combination set (TFCS) configured to MAC by RRC.

6.5.1 Logical Channels and MAC Architecture

The overall MAC architecture is illustrated in Figure 6.6. The figure is applicable to both the network and a terminal with the addition that in the network, several MAC-d entities may exist (one per terminal). The transport services of MAC have been defined by mean logical channels. Five logical channels separated into two types are described:

- Control channels

  - BCCH for transmission of broadcast information. BCCH is unidirectional and exists in downlink only. It is always connected to a BCH transport channel.

  - PCCH for transmission of paging information. PCCH is unidirectional and exists in downlink only. It is always connected to a PCH transport channel.

  - CCCH for transmission of control data when an RRC connection to the controlling RNC of the accessed cell does not or may not exist. In downlink, CCCH is mapped to an FACH, and in uplink to an RACH transport channel. Addressing on CCCH is provided by RRC.

  - DCCH for transmission of control information when an RRC connection exists. There are two DCCHs per terminal: one for unacknowledged-mode transmission and one for acknowledged-mode transmission on RLC. These two logical channels may be multiplexed by MAC depending on con-figuration. DCCHs can be mapped either to common or dedicated transport channels. When on common channels, FACH is used in the downlink and RACH in the uplink. Addressing on DCCH is provided by MAC based on an identifier (RNTI) allocated by RRC.

- Traffic channels

  - Dedicated traffic channel (DTCH) for transmission of user plane information. DTCHs are only defined when an RRC connection exists. The number of DTCHs per terminal is dynamic and configured to the MAC by RRC. DTCHs can be mapped either to common or dedicated transport channels. When on common channels, FACH is used in the downlink and RACH in the uplink. For dedicated transport, MAC can multiplex several logical channels into one or more DCHs, depending on configuration.

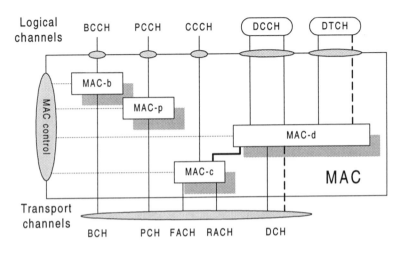

Figure 6.6 MAC architecture.

The operation of MAC is described by means of four functional blocks:

- MAC-b for handling broadcast information;

- MAC-p for handling paging information;

- MAC-c for performing tasks related to transmission on RACH and FACH. There is one MAC-c per UTRA access point (cell);

- MAC-d for performing terminal-specific tasks and controlling transmission on DCHs. A MAC-d entity is set up when an RRC connection is established. All DCCHs and DTCHs are always mapped through MAC-d.

MAC-c and MAC-d are described in more detail in the following sections. A MAC control SAP is defined for describing the configuration operations performed by RRC. The control SAP is connected to all functional entities.

6.5.2   MAC-d

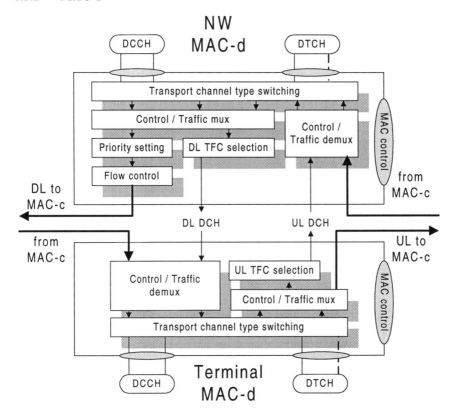

Figure 6.7   MAC-d functional model and connection between the NW and a terminal.

A functional entity known as MAC-d (Figure 6.7) handles mainly terminal-specific functions and provides a connection to dedicated transport channels. The main functions of MAC-d are:

- Dynamic switching between common and dedicated transport channels based on RRC decision. When connected to common channels, data is passed via a MAC-c entity.

- Multiplexing of dedicated logical channels onto one transport channel denoted control/traffic mux. Multiplexing onto each transport channel is defined separately. This function is placed logically after the dynamic switching function to indicate that the multiplexing rules for common and dedicated transport can be different. The multiplexing function on MAC is dynamically configurable by RRC, resulting in a variable-length multiplexing header.

- Selection of TFC for dedicated channels, based on offered load in the logical channel inputs (RLC buffers) within the limits configured by RRC in the TFCS.

- To support a distributed NW architecture, priority setting and flow control functions are added to control transmission between MAC-d and MAC-c in the downlink.

6.5.3 MAC-c

Properties of MAC related to RACH and FACH are described by means of a MAC-c functional entity. The organization of the different MAC-c functions and the relations of network and terminal MAC-c entities are described in Figure 6.8. The network MAC-c receives *downlink* data by two means: DCCH and DTCH data from MAC-d; and CCCH data from a separate SAP connected directly to a transparent-mode RLC. The processing of these two types of data in MAC-c is different.

Data from NW MAC-d first passes through a flow control entity that is the counterpart of a similar entity in MAC-d. The purpose of these entities is to cope with air interface congestion resulting in prolonged queueing and possible overflow in MAC-c. The scheduling and priority handling function is placed before any MAC-c multiplexing to be able to prioritize between all flows. Because MAC has the task to provide terminal identification on DCCH and DTCH, an identifier (RNTI see Section 6.3) is added and the flows from different terminals are multiplexed.

On CCCH there is only control data, and terminal addressing is provided by RRC, thus none of the above multiplexing levels are needed. When CCCH is multiplexed together with DCCH and DTCH, a second multiplexing header, the common/dedicated mux field is added. As FACH is also able to change rate dynamically, the final step before passing the data to L1 is to select a transport format based on the offered traffic. Note that there is no transport format combination as there are not several transport channels that would be supported by the same transport format indicator.

After transmission, a corresponding MAC-c entity in the terminal receives data on FACH. First, the common/dedicated multiplexing header is checked and the data is routed toward either CCCH or DCCH / DTCH accordingly. CCCH data is passed directly to the CCCH SAP. For data on dedicated logical channels, the RNTI is checked and only data destined to this terminal is passed on to MAC-d for further processing.

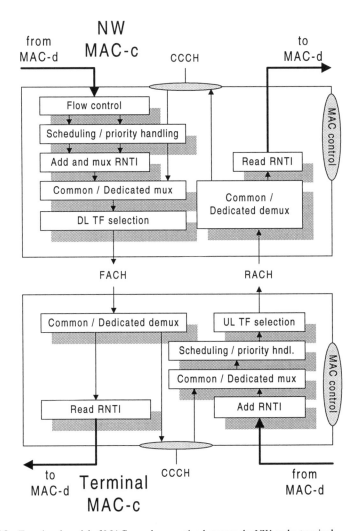

Figure 6.8 Functional model of MAC-c and connection between the NW and a terminal.

Uplink data is also received from two sources, CCCH and MAC-d. Before multiplexing common and dedicated logical channels, RNTI is added to DCCH / DTCH to provide identification of the terminal. For CCCH, terminal identification is provided by RRC. After all multiplexing stages, data is scheduled for transmission according to a priority order. Before transmission, a transport format for the RACH is selected.

When a NW MAC-c entity receives data from RACH, it first checks the common/dedicated multiplexing header. Data destined for CCCH is passed directly to the CCCH SAP. Dedicated logical channel data is passed to a terminal-specific MAC-d entity after checking the RNTI.

MAC control SAPs are provided in both the terminal and the NW to enable RRC to configure the operation of the MAC-c entities.

6.6   RADIO LINK CONTROL (RLC)

The radio link control protocol can be set up to provide unacknowledged or acknowledged data transmission service [4]. An instance of the RLC protocol is defined for each logical channel (Figure 6.1). In the case of a bidirectional logical channel (CCCH, DCCH and possibly DTCH) and an unidirectional RLC (transparent or unacknowledged mode), two RLC entities connect to one logical channel. Each RLC instance is configured by RRC to operate in one of three modes:

Transparent mode: No protocol overhead is added to higher layer data. Error detection is provided by the physical layer and erroneous protocol data units (PDU) can be discarded or marked erroneous. The service can be either streaming-type of transmission, where higher layer segmentation is not maintained, or in special cases limited segmentation capability can be accomplished. For segmentation in transparent mode, the following conditions have to be met:

- Higher layer PDU sizes have to be known at the time of setup;

- A unique transport format with exact rate matching to the higher layer PDU must be configurable for each size;

- Higher layer PDU size may not exceed the maximum amount of data that can be transmitted within a transmission time interval.

Unacknowledged mode: No retransmission protocol is in use and delivery is not guaranteed. Erroneous data is detected and either marked or discarded depending on configuration. The PDU structure includes sequence numbers so that integrity of higher layer PDUs can be observed. Segmentation and concatenation is provided by means of header fields added to the data. An RLC entity in unacknowledged mode is defined as unidirectional, because no association between uplink and downlink is needed.

Acknowledged mode: In this mode, an ARQ mechanism is used for error correction. The quality versus delay performance of the RLC can be controlled by RRC through configuration of the number of retransmissions provided by RLC. If RLC is unable to deliver the data correctly, the upper layer is notified. An acknowledged-mode RLC entity is bidirectional and capable of piggybacking user data with a status indication of the opposite link direction. RLC can be configured for both in-sequence and out-of-sequence delivery. With in-sequence delivery the order of higher layer PDUs is maintained, whereas out-of-sequence delivery forwards higher layer PDUs as soon as they are completely received. Acknowledged mode supports RLC peer-to-peer signaling exchanges, and in addition to status reports on RLC PDU delivery a reset can be signaled between the peer entities.

RLC, like MAC and the physical layer, is entirely configured by RRC. CRLC-Config request primitives are used for the purpose. When operating in transparent or unacknowledged mode, the resulting state representation is extremely simple. Both transparent and unacknowledged modes are illustrated in Figure 6.9. RRC switches the mode of an RLC entity between the null state and either transparent or unacknowledged data transfer ready by changing parameters of the primitive.

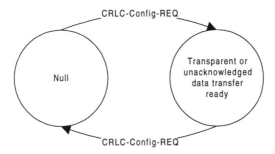

Figure 6.9   State representation of transparent and unacknowledged mode.

Acknowledged-mode RLC operation is more complex due to the necessity of managing retransmissions and buffering in both the transmitter and the receiver. It is possible that the state variables in the transmitter and the receiver get out of sync due to undetected errors on the transmission path. To solve cases with inconsistent state variables and proceed transmission, the RLC has a reset procedure that resets the protocol machines in the terminal and the network to a known state. This operation gives rise to a third state, reset pending, which is illustrated in Figure 6.10.

Upon transmitting a RESET PDU, an RLC entity moves to the reset pending state. When the peer entity receives the RESET, it triggers a RESET ACK and moves back to acknowledged data transmission ready state. A received RESET ACK moves the first RLC back to ready state. Additionally, a configuration primitive from RRC can release the reset pending state to null.

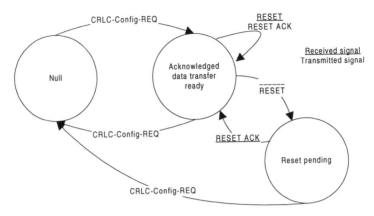

Figure 6.10 State representation of acknowledged mode.

6.7 DATA FLOW

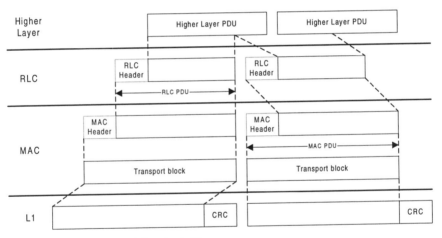

Figure 6.11 Data flow from RLC to physical layer.

The data flow from a higher layer PDU to the physical layer is shown in Figure 6.11 [1]. Summarizing the impacts of what has been described for the different sublayers in the previous sections, a number of operations take place when data is passed through the radio interface protocol L2.

RLC takes higher layer PDUs as input and, depending on operation mode, segments and concatenates the data into blocks of defined size. The size of the RLC-PDU and the number of RLC-PDUs in a transmission time interval can be derived from the transport format selected by MAC, as one RLC-PDU always maps into one transport block. The format of the RLC header also depends on the mode of the RLC. In transparent mode there is no header, for unacknowledged and acknowledged modes different types of headers have been defined.

Depending on multiplexing definitions and mapping to either common or dedicated channels, MAC may add header fields to the RLC-PDU. If several logical channels are mapped onto one transport channel, a multiplexing header field is added to identify the logical channel to the receiver. If a dedicated logical channel is mapped onto a common transport channel, MAC adds an RNTI to identify the receiver.

A complete MAC-PDU corresponds to a transport block, which is delivered to the physical layer for transmission. Error detection has been defined as a task for the physical layer. As an example solution, the addition of a CRC is illustrated. The *receiver* executes reverse operations.

The physical layer performs detection and passes the check result up to MAC and RLC.

If no error is indicated, the receiving MAC reads the header, if any, and forwards the PDU to the corresponding RLC entity. As both the RLC-PDU and the MAC header are protected by the same CRC, a CRC error on a transport channel utilizing MAC multiplexing leads to the PDU being discarded by MAC, because there is no certainty of a destination RLC.

Upon receiving its PDU, RLC acts according to the mode in which it has been set to work. In case of a CRC error, the PDU may either be discarded or forwarded with an error notification, depending on configuration.

6.8 RADIO RESOURCE CONTROL (RRC)

In UTRA, all transport control signaling, with the exception of the TFCI in the physical layer, is handled by RRC. As shown in Figure 6.1, RRC is attached to all logical channels that transfer control information (BCCH, PCCH, CCCH, and DCCH). Terminal-specific control signaling during an RRC connection is conveyed on the DCCH. The messages can use either acknowledged-mode or unacknowledged-mode transmission on the RLC level, which is why two DCCHs used by the RLC entities are defined. The RRC protocol handles a large number of signaling tasks, the most important ones being [5]:

- System information broadcasting, including both data originating from the UTRAN and core network data;

- Establishing, maintaining, and releasing an RRC connection between the terminal and the UTRAN;

- Establishing, reconfiguring, and releasing radio bearers (described in Section 6.8.1);

- Transport control functions, including assignment and release of radio resources and the associated signaling of physical channel, transport channel, logical channel, and RLC parameters between the terminal and the UTRAN;

- Mobility procedures, including different types of handover and cell location update procedures;

- Reporting of terminal measurements and control of the reporting;

- Outer loop power control to signal the target setting of the closed loop power control;

- Transparent transmission and routing of higher layer signaling messages;

- Handling of paging messages, which may be requested by and include data from the core network.

Additionally, the RRC layer takes care of configuring all lower radio layers based on the parameters that are exchanged between peer RRC protocols.

To maintain focus, this presentation concentrates on the functions that directly relate to setting up and controlling data transmission and configuring the lower layers. Additionally, a short description of RRC connection and related mobility handling is found in Section 6.3.

6.8.1 Radio Bearer Related Procedures

There are three radio-bearer-related procedures, namely, radio bearer establishment, radio bearer release, and radio bearer configuration. These three procedures are discussed below.

6.8.1.1 Radio Bearer Establishment

Figure 6.12 Radio bearer setup procedure.

A new radio bearer is established through this procedure (Figure 6.12). Traffic and QoS attributes of the radio bearer are mapped to RLC parameters, DTCH multiplexing priority, DCH scheduling priority, the transport format set of the DCH, and an update of the transport format combination set. An assignment of one or several physical channels and a change in the applied transport channel types can also be included.

There are a number of different ways the radio bearer setup procedure can be utilized with different effects to the connection between the terminal and the network:

- The radio bearer establishment can include the *activation of a new dedicated physical channel*, in which case the RADIO BEARER SETUP message includes the parameters for the new physical channel. This is done if either no dedicated physical channel has been previously allocated, or if a new physical channel in parallel to old ones is to be established. The timing of the procedure follows the detection of the new physical channel and no additional timing information on higher layers is needed.

- As multicarrier transmission should be avoided, especially in the uplink, usually a new radio bearer is established with *synchronized dedicated physical channel modification*. The parameters of an existing physical channel are modified in such a way that the old configuration and the resulting new configuration are incompatible, so that the change of configuration has to be exactly time-aligned in both the terminal and the network. For this purpose, an activation time for executing the modification is set and signaled by the network RRC.

- In special cases, the old and new configuration can be compatible, if transport formats are added or removed without affecting the interpretation of transport formats for any other service. In such cases, the procedure can be executed without an explicit activation time resulting in *unsynchronized dedicated physical channel modification*. This procedure should normally be faster than the synchronized modification, because the new configuration can be used immediately after it has been confirmed. In the more general synchronized case, the activation time has to be set according to worst-time estimation, taking into account possible errors in signaling.

- A radio bearer can also be set up *without dedicated physical channel*. If no dedicated physical channels are readily allocated, and the QoS characteristics of the new bearer allow delays long enough so that common transport channels can be utilized, the default way to set up a radio bearer is to start on common channels and set up a dedicated physical channel when there is more data than what can be efficiently sent on common channels. Alternatively, the new radio bearer can also be multiplexed together with an existing DCH transport channel on MAC without requiring any immediate changes in the physical channels or TFCS.

6.8.1.2   Radio Bearer Release

As a radio bearer is released, all the related resources including the associated RLC entity are also released. Additionally, the procedure (Figure 6.13) may release one or more physical channels and change the transport channel type used by RRC from dedicated to common.

As in the establishment procedure, different parameterizations result in slightly different operation. The release can be either *synchronized* or *unsynchronized*, depending on whether the changes to the physical channel structure, including the TFCI, cause changes to the signaling of other radio bearers. The release can also be executed without any modification to the existing dedicated physical channels, in which case the radio bearer was either multiplexed with another radio bearer or routed through common transport channels (RACH, FACH).

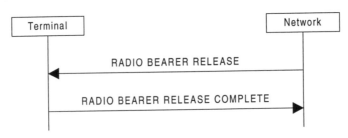

Figure 6.13 Radio bearer release procedure.

6.8.1.3 Radio Bearer Reconfiguration

The radio bearer reconfiguration procedure (Figure 6.14) can be used for reconfiguring parameters to reflect a change in QoS. Changes in RLC parameters, multiplexing priority for DTCH/DCCH, DCH scheduling priority, TFS and TFCS for DCH, assignments and releases of one or more physical channels, and applied transport channel types can be included.

As with other procedures that involve changes in radio bearers, the procedure can be executed both as *synchronized* and as *unsynchronized*. For the signaling radio bearer, only synchronized reconfiguration is possible, because the connection between the terminal and the network has to be usable at all times.

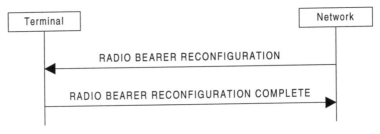

Figure 6.14 Radio bearer reconfiguration procedure.

6.8.2   Transport Channel Reconfiguration

The transport channel reconfiguration procedure (Figure 6.15) can alter the configuration of transport channels (TFS, TFCS) and the resources below it (i.e., the physical channels). The reconfiguration can be synchronized or unsynchronized.

Figure 6.15 Transport channel reconfiguration procedure.

Additionally, a transport channel can also be preconfigured before a corresponding physical channel exists. This approach can be used prior to transport channel switching when the radio bearer is assigned to common channels. A physical channel reconfiguration can be used later to connect the transport channel to a dedicated physical channel.

6.8.3   Transport Format Combination Control

The transport format combination control procedure (Figure 6.16) can be used to control the TFCS of the terminal in the uplink. It can be used by the network as a radio resource management tool (e.g., restricting the TFCS temporarily in case of uplink congestion or controlling the time-multiplexing of high data rate best-effort radio bearers of different terminals in the uplink).

Figure 6.16 Transport format combination control procedure.

6.8.4   Physical Channel Reconfiguration

The physical channel reconfiguration procedure (Figure 6.17) may assign, replace, or release a set of physical channels used by a terminal, possibly influencing also the applied transport channel type and the state of the RRC. Several different motivations for using the procedure are identified:

- Assignment of a dedicated physical channel to switch from common channels to a dedicated physical channel;

- Release of a dedicated physical channel to switch from a dedicated physical channel to common channels;

- A synchronized replacement of a dedicated physical channel. This type of procedure can be used, for example, for reorganization of the downlink code tree to avoid fragmentation.

Figure 6.17 Physical channel reconfiguration procedure.

6.9   CONCLUSIONS

The structure presented herein and its predecessors have been subject to extensive studies in FRAMES RN2 and RN3, ETSI SMG2 L23 Expert Group, and its successor, the 3GPP TSG RAN WG2. Many parts have been through several steps of evolution since the completion of FRAMES issue 1 deliverables on these topics. The development has been logical and focused on building a coherent and well-defined model to describe the essential parts of UTRA transport control.

The way that logical channels were defined by ITU [6], through the type of the information content, led to problems when attempting to place logical channels above the physical layer, as in second generation systems. As the intention was to transmit several types of information such as user and control data on the same physical transport resource, the definition was leading towards a large number of logical channels for no apparent reason, because the transport service from the physical layer was the same, irrespective of the information content. To solve the contradiction, transport channels were introduced to define physical layer transmission characteristics and logical channels were pushed one step higher, to detach them from the differences in transport mechanisms and be able to create definitions based on content only, conforming to the original ITU definition.

In architectural studies, it became apparent that efficient support of macrodiversity in the network required the termination of the lowest retransmission layer above the macrodiversity combining point. This together with the requirement that UTRAN should be able to provide lossless handovers, lead to the removal of a separate logical link control (LLC) protocol, and RLC remained as the only retransmission protocol in UTRAN. Link simulation work confirmed that with fast power control the best radio interface efficiency can be achieved with a small retransmission unit size (in the order of 300 – 600 bits) and a relatively high block error rate (5%-10%). This influenced the later development of RLC PDUs towards a highly optimized lightweight structure that still incorporates a long sequence number field to cope with small PDU size and high data rate.

After FRAMES RN2 and RN3 had separately worked on radio resource allocation mechanisms and related signaling, the joint work in ETSI and 3GPP led into a unification of mechanisms. Peer-to-peer MAC signaling, which was present in FRAMES work, was moved up to RRC to utilize the services of RLC for the signaling. The radio resource allocation signaling of MAC was reduced to the selection of a transport format combination from a set provided by RRC to be conveyed as a physical layer field.

Refinements in UTRAN architecture and the need to be able to address both channel-specific and terminal-specific functions required a more precise modeling of the separation of MAC functions. The model has been developed to describe MAC associated either to common channels or to terminal-specific dedicated channels. This split assists in defining interfaces in a distributed architecture. Even if the evolution on all layers has been strong, the most important structures and the key drivers have proved their competence also on the wider research arena of global standardization. The ability to efficiently transfer multiple radio access bearers through a common radio interface is still the core of the protocol structure.

References

[1]   3GPP TS 25.301: *Radio Interface Protocol Architecture*, v. 3.1.0, June 1999, p. 47.

[2]   3GPP TS 25.302: *Services Provided by The Physical Layer*, v. 3.0.0, August 1999, p. 39.

[3]   3GPP TS 25.321: *MAC Protocol Specification*, v. 3.0.0, June 1999, p. 35.

[4]   3GPP TS 25.322: *RLC Protocol Specification*, v. 1.1.1, July 1999, p. 62.

[5]   3GPP TS 25.331: *RRC Protocol Specification*, v. 1.3.0, August 1999, p. 148.

[6]   Recommendation ITU-R M.1224: *Vocabulary of Terms for Future Public Land Mobile Telecommunication Systems (FPLMTS)*, 1997, p. 50.

Chapter 7

Radio Resource Management

As new wireless systems evolve to complement and replace current second generation wireless access systems (e.g., GSM), a distinct shift in design criteria can be noted. From being primarily systems for voice communication, future wireless systems will deal with data and will in particular be tailored to different multimedia applications. Due to the large impact of the web, and the web browser as the common software platform for various IT applications, provisioning Internet services has become the main design paradigm in defining third and subsequent generations of wireless access systems, UMTS wide-area access systems. Comparing market estimates for wireless personal communication, and considering recent proposals for wideband multimedia services with the existing spectrum allocations for these types of systems, show that spectrum resource management remains an important topic in the near and distant future. Resource management takes on new dimensions and can no longer be restricted to be a matter of spectrum utilization only. Other important components are mobile equipment power management and infrastructure deployment and cost structure.

In the following, we outline some of the new distinctive features of the UMTS wireless access systems and identify what impact these have on the resource management schemes [1-31].

High Bandwidth

As was noted in the introduction, future systems are expected to require much higher data rates than current systems. Since most of the current resource management schemes are not directly tied to any specific data rate, this fact per se motivates the development of new methods. However, data rates in personal communication systems will certainly in many cases be limited by propagation conditions such as distance loss, multipath and so forth. The primary constraining factor is the link budget. Since the required transmitter power increases linearly with the bandwidth, high speed wireless access will have but a very limited range. This will increase the number of required wireless access points and thereby increase the complexity of the resource management schemes.

Multiple QoS Requirements

If the bandwidth as such is not that important to the design and performance of RRM algorithms, the traffic characteristics are. The key resource management problems in "multimedia"-type systems are related to the data rates, and delay constraints traffic in small cell environments will exhibit very large peak-to-average capacity demands. Video users with absolute delay requirements may require considerable portions of the spectrum that they share with e-mail-type message traffic with no such absolute constraints. Dynamic channel allocation (i.e., statistical multiplexing) will provide even larger capacity gains in these situations than in today's mobile phone scenarios.

On the other hand, data traffic provides us with an extra degree of freedom in the resource allocation procedure, leading to better resource utilization. Circuit switching systems are normally designed to meet absolute delay constraints, whereas the delay for data traffic normally is constrained in the statistical sense (e.g., average delays). The latter type of constraints implies an extra degree of freedom in the resource allocation procedure, leading to better resource utilization. We may trade off blocking for additional delay. This has lead to the design of radio access schemes particularly designed for delay nonsensitive, very "bursty" traffic, the so-called packet radio system. Since messages are short and delay is kept low, there is no time for the exchange of resource allocation information. Instead, random access schemes are utilized in which the terminals compete for the radio resources. This will at times lead to message collisions or conflicts that have to be resolved by special conflict resolution schemes (at a delay penalty).

Systems with intermittent data transmission will also suffer from a different kind of problem. Since there are no continuous transmissions, good link quality estimates cannot be made at will but only when there actually is a transmission in progress. In particular, when the traffic is very bursty, the statistical estimates of the link-quality parameters can degrade considerably since the terminal may move some distance between transmissions. This affects all type of RRM decisions (e.g., channel allocation, power control, and handoff decisions). In these situations, channel allocation decisions and power control have to be made on estimated average link qualities rather than on instantaneous values. In these cases, the concept of a handoff loses it meaning in the physical sense, and one may instead consider different "connectionless" (or multiport) schemes, where any RAP in some area may receive messages from a mobile terminal without the explicit establishment of a logical/physical connection [25]. Another possibility considered (in particular in CDMA-type systems) is to "artificially" maintain a physical link even when there is no data to transmit by prescribing a minimum "idle" power level. These tradeoffs are, of course, the most important, the more rapidly the terminals are allowed to move relative to the duration of these idle periods.

A key problem in a system with several users or user groups with different QoS requirements is that it is mostly not obvious how to combine these quality criteria into a single performance measure or cost function allowing a straightforward mathematical optimization formulation.

Asymmetric Traffic

Current mobile communication systems are mainly geared for speech traffic and operate in symmetric full-duplex fashion. Data rates and other quality of service parameters in these systems are the same in the uplink and downlink. In third generation systems, data traffic, as generated by IP-based information retrieval applications, is expected to dominate. In many of these applications, an increasing fraction of the total offered traffic is expected to load the downlink segment of the system. Traditional symmetric full duplex (e.g., FDD) systems are obviously not very spectrally efficient when faced with handling typical web-browsing traffic with down/uplink average data rate ratios of 10/1 or more.

To provide multimedia services in wireless systems places new and severe demands on efficient radio resource management (RRM). Users will simultaneously require a number of links with various QoS parameter settings. To handle this multitude of end-user services, the UMTS access system will supply a possibly large but finite set of basic bearer services, each providing a particular QoS parameter set. The set of bearer services covers a large variety of bit rates, error performance, availability, and delay characteristics. The physical implementation includes link adaptation (i.e., various modem waveforms) (e.g., spreading codes) that provide the various bitrates. The selection of the transmission time (scheduling of the transmission) affects the delay characteristics. In many potential bearer services, so-called RT services, the user applications are guaranteed certain instantaneous data rates. Choosing how the transmitter power should be controlled to achieve such fixed or minimum rate requirements is a well-researched task. nonreal time (NRT) applications will, however, also be able to use any excess bandwidth that the system can provide in a best-effort fashion as soon as the basic guarantees have been honored. Resource management in the latter situation is the focus of this chapter.

The chapter is organized as follows. Resource management for the TDD and WCDMA modes are described in two sections. In the first section, the main emphasis is on "bunched" systems with (quasi-) centralized resource management. Power and admission control schemes dominate the WCDMA section. The chapter closes with a section on those parts of transmission scheduling that can be treated independently of the physical access schemes.

7.1 SOME FUNDAMENTAL LIMITS AND PROPERTIES

In this section, we establish a more fundamental problem definition for the wireless access resource management problem, with special emphasis on multirate best-effort NRT resource management. We consider a general wireless network scenario with a number of transmitters and receivers potentially using the same allocated spectrum to transmit their data messages. The instantaneous interference situation is described in a standard way by means of a gain matrix, which is assumed to be constant over the typical duration of a data message. We assume that all relevant information regarding propagation and interference is available to a (hypothetical) ideal, central resource manager.

By selecting the proper bit rates, transmitter powers, and the transmission schedule in each logical link, we seek to maximize the total throughput defined as the sum of all (average) data rates in all the links currently active. A subproblem is to find the combination of link bit rate that can instantaneously (in the same time slot) be supported at the given quality level and the corresponding transmitter power levels that make this possible.

Consider a "snapshot" of a wireless network consisting of a given set of access points and terminals. At this instance of time, we would like to support a number of links 1,2,3...N between receiver-transmitter pairs. The propagation gain between the transmitter in link (pair) i and the receiver at link j is denoted G_{ij}. The collection of the form the $N \times N$ link gain matrix of the system [24]. We assume that for each link i the emitted power of the transmitter, P_i and the link data rate R_j can be controlled. The rate adaptation mechanism allows data rate

$$R_i \leq f(\Gamma_i) \tag{7.1}$$

where Γ_i is the *signal-to-interference ratio (C/I)* at the receiver of link i, and $f(\gamma)$ is a monotonously increasing modem (waveform set/coding) dependent function. The C/I at the receiver i is given by

$$\Gamma_i = \frac{P_i G_{ii}}{\sum_{j \neq i} P_j G_{ji} + N_i} \tag{7.2}$$

where N_i denotes the power of the external (i.e., interferer independent) noise. Now, let us assume that the transmitter i may use at most transmitter power $\overline{P}_i$. Let $\overline{P}$ denote the vector of maximal powers. We have the following

Definition 1: A rate vector $R(\overline{P}) = (R_1, R_1, \ldots R_N)$ is *instantaneously achievable* if there exists a positive power vector

$$P = (P_1, P_2, \ldots P_N) \leq \overline{P}$$

such that

$$R_i \leq f(\Gamma_i(P)), \quad \forall i$$

Definition 2: A rate vector $R^*(\overline{P}) = (R^*_1, R^*_1, \ldots R^*_N)$ is *achievable* (in the average sense) if it is expressed as

$$R^* = \sum_k \alpha_k R_k$$

where

$$\alpha_k \in [0,1]$$

and where all the R_k s are instantaneously achievable rate vectors.

If a set of rate vectors R_k is instantaneously achievable, we can clearly switch between them, each using a fraction α_k at the time (TDM). Alternatively, we can allocate this fraction of bandwidth to this particular set of rates (FDM), yielding the average rate R^* according to definition 2. As a corollary, we observe that the (average) achievable *rate region* (i.e., the set of all achievable rate vectors in N space for some maximal power vector and gain matrix) is thus the *convex hull* of the *instantaneous rate region* (i.e., the set of all instantaneously achievable rates).

Now, we assume that each link has a bearer service contract that has to be honored. In our case, the contracts require a minimum data rate, denoted ρ_i. For the sake of simplicity we will assume that all ρ_is are nonzero. We will further assume that any excess data rate provided to the user by the system is potentially consumed (and paid for) by the user. The operator therefore has an interest to provide as much excess data rate as possible. If the connection is of the real-time type, we may want to maximize the sum of the provided instantaneous rates. For nonreal-time traffic, we instead have the following optimization problem

$$\max \sum_i R_i^*(\mathbf{P}) \tag{7.3}$$

$$R_i^*(\mathbf{P}) \geq \rho_i \tag{7.4}$$

We make the following observations about this problem:

Observation 1: For any gain matrix, there exist

a) Maximal power vectors $\overline{P}$ sufficiently large (or noise levels N sufficiently small), such that the maximal sum rate in Problem (7.3) is achieved by a singular rate vector (i.e., only one link used at the time) or a combination of such rate vectors.

b) Maximal power vectors $\overline{P}$ sufficiently small (or noise levels N sufficiently large), such that the maximal sum rate in Problem (7.3) is achieved by full vectors (i.e., where all links are used simultaneously) or a combination of such rate vectors.

The proof can be found in [29]. The consequence of this observation is that if the available power is high (i.e., the SNR is high enough), sequential scheduling of transmission (i.e., TDM) provides the highest throughput. Conversely, if the SNR is low, the links will be noise limited and simultaneous (parallel) transmissions in all links provides the best results. This is illustrated by the following simple two-link example. We use the following two sample relations between the data rate and the required C/I:

A. $f(\gamma) = c\gamma$ \qquad (Unlimited bandwidth)

B. $f(\gamma) = c'\log(1+\gamma)$ \qquad (Band-limited)

Case A corresponds to the case where the available bandwidth is unlimited and the same waveform is scaled to achieve a constant E_b/N_0 (e.g., to provide a constant link quality in terms of bit error probability). Case B is simply the Shannon limit for a band-limited channel. Since in both cases there is a one-to-one correspondence (7.1) between the instantaneously achievable data rates and the required link C/I, we begin our investigation by studying the sets of achievable C/I. Simplifying (7.2), we find analytical expressions for the achievable rate region [29]. The results are shown in Figure 7.1 and Figure 7.2 for two bandwidth/quality criteria A and B, respectively. Starting with the unlimited bandwidth case (Figure 7.1), we see that the achievable instantaneous rate regions (as indicated by the region below the solid lines) are not convex, but the average sense rate regions (as indicated by dotted lines) are. The average rate region is in fact given by a set of linear equations (straight lines). The figure also illustrates a minimum rate requirement $R_i > R_{i,min}$ (i.e., within the rectangle in the upper right corner). In this example, we see that for maximal power 5, there are no rate pairs meeting the constraints, neither instantaneously nor in the average sense. For maximal power 10, the rate requirements can be achieved in the average sense but not instantaneously. For the highest maximal power, the rate constraints can be met in both senses. In this case, the sum rate is maximized by time multiplexing singular rate vectors (i.e., individual transmissions).

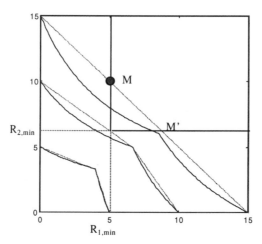

Figure 7.1 Achievable rates for various maximal transmitter powers. No bandwidth limitation (case A). Maximal powers 5, 10, and 15. $A_{12}=0.05$, $A_{21}=0.02$. $N_1=N_2=1$.

Since the average rate region is given by a linear expression, the maximal average sum rate is found in one of the corners of the achievable and feasible rate region (i.e., in either point M or M'). Because the relative noise powers are the same in both links in this particular example (and the maximal singular rates thus become the same) the sum rate happens to be the same anywhere on line M-M'.

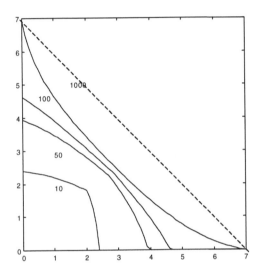

Figure 7.2   Achievable rates for various maximal transmitter powers. Bandwidth limited case (case B). Maximal powers 10, 50, 100, and 1000. $A_{12}=0.05$, $A_{21}=0.02$. $N_1=N_2=1$, $c'=1$.

The bandlimited case B is illustrated in Figure 7.2. Here, in contrast to the previous case, the instantaneous rate regions are convex for low maximal transmitter powers. This means that for these maximum powers, any achievable rate pair can be achieved instantaneously (i.e., by assigning appropriate powers, both transmitters may transmit simultaneously). For higher transmitter powers, the regions become convex, and then higher (average) sum rates can be achieved by time multiplexing the transmissions.

To conclude this section, let us summarize. We have introduced a problem formulation for resource management in wireless data systems with multiple service classes, each looking for maximal data rate allocation but with different minimal guaranteed data rates. The objective from a wireless operator point of view is to provide at least the minimal data rates to the individual users as well as maximizing total throughput. The results show that the maximal total (sum) data rates that can be achieved in the average sense by single transmissions (i.e., reserving the waveform in the entire network), are in many cases (but not always) higher than when the waveform is reused several times in the bunch. In particular, this holds true in a high power (or short range) scenario, where interuser interference is the dominating source of disturbance. The latter statement also applies to the instantaneous sum rate. For lower power levels (e.g., longer ranges), maximal sum rates are mostly reached for simultaneous transmission schemes.

We note that the results are general in the sense that they are not sensitive to the choice of modulation schemes and thus also apply for CDMA-type waveforms. As a practical consequence, the results demonstrate that for short-range, interference-limited systems (e.g., indoor, microcell), coordinated orthogonal multiplexing provides better performance than simultaneous transmissions (e.g., CDMA). For longer ranges when noise (i.e., coverage) comes into play, nonorthogonal schemes have advantages.

7.2   RESOURCE MANAGEMENT FOR THE TDD MODE

RRM in bunched systems, interference matrix based centralized RRM, and decentralized resource allocation are discussed in this section.

7.2.1   Resource Management in Bunched Systems

The centralized "bunch" concept assumes that a limited number of *remote antenna units* (RAUs) are connected to a *central unit* (CU). The resulting group of cells is called a *bunch*. Micro- and picocell systems cannot always be arranged in a regular (e.g., hexagonal) cell structure, resulting in a high probability of overlap between cells. To improve the trunking efficiency, the resources are shared between the cells in the bunch. Bunches should be designed to handle areas with high traffic load (urban environments) where high capacity is important, even at the expense of increased algorithm complexity.

Within the bunch, main interference is caused by intrabunch transmissions. To maximize spectrum efficiency, the knowledge of all allocated resources, transmitter powers, and path gains is required together with synchronization between the cells. It is of course inevitable that some bunches overlap each other and thus interbunch interference will arise, especially at the bunch borders. This interbunch interference is handled in a decentralized manner and does not require communication between bunches.

Two types of bunched RRM algorithms for TDD mode are presented. One uses a link gain matrix approach and the other uses an interference matrix approach. The smallest available physical resource (e.g., one slot in the frequency, time, and code domain) that can be assigned in a frame is denoted as RU.

7.2.2   Link Gain Matrix-Based Resource Allocation

Here we present and analyze a centralized resource allocation scheme based on measurements of the link gain matrix. The model for our wireless system is described in [2]. It is based on the concepts initially discussed in [3-5], where a limited number of RAUs are connected to a functional entity called a hub or CU. We call this group of cells a *bunch*. This is a similar concept as in GSM, where a number of BTSs are connected to a BSC. The main difference is that all intelligence and a significant part of the signal processing are located in the CU. In our case, the RAUs may even be physical antennas with their corresponding BTSs colocated at the CU site.

A lot of overlap between cells is quite common in micro- and picocells. This overlap makes it difficult to predict interference between cells, and thus a lot of resources are usually wasted when channels are assigned. To improve the trunking efficiency in such a case we must better share the resources between cells. Sharing measurement and status information between RAUs for the purpose of dynamic resource management is not a significant problem if all intelligence is concentrated at the CU. With the proliferation of techniques for high-speed networking, we believe that it is feasible to share information and achieve synchronization among the RAUs locally within a bunch, even if a more distributed implementation is used. Between CUs in different bunches, communications paths of only moderate capacity exist and thus the interference caused by other "bunches" cannot be properly controlled. This problem has not been considered in previous studies (e.g., [3]). Through network planning, the traffic intensity at the border of a bunch, where interference problems are most prominent, can be reduced. The remaining interference can be averaged out, either by time/frequency hopping or by the spreading inherited in UTRA. If only one carrier is available then we cannot use frequency hopping. Time hopping might be a problem in TDD mode, since the number of timeslots to hop on are limited. Our current working assumption is that the spreading will be sufficient when combined with either a reduced load near the border or a segregation strategy.

7.2.3   Intrabunch Resource Allocation

Resource allocation is performed mainly by four subalgorithms:

- RAU selection, which finds the most suitable RAU (e.g., with the lowest path loss);

- RU selection, which selects resources to try (random selection or some heuristic algorithm);

- Feasibility check, which calculates if the selected resources can achieve an acceptable quality signal-to-interference-ratio (SIR) without disturbing the existing users;

- Interactive admission and initial power algorithm.

We assume all assigned RUs to be SIR-balanced at an acceptable C/(I+N), using a distributed slow power control scheme (e.g., distributed constrained power control (DCPC) [6]). By SIR-balancing we mean that some power-control algorithm controls all transmitter powers such that all receivers have their required average SIR level. The SIR averaging is required due to rapid signal level variations as well as the intermittent interbunch interference due to burst packet traffic.

Within the bunch, we base our RRM decisions on the link gain matrix G [7]. It is constructed from downlink path loss measurements on the beacon channel, on which all RAUs broadcast. By proper use of the training sequences in each traffic and access burst, it would also be possible to do uplink gain measurements. An example of G-matrix construction is shown in Figure 7.3. The beacons contain the RAU/bunch identity and it is transmitted so that the mobiles can listen to all RAUs within range. All RAUs in the figure are within the same bunch. Powerful error control coding is used, which permits the decoding even in case of very low SIR. The mobiles measure the signal strength and report this to the CU together with the decoded RAU identity. Ideally, with complete knowledge of the gain matrix and all transmitter powers within the bunch, we would be capable of determining if a new allocation is feasible or not. An allocation is feasible if all cochannel users can achieve their SIR target.

For the links that cannot be measured on a regular basis, we use a static or semistatic value from what we call the *structure matrix*. The structure matrix contains averaged or otherwise filtered gain values from RAUs to zones, and in this study we have defined a zone as the area in which all mobiles are served by the same RAU (i.e., a cell). For instance, if mobile i in cell j cannot measure on the beacon from RAU k, it uses a filtered value from RAU k to cell j instead. By filtering we mean the average (or some percentile) over all the mobiles in cell j that have been able to measure on RAU k. It is also possible to measure the link gain from RAU to RAU and use this as the default gain until any measurements can be made.

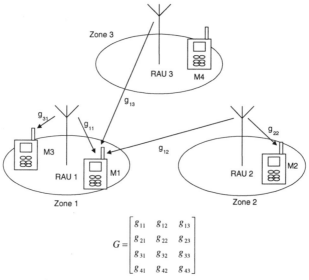

$$G = \begin{bmatrix} g_{11} & g_{12} & g_{13} \\ g_{21} & g_{22} & g_{23} \\ g_{31} & g_{32} & g_{33} \\ g_{41} & g_{42} & g_{43} \end{bmatrix}$$

Figure 7.3 The gain matrix G.

7.2.4   Interbunch Interference Handling

Due to computational complexity, signaling load, and synchronization problems, one single bunch might not be possible if we want to cover a large area. In order to handle the interference between different bunches, we use interference averaging by TH/FH for a TDMA system. Results from [15] when TH/FH is employed show that even if all users have error correction coding with rate 0.5 (and thus need two RUs each), performance is still better than for the best FCA system. In reality only some of the mobiles will need much coding since the ones far from the bunch border will only experience controlled intrabunch interference. When spreading (TD/CDMA) is used as in the current study, hopping is not necessary, since the spreading codes give a reasonable amount of interference averaging. With spreading, there are usually fewer slots to hop on than without spreading. This means that we might need to reduce the load near the border or use a segregation strategy to handle the interference. A simple and effective but not necessarily optimal solution is to try the channels with the least total interference first. This will make the bunch "nicer" to its neighbors and makes the system look decentralized on a global scale (between bunches). Another advantage is that it makes the system more robust to measurement errors, especially at low to medium loads. Communication between bunches will not be needed anyway for RRM purposes, except for interbunch handovers.

7.2.5   Dynamic Link Asymmetry

Speech users usually generate an equal amount of traffic in uplink and downlink, respectively. Data users on the other hand, who are generating an increasing fraction of the offered traffic, frequently generate more traffic in downlink than in uplink or vice versa. This kind of traffic can be handled more efficiently if the system provides asymmetrical links (i.e., different effective data rates in uplink and downlink, respectively).

TDMA/TDD facilitates the implementation of asymmetrical links. In principle, any number of the timeslots in the TDMA frame (see Figure 7.4) can be used for uplink or downlink. This freedom is in practice severely limited. If adjacent cells do not agree on which timeslots to use in which direction, uplinks of one cell will interfere with downlinks of other cells, and vice versa [10, 11]. This interference is difficult to predict or avoid. Nevertheless, substantial capacity gain can be achieved by allowing for asymmetrical links. It is desirable to have a dynamic system that tracks the asymmetry demand. Cellular systems with partially centralized radio resource management, such as bunched systems, are particularly suited for providing dynamic link asymmetry. The more information available to a centralized resource manager, compared to that of a decentralized resource manager, makes it less difficult to avoid cotimeslot interference, though it is still difficult.

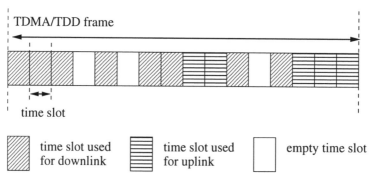

Figure 7.4 TDMA/TDD frame.

Not all transmissions can be synchronized at all stations, BSs, and MSs, propagation delays will cause adjacent timeslot interference between cells. A study of propagation delays and interference in the UTRA-TDD mode, which was performed within this project, concludes that the short distances in indoor and pedestrian environments render adjacent timeslot interference negligible for the UTRA-TDD mode. Another conclusion is that cotimeslot interference should be avoided by means of sharing the same timeslot assignment throughout the system. Due to the behavior of pairs of users engaged in conversation (i.e., not talking simultaneously), considerable statistical multiplexing gains are already obtained for a small population of users. In [12], a scheme where uplinks and downlinks are sharing timeslots is shown to give considerable capacity gains compared to traditional TDD, where timeslots are statically designated for either uplink or downlink traffic. A dynamic channel assignment algorithm with an adjustable boundary or switching point between uplink and downlink is proposed and evaluated in [11]. Our study focuses on possible performance gains from providing dynamic link asymmetry in bunched cellular systems with asymmetrical uplink and downlink capacity demands. A simple scheme with a soft switching point, very similar to that of [11], is described and evaluated.

7.2.5.1   Resource Allocation Algorithm for Dynamic Link Asymmetry

In the centralized bunch concept, all link gains between MSs and BSs are assumed known through intelligent measuring. Assuming the cross base station link (CBL) and cross mobile station link (CML) gains are also known, the CU can provide dynamic link asymmetry. Compared to static link asymmetry, substantial capacity gains can then be achieved.

Though it may be tractable to measure CBL gains, measuring CML gains is usually considered infeasible due to the large number of MS. Heavily relying on the knowledge of CML gains will make the system sensitive to measurement errors. Thus, we show that even with a very simple scheme only considering the CBL gains, capacity gains are achieved. Assuming the link gains between MSs and

BSs and between pairs of BSs are known, the resource allocation algorithm goes as follows:

---

1. Initialize:
   Number of active users: N
   Set of active users: $M = \{M_1, M_2, ..., M_N\}$
   Number of admitted users $k = 0$
   Set of admitted users: $U = \{\}$
2. Test if users $\{U, M_{k+1}\}$ can all be supported.
   i) Temporarily assign a timeslot to user M_{k+1}. If a downlink slot is requested, take the first free slot starting from the beginning of the frame. If an uplink slot is requested, take the first free slot starting from the end of the frame.
   ii) Calculate/estimate received SIR for all links considering CBL gains, but neglecting CML gains.
   iii) Compare SIRs to required SIR threshold. If all SIR are greater than the threshold, all users are considered to be supported. If not, then temporarily assign next free timeslot to user M_{k+1} and repeat from ii). Abort when there are no more slots to try.
3. If YES, then add user M_{k+1} to the set of admitted users U.
4. Increase k by 1.
5. If $k<=N$, then repeat from 2.

7.2.5.2   Performance Evaluation

Our performance measure is the user assignment failure rate, v_u, which is defined as the fraction of users that either do not get a channel or get a channel that is unusable due to bad quality (SIR). The assignment failure is plotted against the relative traffic load ϖ_c , which is the number of users in the system divided by the number of cells and the total number of channels in the system. Statistics are collected from all mobiles in the area. Capacity is measured as the load where the assignment failure rate equals 2%.

Indoor Office Environment

The models used to investigate the bunch concept in an indoor environment are the same as the ones given by ETSI [1] to evaluate different UMTS proposals. The scenario is a multistory building with offices and corridors. The basis of the performance evaluation is a series of snapshot simulations where only the downlink is considered. The mobiles are static and are distributed evenly over the area, with the exception that the probability is 0.85 of being located in the office and 0.15 in the corridor. The path loss model is based on the COST 231 model and is simplified as:

$$L\text{ [dB]} = 37 + 30\text{Log}_{10}(R) + 18.3 \cdot n^{((n+2)/(n+1) - 0.46)}$$

where R is the transmitter-receiver separation given in meters and n is the number of floors in the path. Shadow fading is added and it has a lognormal distribution with a standard deviation of 12 dB. There is no outdoor to indoor interference present at this stage of the evaluation. The propagation delay is not taken into account because of the short distances in the environment. We also assume perfect knowledge of the gain matrix (i.e., all the path losses between the base stations and mobiles are known).

When we evaluate the performance of the bunch concept, we compare it with fixed channel allocation (FCA). To implement FCA, the feasibility check is ignored when assigning a new user. This means that a channel is used at the selected RAU regardless of the interference the assignment might cause to other users. In an FCA system, the interference can be lowered by splitting the resources between the base stations. In Figure 7.5 we see how the channel groups are divided between the RAUs in a reuse 4 system.

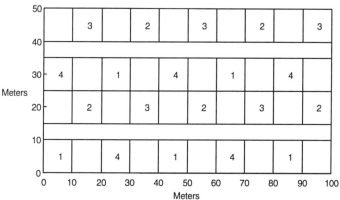

Figure 7.5 Channel group numbering showing the FCA reuse pattern with a cluster size of 4.

Here we have 12 codes and 16 timeslots available, which gives a total of 192 RUs. The channels are picked in a random order when testing for feasibility. In Figure 7.6, we see the performance of a bunch system compared with fixed channel assignment with a cluster size of 1, 2, and 4. The traffic load is always related to the total number of RUs in the system, not the number of RUs assigned to each RAU. The bunch system can carry 2.5 times the load at 2% assignment failure.

We have also looked at the difference between one bunch covering the whole building and one bunch covering one floor. An advantage with one bunch on every floor is that the complexity is lowered. This is due to the fact that fewer cochannel users means a less complex feasibility check. The measurement demand is also decreased with fewer RAUs in a bunch. In order to reduce the simulation time, the number of timeslots used is 4 instead of 16.

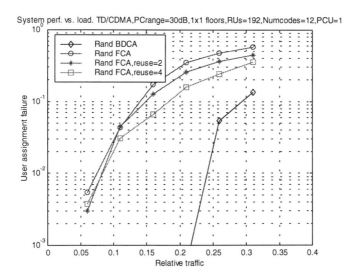

Figure 7.6 Comparison of bunch performance with FCA.

The performance increase with one bunch covering the whole building is shown in Figure 7.7. The load that can be carried at 2% assignment failure is increased from 0.11 to 0.19 (i.e., 2.3 dB). The results show that the performance increase is significant when one bunch is covering the building compared with multiple bunches.

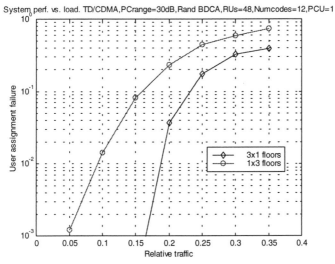

Figure 7.7 Performance for different bunch sizes; 3×1 means that the building has one bunch covering three floors. In the 1×3 case there are three bunches covering a single floor each.

Figure 7.8 shows the results from a simulation with different downtilting angles, Φ_d. The scenario is a three-story building with one bunch covering all three floors. We have 8 codes and 6 timeslots in this snapshot simulation. The RU selection scheme is ordered (i.e., the RUs are tested for feasibility in the same order in all cells). We see that the performance is greatly enhanced when using pattern shaping. When comparing the omnidirectional case (no downtilt) with the case where the pattern is downtilted 8 degrees, we see that the load can be increased from 0.14 to 0.27 with 2% assignment failure. We can almost double the capacity with this interference-reduction method.

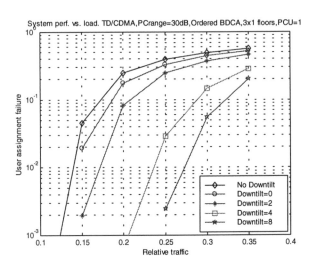

Figure 7.8 Effect of different antenna downtilting angles.

Manhattan Environment

We evaluated the bunch concept in the downlink direction with snapshot simulations on isolated islands of cells (no wraparound). The Manhattan grid for pedestrians from [1] is used together with the recursive path-loss model. Only outdoor users are considered. For line-of-sight (LOS) paths, this model has a path loss proportional to the squared distance (r^2) up to 300m from the base station and r^4 thereafter. The shadow fading is lognormal with a standard deviation of 10 dB and correlated in time but not in space. Fast fading is not modeled. A mobile always connects to the strongest RAU. Unless otherwise mentioned, the simulation area is 12×12 blocks of the Manhattan grid, which means $12^2 / 2 = 72$ RAUs. All RAUs are omnidirectional and the power control dynamic range is 20 dB. The SIR target is set to 10 dB. The multiple access mode is hybrid TDMA/CDMA (TD/CDMA) with joint detection of intracell users, processing gain (PG) equals 16, and 12 codes per timeslot are used in the downlink.

For a system with four bunches, we studied the influence on performance from the channel search method before any updates of the measurement-based power control. This corresponds to the pessimistic and unlikely case where all users are allocated within one power control period (about 200 ms). The result is shown in Figure 7.9 where we see that the least-external-interference method performs best, closely followed by the least-total-interference.

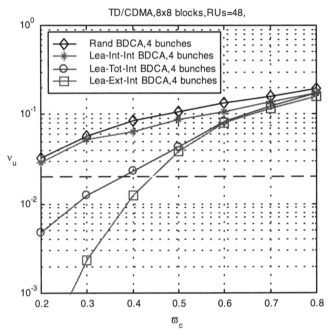

Figure 7.9   The effect of different channel search methods for a system with four bunches covering an area with 72 RAUs. The methods used are random, least-internal-interference, least-total-interference, and least-external-interference. Performance is measured before any power control updates.

Since the least total interference is based only on measurements, we also use it in the following results for the Manhattan environment. To get the external interference, we subtract the interbunch interference from the total interference. After a few (5-10) updates of the power control, the performance differences between the different methods diminish, and they all allow a load of approximately 0.7 at the 2% assignment failure level. We also compare the performance for different number of bunches covering a fixed area. The result is shown in Figure 7.10 where we see that there is a large gap in performance between the single and multiple bunch cases. The system with 72 bunches can be considered as a totally decentralized system since it has one RAU per bunch. The results show clearly that the bigger the bunch, the better the performance.

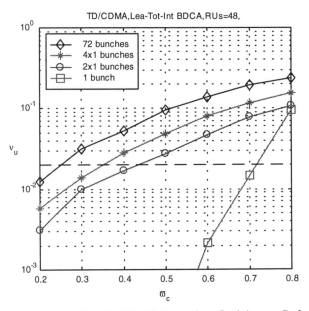

Figure 7.10 Different number of bunches (72, 4, 2, 1) covering a fixed size area. Performance is measured before any updates of the power control.

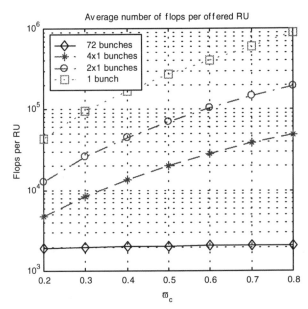

Figure 7.11 Computational complexity comparison between the different bunch sizes covering an area with 72 RAUs (12×12 blocks).

If we collect the performance after about 10 updates of the power control, the performance difference is very small, and in all the tested cases, we allow a load exceeding 0.7 at the 2% assignment failure level. If we are more concerned with computational complexity than with performance, we instead keep the bunches as small as possible, as the results in Figure 7.11 show. The average number of floating point operations (add, multiply, etc.) per RU allocation is plotted against the relative traffic load. As has been shown, the advantage with having large bunches is primarily that the initial SIR is better than in the multiple bunch cases, which means less disturbance of existing connections. The multiple bunch systems can also reach a high capacity, but they are more dependent on the measurement-based power control that occurs about every 200 ms.

7.2.6   Limited Measurements and Signaling

Earlier in our studies, we assumed that a mobile could measure the link gain (inverse path loss) from all RAUs in the bunch. Some results with this assumption are found in [2, 8]. In the current implementation of the bunch concept, we use the gain values between all pairs of mobiles and RAUs to determine if a new allocation is safe for the existing users. Since it is not realistic to measure on all RAUs in large bunches, we wanted to see how much the performance degraded when only a small number of RAU beacons (the strongest ones) could be received. We ran the simulations for 1,2,3,4 and an infinite (Inf) number of measurable beacons. To receive one beacon means to receive the beacon from the serving RAU, since a mobile always connects to the strongest beacon in this study. With two beacons, the mobile also receives the next strongest beacon not necessarily the strongest interferer, since that base station might not transmit on the same traffic channel. Beacons = Inf means that the mobile can receive all the beacons in the bunch.

For the links that cannot be measured on a regular basis, we use a static or semistatic value from what we call the structure matrix. The structure matrix contains averaged or otherwise filtered gain values from RAUs to zones, and in this study we define a zone as the area in which all mobiles are served by the same RAU (i.e., a cell). For instance, if mobile i in cell j cannot measure on the beacon from RAU k, it uses an average or filtered value instead. By "filtering," we actually mean some percentile value over all the mobiles in cell j able to measure on RAU k. The reason for using a percentile instead of the arithmetic mean is that the arithmetic mean overestimates the cell-to-cell gain between cells on perpendicular streets due to the street crossings in the Manhattan environment. This overestimation leads to a high blocking. Instead of a percentile, it would be possible to measure the link gain from RAU to RAU and use this as the default gain until any other measurements can be made. This gives similar performance figures as when the median value is used. The results of a snapshot simulation with the median (50-percentile) method and before any updates of the power control algorithm are shown in Figure 7.12. We see that the performance in the case of four beacons is almost as good as when the mobile is allowed to measure on all beacons. The limiting factor for the case with four beacons is that the median method gives a too-timid allocation. A lower percentile (around 40) would increase the performance in this case, while a higher percentile (around 60) would increase performance for the single beacon case, which is outage-limited.

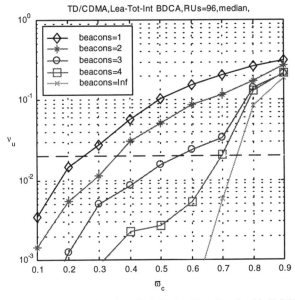

Figure 7.12 Assignment failure as a function of relative load for a bunch with 72 RAUs when the number of receivable beacons is varied between one and all (Inf). The measurement-based power control is not used.

Further capacity improvements should be possible by tuning the gain margins, percentile, number of iterations in the feasibility check, and so forth. The results so far show that the gain measurements are manageable at least in the studied environment. Measurement errors are handled by increased SIR-margins at the cost of a slightly lower capacity.

7.2.6.1   Asymmetrical Links

When evaluating asymmetrical links, we use constant received power control. The multiple access mode is TDMA/TDD, with 16 timeslots per frame. Considering the nature of client-server data applications and voice communications (it is impolite to interrupt), it is assumed that active users do not simultaneously use uplink and downlink. That is, at a certain instance of time they use either uplink or downlink, not both. Two asymmetry scenarios are considered: one where an active user is using downlink and uplink with probabilities 0.7 and 0.3, respectively, and another where the probabilities are 0.5 and 0.5.

In addition to the algorithm described above, curves are also calculated for two reference systems. The first reference system is a static TDD system where timeslots are reserved for either uplink or downlink data. The fraction of timeslots reserved for the different link directions are set equal to the average fraction of traffic offered to that link direction. That is, in asymmetry scenario one, 70% of the timeslots are reserved for downlink traffic, and in scenario two the corresponding figure is 50%. In the second reference system, full-link gain knowledge is assumed (i.e., link gains between MSs are known). This is included as an upper bound on the performance.

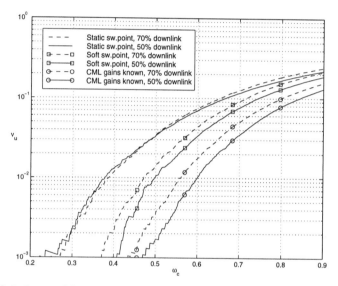

Figure 7.13 Assignment failure rate, v_u, as a function of relative load, ω_c, for three resource allocation algorithms in a bunch with 55 RAUs. Solid lines represent a scenario where 50% of the traffic is offered to the downlink and 50% to the uplink. Dashed lines correspond to 70% downlink and 30% uplink traffic.

From Figure 7.13, we readily see that at 2% user assignment failure, our simple resource allocation algorithm performs noticeably better than the static reference. For higher assignment failure rates, it approaches the performance of the static system.

7.2.6.2   Implementation Aspects

Previously, we primarily considered a bunch to be a collection of cells, but in some cases there are advantages in seeing the bunch as a single cell. If the bunch is a single cell then all the RAUs in the bunch will transmit the same ID and thus the mobile cannot tell to which RAU it is listening. This means that we cannot create the gain matrix by letting the mobile measure on the beacons. The main advantage with the single-cell approach is that we can implement an adaptive antenna selection (AAS) strategy ([4, 5]), where a mobile can perform handover between RAUs without actually noticing the event.

With AAS, the CU in the bunch can switch the mobile's traffic transparently from one RAU to another if the used channels are available and feasible in the new RAU. If the channels were not available in the new RAU, then we would of course have to reallocate the MS to some other channels. The AAS type of handover can be done very quickly and safely, since it does not rely on uplink signaling, which means that the dropping probability will be low. Another advantage is that macrodiversity is easy to implement; we can have several RAUs transmitting the same data and several RAUs receiving the data from one MS. The case with several RAUs transmitting is of course limited to pico- and microcells, where the receiver's equalizer can resolve the different paths and combine them in to one stronger signal.

As stated earlier, the disadvantage of the single-cell approach is that we can no longer measure the gain matrix in the downlink since there is no way for the MS to identify which beacon belongs to which RAU. Therefore, we have to measure all the gain values in the uplink. This could be difficult to perform regularly if the MS does not have so much uplink traffic, but all mobiles need at least some uplink control channel for quality measurements, acknowledgments, and so forth. Furthermore, we can estimate the gain for the serving link by measuring signal strength on the downlink traffic channels as long as macrodiversity is not used. Future studies will show if the advantages of the single-cell approach outweigh the difficulties concerning the gain matrix measurements.

7.2.7 Interference Matrix-Based Centralized RRM

The architecture of this wireless system is originally presented in [22], where a locally centralized approach to perform dynamic resource allocation is proposed. We use the concept of a bunch of cells as a group of RAU that are strongly interconnected. All resources available to a bunch are dynamically allocated by a CU. Previously studied for a WTDMA scheme [15, 16], this system architecture is now updated and its performance is being studied in relation to TD-CDMA access scheme.

In classical systems, the cell is seen as an entity. The limit between two cells generally consists of small areas where a mobile could be received by both base stations. Micro- and picocells in urban areas or in building, which we study here, do not easily integrate a regular pattern. The given environment results in a complex propagation situation. Base stations are likely to be located as close as necessary to guarantee a continuous coverage, so that optimal planning of base station location becomes impracticable. This leads to important overlapping where the mobile station is received by more than one base station.

We account for the cells overlapping by using the concept of coverage zones. A *zone* is an area that is covered homogeneously by one or several RAUs [15]. By homogeneously, we mean that the whole zone is covered by the same number of RAUs (Figure 7.14).

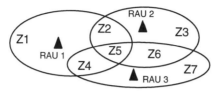

Figure 7.14 The zone concept.

7.2.7.1 C/I Aspects for Intercell Interference

If we consider reuse 1 in a system, the C/I is roughly 0 dB in the worst case (i.e., at the cells' common border) if a single code per timeslot is used. The use of several codes per timeslot (at most 8) implies that the C/I decreases roughly by 9dB (worst case), compared to the single code per timeslot case. To avoid that kind of situation and to enable the deployment of overlapping cells of different sizes, a modified version of the dynamic resource allocation (DRA) algorithm proposed previously in [23] is further considered. This improved scheme is adapted to the use of the CDMA component in a TDMA frame and to cope with TDD aspects. Note that I is only the intercell interference. It is assumed that the intracell interference is eliminated by the multiuser detector [17].

There are several aspects that should be taken into account. Let us consider the case with two RAUs with overlapping coverage (Figure 7.15). As defined previously, a *zone* is homogeneously covered by the same number of RAUs. In Figure 7.15 we have two MSs in the central zone covered by RAU_1 and RAU_2. In the TD-CDMA system, a family of orthogonal codes is associated to an RAU. As a result of this, we could associate the cell coverage to a family of codes. So the resource allocation is in some way RAU oriented. The families of codes in neighboring cells are not orthogonal. The macrodiversity is not required for the TDD mode. Moreover, in our case even if a zone is covered by several RAUs, only one of them will be used for resource allocation in a time slot.

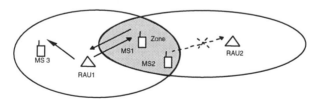

Figure 7.15 Allocation in a zone is RAU oriented.

For example, the MS_1 and MS_3 use RAU_1 (e.g., the same timeslot but different orthogonal codes from the same code family). To avoid increased intercell interference, a new mobile, MS_2 will not use RAU_2 in the same timeslot. A code used with RAU_2 is not necessarily orthogonal to a code used with RAU_1.

System Matrices

The structure matrix (Figure 7.16), defined as an interference matrix built on path losses between RAU and zones, is the same as in [17].

C: Covered: path loss > threshold L_1
I: Interfered: threshold L_1 < path loss < threshold L_2
NI: Noninterfered: threshold L_2 < path loss

Figure 7.16 Structure matrix.

The threshold values L_1 and L_2 are defined by means of simulations. The status of resources is recorded in a unique *resource management matrix* (Figure 7.17) for uplink and downlink. We define the frequency/timeslot pair as a *soft resource unit* (SRU). In a timeslot we could use from 1 to 8 codes.

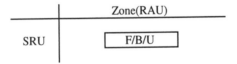

SRU: Soft resource unit
Free (F)
Blocked (B)
Used (U)

Figure 7.17 Resource management matrix.

The state of the SRU in a zone is *free* if the number of codes used in that timeslot is below the maximal permitted number of codes per timeslot. If the maximal number of used codes in a timeslot is reached, this SRU is in a *used* state. The influence of the intercell interference may be considered. The SRU is *blocked* when its allocation would severely disturb another call.

Let us consider the simple case with a zone covered by both RAU_1 and RAU_2. In that case, a row in the resource management matrix is presented in Figure 7.18. For each zone, we take the status of each timeslot for each covering RAU.

Nb of used codes

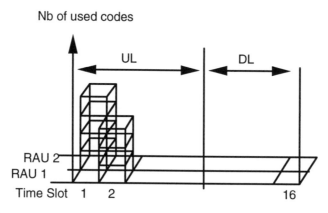

Figure 7.18 Example of a row of resource management matrix.

We see that in the first timeslot, four codes are allocated to RAU_2 in the uplink, and then the slot is blocked for the RAU_1. We use a timeslot only with a single RAU, even if several RAUs cover the zone. The switching point could be fixed and the same for all zones, or dynamic (depending on traffic variations) for a zone or group of adjacent zones.

Unique Switching Point for All Zones

If the switching point is fixed and common for all zones, the resource matrix resembles that found in Figure 7.19.

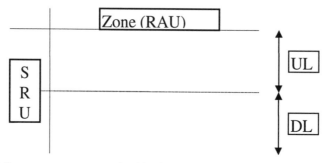

Figure 7.19 Resource management matrix with unique switching point.

In the case of unique switching point for all zones, we make the working assumption that considering only the threshold for coverage and neglecting the interference threshold in the structure matrix, the CIR could be ensured. This hypothesis would be studied by means of simulation.

Zone-Specific Switching Point

If the switching point is mobile and not the same for all zones, the resource matrix is shown in Figure 7.20.

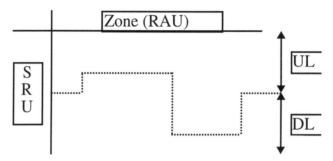

Figure 7.20 Resource management matrix with zone-specific switching point.

As the cross mobile link is the more critical from the interference point of view, the allocation algorithm should try to avoid this situation.

In this case, we have to respect two constraints when we perform allocation. First, each zone has a single switching point, even if it is covered by several RAU. Second, all the zones covered by the same RAU have a unique switching point. We note that these constraints are minimal. When SRU blocking is performed as presented earlier, these constraints are intrinsically fulfilled.

Allocation Algorithm

We consider the association of frequency/timeslot/code as an RU. Two types of policies are defined to avoid an unacceptable increase of intercell interference level:

- *Polite policy*: block an SRU in a zone if its allocation would severely disturb a current call;

- *Aggressive policy*: block an SRU in a zone if its allocation would be severely disturbed by the current call.

In the case of unique switching point for all zones, we make the assumption that considering only the threshold for coverage and neglecting the interference threshold in the structure matrix, the CIR could be ensured. This hypothesis would be studied by means of simulation.

To illustrate how we perform the blocking of SRU when an allocation is carried out, we consider the following configuration in Figure 7.21.

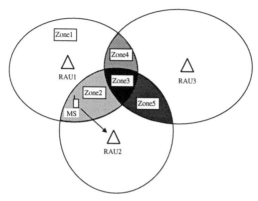

Figure 7.21 Example of SRU blocking for uplink allocation.

We suppose that we have an MS in $Zone_2$ and we choose the RAU_2 for the uplink allocation. According to the aggressive policy, we block the SRU for the pairs (Zone, RAU) that could interfere with our link. These zones are covered by at least two RAUs of which at least one is the current RAU_2. The pairs are $(Zone_2, RAU_1)$, $(Zone_3, RAU_1)$, $(Zone_3, RAU_3)$, and $(Zone_5, RAU_3)$.

According to the polite policy, we block the pairs (Zone, RAU) that can be interfered by our allocation. If the current zone is covered by several RAUs, we block the SRU for all the zones covered by these RAUs, the current RAU excepted. In our example, the pairs are $(Zone_1, RAU_1)$, $(Zone_2, RAU_1)$, $(Zone_3, RAU_1)$, $(Zone_4, RAU_1)$. In the case of downlink allocation, the SRU blocking is performed in a similar manner. In the following, we present the allocation algorithm. There are n_r RUs requested in a zone:

Step 1: The CU preselects all a priori-free SRU n_f.

- A priori-free SRU are those of which the state is free in the resource management matrix.

- If $n_r < n_f$, then the call is blocked and cleared.

Step 2: The score of the n_f resources is evaluated.

- The CU evaluates which zones would be blocked by the allocation of the resource. If at the location the resource is already blocked, we increase the score of that resource. The blocking is performed according to aggressive and/or polite policy. The scoring idea is based on the reblocking principle to maximize the reuse of the same SRU if possible.

Step 3: The n_r SRU with the highest scores are selected.

- The resources are allocated and the resource management matrix is updated, accounting for the additional blocking caused by the new allocation.

- After deallocation, the resource management matrix is also updated.

7.2.7.2   Evaluation of System Performance

The system performance was evaluated through simulations performed in an urban environment. The QoS is defined by considering the new call-blocking probability for a certain C/I ratio.

The simulation model is a Manhattan-like structure considered for the urban environment with a grid of rectangular streets and squared buildings. No wrap-around is assumed. The structure and deployment models are a reduced version of the ones proposed in [1] for UMTS radio transmission technology evaluation. The propagation model is a recursive model (see more details [18]) that calculates the path loss as a sum of LOS and NLOS segments. Note that this propagation model is valid only for microcell coverage with antenna situated below the rooftop. The slow fading (shadowing) is modeled by a lognormal distribution with zero mean and a 10-dB standard deviation. It is constant in time and depends only on the mobile position and the position of the RAU, to which the mobile is connected. A mobile in the same geographical position and connected to the same RAU will experience the same slow fading. We consider that the fast fading is averaged out.

The mobiles are uniformly distributed in the street areas. Mobility is not considered in this study. It is assumed that the MS localization in a zone is perfectly done by the CU. The effects of imperfect localization on the algorithm are for further studies. The simulated system contains 32 RAUs. A simple system configuration with seven timeslots per each link and a maximal number of codes per time slot equal to eight was simulated. Note that from the 16 slots available in the TDMA frame, two slots are used for signaling (i.e., one for the downlink beacon channel and one for the uplink random access).

Circuit Allocation

The traffic model includes only circuit services. Speech calls are generated according to a Poisson process assuming mean call duration of 120s. No voice activity detection is considered. The performance of the DRA algorithm is studied from the interference point of view characterized by the C/I ratio and from the traffic point of view characterized by the new-call-blocking probability. From the traffic point of view, the performance of the algorithm is shown in Figure 7.22. We see that the carried traffic for 1% new-call-blocking probability is about 17 Erlang per RAU.

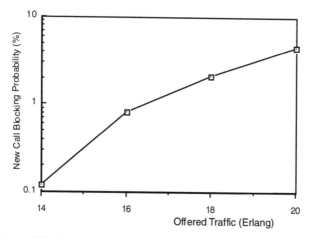

Figure 7.22 New-call-blocking probability versus offered traffic.

In Figure 7.23 we plotted the cumulative distribution function (CDF) for the C/I ratio in the case of the DRA algorithm with the aggressive policy for resource blocking, and in the case of the DRA algorithm with mixed aggressive and polite policies for resource blocking.

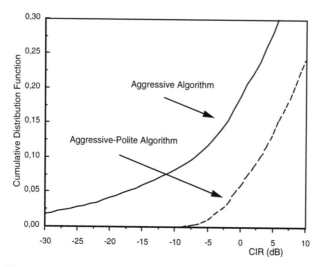

Figure 7.23 CDF comparison (for the C/I ratio) for different allocation policies.

We estimate that the range of CIR target of the TD-CDMA system should be roughly between –15 dB and 0 dB, depending on the type of service. For this CIR range, the allocation algorithm performs well. For instance, the mixed aggressive-polite policy ensures that almost all mobile stations have a CIR inferior to –5 dB.

The DRA algorithm with the aggressive policy performs less well (e.g., less than 10% of the mobiles have a CIR inferior to –5 dB).

Packet Allocation

The traffic model contains only WWW browsing sessions. The model used is the one described in [1]. Only downlink sessions are accepted. The parameters of the session are in Table 7.1.

Table 7.1

Parameters of WWW Session Traffic Model

Packet information types	Average number of packet calls within a session	Average reading times between packet calls [s]	Average number of packets within a packet call	Average interarrival time between packets [s]	Parameters for packet size Pareto distribution
WWW surfing UDD 144 Kbps	5	412	25	0.0277	k=81.1, α=1.1

Our goal is to study the allocation algorithm from the interference point of view. We consider that the queuing discipline decides what packet should be transmitted and the allocation algorithm only takes care of the interference in the system. So, the only QoS parameter considered is the C/I ratio. The study of the queue discipline should take care of the packet delay. In our simulations, only best-effort discipline is used. Depending on the number of available resources, we allocate to a packet 1 to 10 codes per frame. The CDF for the C/I ratio is presented in Figure 7.24. A comparison between the aggressive and aggressive-polite policies is plotted. The number of session arrivals is 0.13 sessions/sec. One packet is discarded if a resource is not available in less than 200 ms.

We see that the aggressive policy performs well in the case of packet allocation. For example, we have less than 5% MSs with a CIR inferior to –10 dB. Two types of multirate are investigated; multicode and multislot. A comparison from the C/I ratio is plotted in Figure 7.25. To avoid the influence of the allocation algorithm, we use a simple random allocation algorithm. A resource is randomly chosen from the set of available resources. Any special care to limit the interference is taken. We see that the multislot allocation outperforms the multicode allocation from the CIR point of view.

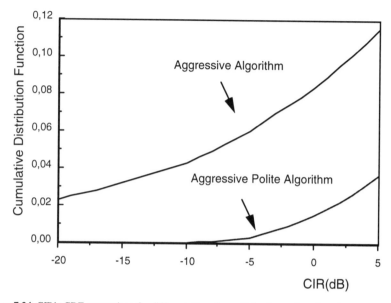

Figure 7.24 CIR's CDF comparison for different allocation policies for packet allocation.

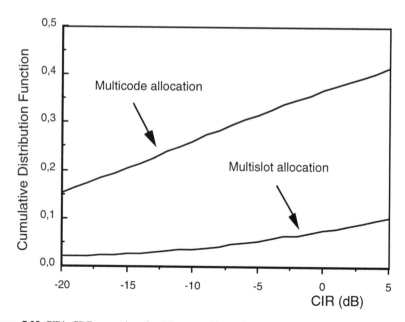

Figure 7.25 CIR's CDF comparison for different multirate allocation.

Mixed Packet-Circuit Allocation

The best packing strategy is evaluated in the case of mixed traffic. Circuit calls are bidirectional, voice calls. Packet calls are of the same type as mentioned previously, unidirectional in the downlink. The number of voice call arrivals and WWW session arrivals is the same (i.e., 0.13 calls/sec). The CIR is plotted in Figure 7.26. We see that circuit calls have a greater C/I ratio than the packet calls. The algorithm copes well with the burstiness of the WWW traffic.

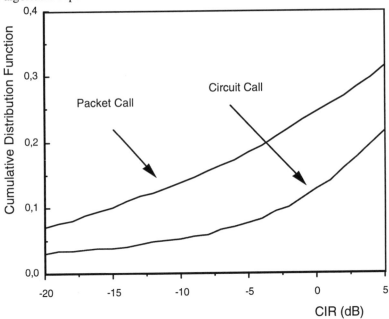

Figure 7.26 CIR's CDF for mixed packet-circuit allocation.

7.2.7.3   Summary

We propose a locally centralized architecture for resource allocation in microcellular environments for the UMTS TD-CDMA access scheme. The performance of this scheme was evaluated through simulations in the case of circuit calls, packet calls, and mixed scenario. The results show a good behavior face to the interference. Two types of allocation policies are proposed.

7.2.8   Decentralized Resource Allocation

Especially in the TDD mode, an allocation method suitable for packet traffic should be investigated, since this mode will predominantly be used in micro- and picocell layers, where the data traffic load is expected to be higher than in macrocell layers. The TDD channel structure (and other features), as defined in the summer of 1998, is used to define the system. Starting points are decentralized methods, which can be adapted to work in a distributed manner. A decentralized channel segregation algorithm is considered. The basic algorithm is presented, and the improvements needed to adapt this algorithm to mixed services allocation are studied. Some simulations results are also presented to validate performance.

Radio Access Scheme

The UTRA TDD mode description is not yet finished/available, but here are the assumptions used throughout this investigation:

- Five MHz carriers, so multicarrier operation not necessary/foreseen;

- Frames of 16 timeslots on one carrier;

- Dynamic switching point for up-and-downlink traffic within the frame;

- Eight or 12 CDMA codes per timeslot, codes 16-bits long;

- Frequency reuse: 1;

- No specific arrangements for NRT/packet channels.

Traffic and Area Models

From [1], the area and propagation models are taken. The Manhattan area model from Section 7.2.5 is used, with the accompanying path-loss model. The traffic model has up to now been confined to speech, with a single duplex link requirement, due to simulator problems.

Layer 2 Assumptions and Design

To make the separate functions transparent in Layer 2, algorithms for channel allocation and for the scheduler should be separated. This way, the allocation algorithm only needs to multiplex data on to physical channels, while the scheduler only needs to know how many resources are available (Figure 7.27).

However, while investigating channel allocation algorithms keep in mind that the incoming traffic would need scheduling (and possibly load control).

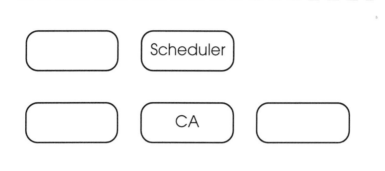

Figure 7.27 Layer 2 algorithm relations.

7.2.9   Channel Allocation Algorithm (Segregation) Description

The resource allocation studied in this part is based on the DCA algorithm proposed in the RACE/ATDMA project [21]. The main advantage of using a DCA concept is to avoid any resources planning, and to adapt the resource allocation to a channel quality criterion or to a traffic parameter. Some studies have been carried out in the ATDMA project, and the segregation channel algorithm has been chosen for its self-adaptive learning capability and its adaptability to packet-switched schemes. Since the TDD mode will be used for asymmetric traffic and high-bit-rate data traffic, this DCA scheme seems worth considering.

The segregation channel algorithm is based on channel quality, which is evaluated considering the channel interference level. Each base station has a priority list of preferred timeslots. The basic algorithm is implemented as follows:

- On a frame basis, each base station defines a priority value of each timeslot based on interference level experienced by the different users on the slot. A priority function gives the relationship between the priority value and the interference level of a slot.

- The priority list of the timeslot is updated in each base station considering a geometrical window formula that considers the new priority value, and also old priority values. Each priority value is weighted to take into account the measurement time of this value, so the values' influence decreases with time.

- For a resource allocation, the base station takes resources on the slot with the highest priority.

Associated with the channel allocation algorithm, a channel reallocation (intracell handover) is also implemented. The decision is based on quality criteria.

Considering this basic algorithm, some points should be studied to adapt this algorithm to a TDD mode and to mixed traffic services:

- The impact of data services using multicodes and/or multislots resource allocation on the priority list evaluation;

- The impact of NRT services with variable bit rate;

- The allocation strategy for mixed services, especially for the cases mixed circuit low/high bit rate users and mixed circuit/packet services (e.g., the necessity to keep free resources for high bit rate services).

The basic algorithm chosen is an isolated concept where each base station makes its own decisions based on local measurements, and no communications between base stations are required. For the downlink, the mobile performs measurements of the interference level on the downlink traffic channel of the serving cell. These measurements are preprocessed and transmitted periodically to the base station. For the uplink, the base station performs measurements of the interference level on the uplink traffic channels (for each user connected to the base station). With these measurements, the base station constructs its priority list of resources.

Simulation Description

The investigation is based on dynamic system simulations performed at the Research and Development Centre of France Telecom (CNET). This radio mobile network simulation tool was designed to provide long-term and wide-scale performance statistics for different radio access systems, and to evaluate the capacity of these systems under various environments and configurations. The simulator has been adapted to the TDD mode with the following issues:

- Modeling of intracell and intercell interference;

- Modeling of a variable switching point;

- Path loss and interference calculation for MS-MS and BS-BS situations;

- Adaptation of a DCA algorithm to a TDD mode.

The simulations consider scenarios defined in [1]. The test environments, propagation models, shadowing model, mobility models, and traffic models are those described in this book. The scenario chosen (speech service and Manhattan environment) allows us to study the general mechanisms of the algorithm and to make some preliminary remarks.

Figure 7.28 shows the users' distribution among the slots of each base station. For the case without DCA, users are uniformly distributed on each slot. On the contrary, for the DCA case, for base stations in direct visibility (e.g., base stations 9 and 10 or 20 and 21), some sort of cluster of slots appears due to the segregation algorithm.

- For low interference situations (base stations without neighbour cells in direct visibility), the channel segregation algorithm shows little differences compared to the random case. In fact, with low interference level, the channel segregation is not really needed. The users are quasiuniformly distributed on each timeslot (base station identities from 11 to 15).

- For high interference situations (base stations with neighbour cells in direct visibility), the segregation is very efficient. Results show a sort of cluster of the timeslots between neighbouring cells. This is illustrated by the example of the base stations 9 and 10. Base station 9 uses slots 2, 3, 6, 7 in particular; on the contrary, base station 19 uses slots 1 and 4.

- For high traffic load, the basic DCA algorithm is not sufficient. In fact, the figures presented have a traffic load of 45%. For higher traffic load, the cluster of timeslots becomes less visible since most of the resources are needed. Furthermore, since more resources are needed, even timeslots with very low priority (i.e., high interference level) are allocated. This leads to a rapid degradation of the quality of these slots. Some sort of admission control should be added to this algorithm to avoid the allocation of a resource with a too-low quality. This admission control decreases the system capacity but protects user quality, especially for high bit rate services.

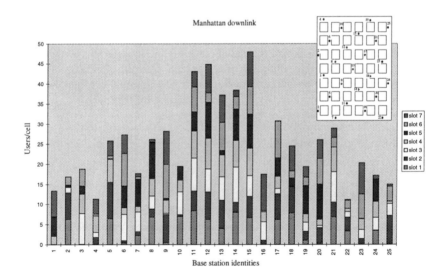

Figure 7.28 User distribution among slots for speech service.

For evaluation of the capacity, the definition described in [1] is taken into account. The spectrum efficiency is defined as the system load where there are exactly 98% of satisfied users. A user is satisfied if all three of the following constraints are fulfilled:

- The user does not get blocked when arriving at the system.

- The user has sufficiently good quality more than 95% of the session time. We point out that only the active session time is considered in the evaluation of the percentage of bad quality.

- The user does not get dropped. A speech user is dropped if it experiences bad quality for more than 5 seconds.

First capacity results show (Figure 7.29) that for low to medium traffic situations, DCA improves the quality, and this quality could be further improved by applying intracell handover. However, for high traffic load (> 65%), since many of the slots are used, the capacity reached is about the same with or without DCA. The capacity obtained for speech service in a Manhattan environment is 92 Kbps/cell/MHz.

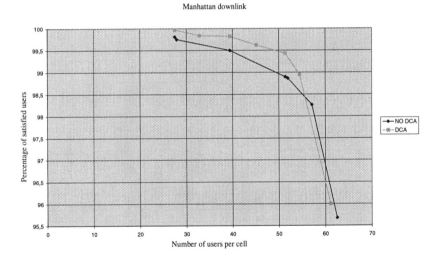

Figure 7.29 Satisfied users for systems with and without DCA for speech service.

This first study has shown the efficiency of the channel segregation algorithm. This point becomes even more important considering higher bit rate services than speech. In fact, for speech service, capacity is reached for a high traffic load, but for higher bit rate services, the efficiency of the segregation is more important to avoid the reuse of a slot in neighboring cells. Considering a 144 Kbps user, which needs nine codes allocated on one slot, this slot becomes too loaded to be reused in adjacent cells. The channel segregation algorithm allows you to avoid such problems without needing any frequency or time reuse.

Results for LCD144 Service

To verify the efficiency of the segregation algorithm for higher bit rate services than speech, simulations have been performed for the LCD144 service. The same scenario is chosen to directly compare the algorithm efficiency for different services. The LCD144 service assumes the allocation of one slot and nine codes on that slot. Since the maximum number of codes per slot is 12, one slot can support only one LCD144 user.

As said above, a slot supporting an LCD144 user is too loaded to be reused in neighboring cells, and a time reuse needs to be implemented. The channel segregation algorithm allows you to have this time reused dynamically and avoids any planning.

First results obtained for the LCD144 service show the efficiency of the algorithm; the cluster of the timeslots appears in the same manner as for speech service, see Figure 7.30.

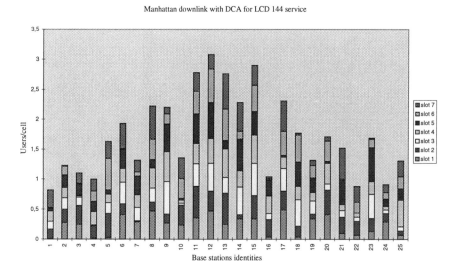

Figure 7.30 User distribution among slots for LCD144 service.

Furthermore, the capacity evaluation shows a gain of 25% of the capacity when the DCA is introduced. Without DCA, the capacity is reached for two users per cell (i.e., 115.2 Kbps/cell/MHz). With DCA, this capacity is increased to 2.5 users per cell (144 Kbps/cell/MHz).

7.2.10 Further Studies

The channel segregation has proven its efficiency. However, to support mixed services (speech and LCD144 in our case), changes in the allocation strategy are needed:

- Speech users should be multiplexed on the same slot in order to keep free resources for high bit rate. Results for speech services have shown that for a maximum number of seven codes per slot, the slot can be used in neighboring cells.

- LCD144 users are allocated on a complete slot. This slot cannot be used in neighboring cells.

- Some kind of reshuffling of resources may be required to keep this type of allocation (above all for speech users).

- Intracell handover is also necessary to avoid problems caused by the allocation of a slot already used by a neighboring cell.

This automatic allocation procedure needs to be carefully studied to achieve such an arrangement. This allocation will be useful for mixed services and also to allow base stations to have variable switching point. The reshuffling of users will be able to free a timeslot for a direction change.

Thanks to this procedure, the TDD mode could present a flexible boundary between uplink and downlink slots as well as an adaptation of the asymmetric traffic variations.

7.2.11 Summary

Results show the efficiency of the channel segregation algorithm and the sort of cluster created by the algorithm. Furthermore, for the LCD144, a capacity gain brought by the introduction of this DCA was demonstrated. Other points need to be studied to improve the efficiency of the algorithm for mixed services and packet data services.

7.3 RESOURCE MANAGEMENT FOR THE WCDMA MODE

This section describes the RRM schemes for UTRA FDD proposed by the FRAMES RN workpackage. UTRA FDD uses wideband DS-CDMA technology and is also referred to as WCDMA. The main idea of the proposed schemes is to ensure the existence of a simple and robust scheme for RRM for the UTRA FDD mode. The cornerstone of the presented scheme is interference control by using fast power control both in uplink and downlink, as well as soft handover. The stability of the system will be ensured by admission and congestion control. Due to the reuse of one, all the channels (code) of a given cell in the system experience almost equal interference conditions. This greatly simplifies the RRM algorithms.

The algorithms presented are PC, admission control (AC), congestion control (CC), and handover (HO).

7.3.1 Basic Principles

The traffic served by the system can be divided into two classes (i.e., NRT and RT traffic). The RRM algorithms are similar for both these classes. However, the UTRA FDD concept treats short packets and long packets differently. For NRT traffic, a (fast) load control is provided by the AC and CC schemes mainly based on the interference level (uplink) and used output power (downlink). These schemes have to be applied in the UTRA FDD system because the number of channels is (virtually) unlimited. Short packets use an open loop power control and longer packets (allocated a return channel) use fast-closed loop PC. The handover for NRT traffic may be based on the service demand (e.g., UL/DL capacity requirements).

For RT traffic, AC and CC algorithms are also applied but with lower constraints on the execution delay. The admission control algorithm is almost the same as for NRT bearers but include bearer prioritizing functionality. RT bearers also use the fast-closed loop PC (i.e., SIR-based) and an outer PC loop that sets the SIR target according to the radio environment (channel).

The proposed solutions may seem simple, but they are expected to be effective. They will also support joint detection and adaptive antenna techniques.

7.3.2 Overview of the RRM Algorithms

The power control algorithms are divided into closed loop PC (fast PC), outer loop PC (slow PC), and open loop PC. The fast PC will be implemented as a closed loop in which the receiving part returns up/down commands on a return channel, adjusting the received SIR toward a target value. The desired SIR target value will have to be adjusted to satisfy the received quality due to channel condition variations. Therefore, an outer loop is required in which updates of the target SIR value are sent depending on the received quality. Furthermore, an open loop PC, C, or SIR-based, is required in the uplink when closed loop commands are missing (e.g., for random accesses and initial transmission on a new channel).

The load control algorithms are divided into an AC algorithm and a CC algorithm. The AC algorithm will ensure that the users requesting access into the system can be granted their QoS requirements. The uplink AC algorithm is based on interference or uplink SIR threshold measurements. The purpose of CC is to maximize the achieved bitrate over the network without causing an overload situation. The CC control algorithm strongly depends on the fast SIR-based power control algorithm and uses interfrequency HO (move a user to another frequency with lower interference).

The HO algorithm can be located in the MS or in the RNC. The MS reporting (path loss of active set and candidate set pilots) to the RNC is mainly event triggered. The HO algorithm supports soft/softer and interfrequency HO and can be purely path loss or service and load dependent. One additional measure when performing an interfrequency HO is the MS speed.

7.3.3 Power Control

The uplink power control (PC) is essential for a CDMA system to combat the uplink near-far problem (i.e., that interferers located near the base stations overwhelm the signals from more distant mobile stations). The uplink power control must be fast enough to track (slow) Rayleigh fading and other propagation loss variations. The same properties are desired for the downlink PC, although the near-far problem is smaller. Users with low velocities in the outer parts of the cells especially benefit from the downlink PC. Moreover, when employing adaptive antennas and transmitting with equal Tx downlink power to all bearers, the relation between the desired signal and the intracell interference will vary between users within the same cell. A SIR-based PC changes that situation and improves the performance. Equivalently, the use of adaptive antennas in the uplink makes it important to control the SIR level rather than the Rx level, since the uplink interference is no longer identical for all users within a cell.

Figure 7.31 shows the C/I and SIR definition. R_c is the chip rate and R_b is the information bit rate. PG is defined as R_c/R_b leading to PG = 512 (27 dB) for an 8-Kbps voice service. Further, SIR can be seen as C/I times PG.

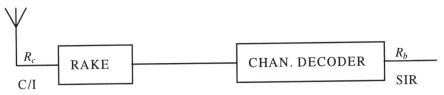

Figure 7.31 C/I and SIR definition in this chapter.

The fast PC is implemented as a closed loop in which the receiving part returns up/down commands on a return channel, adjusting the received SIR toward a target value. The desired SIR target value must be adjusted to satisfy the received quality due to channel condition variations. Therefore, an outer loop is required in which updates of the target SIR value are sent depending on the received quality. Furthermore, an open loop PC is required in the uplink when closed loop commands are missing (e.g., for random accesses and initial transmission on a new channel). In the open loop, the transmit power is determined from a path loss estimate, which in turn is derived from measurements of a reference signal transmitted at a known power level.

The power control consists of three components: closed loop PC (fast PC), outer loop PC (slow PC), and open loop PC, where the outer loop in a sense controls the fast closed loop.

Closed Loop

To achieve a closed loop, the data channel needs an associated return channel for the PC commands. Such a return channel will always be available for traditional circuit switched bidirectional services like speech. However, for packet services, this will depend on if a dedicated connection is set up for the packets, or if a common channel is used. A dedicated connection needs a return channel for other control signaling (e.g., synchronization), which also can be used for the PC commands. When a common channel is used for the packet transmission, there is no dedicated return channel for the PC commands, and accordingly the closed loop PC is not applicable.

The closed loop must be fast enough to track the (slow) Rayleigh fading, which means that the PC commands must be frequent enough and the execution time of the up/down corrections must be short. The price to pay for a high PC command frequency is a high signaling load over the air interface. The required command frequency depends on the present channel condition, and therefore it is desirable to adapt the command rate to the channel variations or at least to the specific environment. The power step should also be adaptive. In case of a too-large step size, the *Tx* level will ripple with large variations around the desired level. Too-small steps on the other hand, make it hard to track the fading unless the time step is made smaller.

A large dynamic range is required for the uplink (about 80 dB) to overcome the near-far problem. For the downlink, the dynamic range requirement is smaller (about 30 dB) since there are less near-far problems. However, it is important to take the future use of adaptive antennas into account when specifying the downlink dynamic range. Further, the large dynamics in data rates among bearers has to be accounted for, since a larger data rate requires larger power and a bearer's data rate may alter very quickly.

In soft handover, the MS can receive different power commands from different BSs (i.e., one BS is giving power-up commands, while the other is giving power-down commands). To deal with this ambiguity and to minimize the interference, the MS only increases its power if all BSs it is in soft handover with are giving power-up commands. In the downlink, the needed transmitted power is split equally between the BSs when the MS enters macrodiversity. The longer the MS stays in soft handover, the more likely the powers, from different BSs, are to drift apart due to errors of the power control commands. This problem can be solved by forwarding the PC commands up to the RNC, which then has to correct these errors and order identical commands to the participating BSs. However, the feasibility of such a solution depends on the signaling load between the RNC and the BSs and possibly extra delay of the PC commands. For MSs in softer handover (handover between sectors located at the same site), the ambiguity will not occur for neither uplink nor downlink if PC commands are demodulated after (RAKE) combining of the macrodiversity branches. The same applies for the uplink PC of MSs in ordinary soft handover between different BSs.

The closed loop PC requires a reliable SIR measurement method with high precision, low bias, as well as low delay properties. A physical channel suitable to measure on is of course required. Both the power of received desired signal and the interference power can for instance be measured on pilot symbols.

Outer Loop

The purpose of the outer loop is to adjust the target SIR of the (fast) closed loop PC that under the current conditions satisfies the required reception quality. The quality has to be measured in some way (e.g., as average raw BER or Frame Erasure Rate (FER)). Typically, FER estimation is slower than raw BER estimation since the required averaging time is longer. The time scale of the outer loop is typically 1 to 5 seconds. NRT bearers are assumed to often transmit their data with a high bit rate, hence only occupying the channels during a short amount of time. Thus, no outer loop updates can be performed. Therefore, the SIR target has to be set on a session (service and radio environment dependent) basis. The outer loop can be used for this task if we collect cell statistics in order to set this SIR target. Perhaps the mean Block Error Rate (BLER) can be used as a quality measure for the outer loop algorithm for these NRT bearers.

In the uplink there is a question of where the quality estimation shall take place. When in soft handoff the most accurate quality estimate is obtained after diversity combining in the RNC.

Open Loop

In the uplink, the open loop PC is needed for determining the appropriate transmit power when PC commands are missing. Before transmitting a random access, a packet on a common channel or the initial transmission on a dedicated channel, a proper transmit power is determined by estimating the path loss to the receiving BS and the interference level at the BS. If necessary, an open loop PC can also be used during a call to provide fast recovery from rapid and large changes, in that case operating in combination with the closed loop. The necessity of such functionality depends on the ability of the closed loop to manage fast changes.

For the path-loss estimation, measurements are done on a reference signal transmitted from the BS at a known power level. The reference signal could be a pilot tone or any other control channel whose transmitted power level is broadcast by the BS. Measuring a reference signal gives a better path-loss estimate than measuring the total in-band power, since the load does not affect it. One should also remember that the Rayleigh fading is different in the uplink and downlink. Hence, filtering is needed to achieve an accurate estimate of the uplink path loss.

Open loop PC cannot be used in the downlink in the same way as in the uplink. In that case, a reference signal from all MSs is required. Initial downlink transmissions can only be power controlled if there are preceding uplink transmissions carrying downlink measurement results. Therefore, the random access should contain a path loss estimate (or a corresponding measurement) from the uplink open loop, which can be used for setting the initial downlink transmission power.

7.3.4   Power Control Algorithms

To have the possibility of dealing with the instability (overload) problem, the following uplink closed loop PC algorithm is investigated. The algorithm is defined to give power-up command if:

$$\frac{SIR_m}{SIR'}\left(\frac{Z_r}{Z'}\right)^n < 1$$

Here, SIR_m is the measured SIR, SIR' is the target SIR, Z_r is the total received power, Z' is a chosen threshold, and n is a chosen coefficient which could look like:

$$n = \begin{cases} 0 & if \quad Z_r < Z' \\ 0.25 & if \quad Z_r \geq Z' \end{cases}$$

A typical step size for the power-up and power-down commands is 0.5 dB, but other step sizes are possible. The maximum allowed uplink output power is terminal dependent.

In the downlink, the instability problem is not as severe as in the uplink, since it often stays local. We suggest that a pure SIR-based power control is employed. The downlink algorithm is defined to give a power-up command if:

$$\frac{SIR_m}{SIR'} < 1$$

The maximum downlink power is limited by the maximum allowed (total) output power from the BS. As previously mentioned, the downlink powers assigned to a bearer in macrodiversity might drift apart. Hence, communication with the RNC is needed to counteract this drift. The communication between the BSs and the RNC is decreased by lowering the power control bit rate during macrodiversity. At the same time retransmit the same power control commands from the MS a couple of times to achieve a higher accuracy in the power control signaling. The slow power control will alter the SIR', very similar to the scheme used in CODIT. The method used to measure the end quality is performed in the same way for both uplink and downlink. However, different methods are used for speech and data services. Note that slow power control must be performed on each bearer, while fast power control is performed on a user basis. The method used for each service is based on which channel coding and interleaving scheme is used for that particular service. Calculating raw BER (reencode and compare), BLER for RLC blocks, or the FER measures the quality. The open loop depends on whether the information about the interference is available. If the information is not available, a gain-based algorithm is used to set output power to

$$\frac{C'}{g}$$

Here, C' is a signal strength target, and g is the gain between the MS and the BS. If interference measurements are available, a different algorithm that aims at SIR target is used to set output power to

$$\frac{SIR' \cdot I_m}{g \cdot PG}$$

Here, I_m is the interference measurement. If the desired quality is not achieved by using the open loop, the output power is increased in steps, where the step size is the same as for the closed loop. However, only a limited number of increased power attempts are allowed.

7.3.5 Measuring and Signaling Requirements

For the closed loop uplink power control, the received SIR as well as the total received power at the BS must be measured (Table 7.2) for each cycle (5-20 times per frame). For the downlink, on the other hand, only SIR measurements are needed. However, these measurements must be made after macrodiversity combining, if the MS is in soft handover.

Table 7.2

PC Measurements

Measurement	Measuring entity	Range	Measurement frequency
uplink_sir	BS (closed loop)	Performed at all BSs	1 or 2 kHz
uplink_interf	BS (closed loop)	Performed at all BSs	1 or 2 kHz
Downlink_sir	MS (closed loop)	One meas. per user (MS)	0.5 or 1 kHz
Path loss	MS (open loop)	Number of candidate BSs	As often as possible
Uplink_quality (BER, ...)	RNC (outer loop)	One meas. per service	As often as possible
Downlink_quality (BER,..)	MS (outer loop)	One meas. per service	As often as possible

The parameters of the PC algorithm are given in Table 7.3. Note that the values of these parameters can be different for different directions (UL/DL) of transmission.

Table 7.3

PC Algorithm Parameters

Parameter	PC algorithm	Description
OL_PC_PWR_STEP	Open loop	Power increment when access failed
OL_PC_TRIALS	Open loop	Maximum number of trials
CL_PC_REP_INTERVAL	Closed loop	Repetition time interval
CL_PC_PWR_STEP	Closed loop	Power step of the closed loop

7.3.6 Interaction With Other Algorithms or Entities

Stability is a critical issue for an SIR-based PC, in which the so-called party effect occurs for high loads. The transmit power is raised continuously for all bearers to overcome the ever-increasing interference. Such effects can be avoided by efficient admission control, so that the load is kept below the maximum allowed level. The admission control would then rather interact with the outer loop, which in turn controls the closed loop. On occasions where the admission control does not manage to avoid the party effect, a congestion control mechanism is required to stabilize the system. Some kind of built-in stabilizing mechanisms in the closed loop PC is desirable (i.e., the n coefficient in the uplink closed-loop algorithm).

The outer loop has a close connection to the admission and congestion control, which should be able to control the offered quality for different loads and different bearer requirements. It is the outer loop that together with link adaptation (changes coding) and channel assignment (multicode) realizes the admission control decisions. Note the bearer-dependent link adaptation (i.e., the coding etc., only will be set at call setup). If the channel of a specific bearer is varying during a call, it is a task for the outer loop to adjust the SIR target to compensate for these channel variations.

7.3.7 Handover

This section deals with handover for UTRA FDD. Three handover algorithms are proposed: gain-based, and two that aim at optimizing the UL or DL, depending on the specific bearer. These algorithms aim to select the base station that can provide the required quality with the lowest possible output power of the MS/BS. Quality is estimated separately for each candidate BSs before the actual handover occurs.

Definitions:

- **Sector:** Smallest service area. Each sector has a unique identity (pilot code) that is broadcasted on the downlink. In this section, a sector is equivalent to a cell.

- **Base station:** A base station serves several sectors. In this section, a base station is equivalent to a site station.

- **Cell layer:** Each cell layer has a different set of frequency carriers in order to avoid severe interference between the layers, that is *true* hierarchical cell structure (HCS).

- **Active set (AS):** The set of sectors to which a user is connected. A connection is defined as a dedicated channel.

- **Candidate set (CS):** The set of the candidate (mainly neighboring) sectors.

7.3.7.1 Idle Mode Handover and Cell Selection

The MS makes measurements on the current sector, and the control channels (PCH and BCCH) associated with the sector (base station) are also monitored. The MS starts to measure on pilot channels of other sectors when the current sector is below a pre-defined threshold (i.e., the MS has moved from the coverage area of the sector). The neighbor set defines which pilot channels will be measured by the MS. The mobile station can be in two different modes: idle or active mode. Perhaps there is a need for a third state of the mobile station. This state may be called the *access* or *standby* state and it means that the mobile station also measures the neighbor set pilot channels. This may improve the cell selection performance, that is, this may be needed when a NRT bearer sends a short (nonpower controlled) data packet.

The estimated path-loss(es), measured by the MS, will be included in the random access burst to improve the cell selection algorithm and to simplify the initial downlink power settings. However, the following sections emphasize handover (i.e., when the user already has one (or more) dedicated channels).

7.3.7.2 Requirements for the HO

If a handover algorithm that is independent of the UL/DL load in the system is desired, we can simply use the path loss, estimated by the MS, as the only HO decision variable. On the other hand, if we believe that the capacity can be improved by making handover decisions bearer and load-dependent, more measurements are required. For the latter approach, all BSs continuously measure their average total output transmission power (TXP). The average TXP is calculated so that the BS add together transmission power of each channel (code). Each BS also measures the total received power it experiences in the uplink direction. The BS transmits this average value (an estimate of the interference level) on a broadcast channel.

The MS measures the path loss of all base stations in active and candidate sets, and these measurements are reported to the BS. The information of the uplink interference level is also received from all BSs within the active and the candidate set.

For the uplink, the goal is to minimize the output powers but still meet the SIR requirement. Therefore, it is natural to let the UL decision variable be based on the UL SIR.

The soft handover algorithm for RT bearers is quite straightforward. However, when we consider NRT bearers, two problems arise:

- The termination of the ARQ protocol will influence the handover algorithm, and it is not obvious that we gain anything when using soft handover together for NRT bearers. The transmission load and the complexity increases with soft handover.

- The complexity of the packet scheduling at the base station (downlink) may increase if soft handover is used.

A possibility is therefore to set the AS_MAX_SIZE to one if only NRT bearers are used (i.e., not using soft handover for NRT bearers). If RT and NRT bearers are used simultaneously, the same active set as for the RT bearers is used.

The softer handover algorithm is similar to the soft handover algorithm, but it is possible to identify if some sectors are associated with the same base station. This means that it is possible to use different handover thresholds for soft and softer handover. Note that for (pure) softer handover, there is no problem with the NRT bearers because the user is only connected to one base station.

7.3.7.3 Handover Initiation Algorithms

The (main) parameters needed by the algorithm are shown below. These parameters are sent on a downlink control channel (L3 messages). The receiver of the handover messages may be a sector or a specific bearer.

The basic behavior of the algorithm (i.e., the adding, replacing, and removal of a sector), is shown in Figure 7.32.

The simplest handover algorithm is to make the handover decision path loss dependent (i.e., connect to the nearest BS).

$$HO\_quality\_estimate = \frac{1}{Path\ loss}$$

If we want to maximize the capacity in the uplink, the handover algorithm will be SIR dependent, since we want to minimize the MSs' output powers.

$$SIR_{UL} = \frac{P_{MS}}{Path\ loss \cdot I_{UL}}$$

Hence, we should try to minimize the *Path loss·I_{UL}* factor. I_{UL} can be broadcasted from the base station.

$$HO\_quality\_estimate = \frac{1}{Path\ loss \cdot I_{UL}}$$

On the other hand, if the downlink is considered then we do not want a BS to transmit at its maximum power and still not meet the required SIR. One way around this, which aims at lowering the probability of having a BS reaching its maximum output power, is to have a *HO_quality_estimate* like this:

$$HO\_quality\_estimate = \frac{1}{Path\ loss \cdot (P_i + P_{i,j})}$$

P_i is the output power from BS i, and $P_{i,j}$ is the required power from BS i to serve MS j. The mobile can be informed about the current output power from the base station either by having the power level broadcasted, or by changing the base station's pilot strength depending on the output power level.

The softer handover algorithm is similar to the soft handover algorithm, but it should be possible to identify if some sectors are associated with the same base station. This means that it is possible to use different handover thresholds for soft and softer handover.

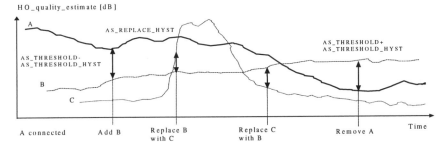

Figure 7.32 The handover algorithm during soft/softer handover (RT services). The scheme shows the pilot strength of three sectors (A, B, and C). The maximum active set size is two.

7.3.7.4 Measurement and Signaling Requirements

The measurements (Table 7.4) needed for the handover algorithms are:

- The desired signal (carrier) strength of the downlink pilot channel(s) (MS measurement);
- The mobile station's speed (MS measurement used for HCS, see the section on HCS);
- The average total received power in the uplink (BS measurement);
- The average total transmitted power in the downlink (BS measurement).

These MAHO measurements can be sent on the uplink:

- On an event-triggered demand;
- Periodically;
- On network demand.

The working assumption is that the mobile station mainly sends the measurement reports on an event-triggered basis, or on a network demand in order to minimize the signaling.

The mobile station measures the path loss to all candidate BSs and sends path loss measurement messages to the BS. Measurement messages are sent when the path loss changes or when base station sends a handover measurement order to the mobile station. Information about I_{avg} level is broadcast to the MS periodically or when the situation changes.

Table 7.4

HO Measurement Requirements

Measurement	Measuring entity	Range	Measurement frequency
Path loss	MS	Performed for all BSs	As often as possible
Average_downlink txp	BS	Performed at all BSs	About once per frame
Average_uplink_interf	BS	Performed at all BSs	About once per frame

Whether RNC or the MS performs the handover calculations, depends on the bearer. This means that the HO algorithm can be located in the RNC or in the MS. However, the network (RNC) can always control the MS by means of the HO algorithm parameters (Table 7.5).

7.3.7.5 HO Layer 3 Signaling

The handover measurement message includes received signal strengths for each candidate set (and active set) base station and an MS speed estimate. The handover measurement order message includes the new candidate set. The handover direction message includes all the required data that the mobile station needs to communicate with the new base station(s).

7.3.7.6 Interaction With Other Algorithms or Entities

There is a mutual dependence between handover and power control. Which handover algorithm works best depends on the power control algorithm that the system employs. Further, the power control for an MS in soft handover will be different compared to an MS that only is connected to one BS. The soft handover algorithm also has an impact on the choice of admission and congestion control algorithms.

Table 7.5

HO Algorithm Parameters

Parameter	Typical value	Description
AS_THRESHOLD	4 dB	The active set threshold. All sectors within the threshold are added (at most AS_MAX_SIZE sectors).
AS_THRESHOLD_HYST	1 dB	A hysteresis of the active set threshold in order to avoid ping-pong. Add sector(s) when the HO_quality_estimate are within AS_THRESHOLD - AS_THRESHOLD_HYST and remove sector(s) when outside AS_THRESHOLD + AS_THRESHOLD_HYST.
AS_REPLACE_HYST	3 dB	The worst sector is replaced when a candidate set BSs is AS_REPLACE_HYST above (the active set size is equal to AS_MAX_SIZE).
PS_MINIMUM	-	The minimum pilot strength of an active set pilot.
AS_MAX_SIZE	2	The maximum active set size.
HCS_LAYER_UP/DOWN	{-}	These thresholds select the cell layer. Each layer has this set of thresholds describing when to select the upper or the lower cell layer. The mobile station measures all active and neighbor set pilot channels.
HCS_LAYER_REPLACE	-	This threshold compares the current active set with the strongest candidate sectors transmitted on other frequencies (layers).
HO_TYPE	Path loss	"Path loss," "UL," or "DL"- based

7.3.8 Admission and Congestion Control

In CDMA systems, the uplink capacity depends on the intracell interference as well as the intercell interference. In the downlink, the capacity is mainly limited by the intercell interference. If the system capacity is exceeded, the fast SIR-based power control will generate party effects, which usually results in extensive dropping. Dropping is assumed to be much worse than blocking. The purpose of admission control is to guarantee system stability and high capacity by denying an arriving bearer service access to the system if the load is already high. However, unstable states may still occur, since the users move, and can give rise to traffic concentration in a certain part of the system. In order to handle these unstable events, congestion control is needed. The strategy behind congestion control is to lower the interference level in the system, hence forcing the system back to a stable state. This can be done by lowering the bit rates for unconstrained NRT bearers by putting them in a queue and allowing them to reenter the system when the system's load is lower. The bit rates of NRT bearers can also be increased if the current load is low. In this chapter, two different admission and congestion (load) control approaches are presented.

Table 7.6

HO Messages

Message name	When	Description
Measurement message	Mobile station sends measurement messages when it receives measurement or when path loss or speed situation changes.	Path loss information of all the candidate base stations and MS speed information.
Measurement order	Base station sends measurement order message for the mobile station whenever the base station wants to update its path loss or MS speed data to the mobile station.	Identification information for all the candidate set base stations.
Handover direction	When handover occurs, the base station sends handover direction message to the mobile station.	All the required information on the new base station(s) the mobile station needs to communicate with the new base station(s).

7.3.8.1 Requirements to the Admission Control

WCDMA systems are dependent on the admission control algorithm, as the system might become unstable if too many bearers are admitted. In addition, the uplink and the downlink are more different than in a TDMA system due to the orthogonal downlink in CDMA systems. Therefore, different admission control algorithms are used in the uplink and downlink, respectively. Hence, a new bearer service should only be admitted if it passes both the downlink and uplink admission algorithm.

Admission control algorithms, for the uplink, in CDMA systems have been proposed in several papers ([19, 20]). Simulations in [20] indicate that an uplink algorithm based on total received power at the own BS (where the new user seeks access) has a good performance. Since this algorithm has only been studied in a single bearer service system, a modification of the algorithm so that it can handle different bearer services is proposed in approach 1 below. The algorithm in approach 2 is quite similar.

Very little research has been conducted in the area of downlink admission control. The downlink can be either code or interference-limited. This means that two different admission algorithms are needed, one for each case. In the code-limited case, an algorithm where one simply calculates the number of used channels at the BS is proposed in approach 1. For the interference-limited case, approach 1 uses an algorithm that considers the transmitted power from several BSs. The reason for having a global algorithm is due to the orthogonality in the downlink. Approach 2 also uses two algorithms, where the first one is similar to the algorithm used in the uplink. Congestion control, in both the uplink and the downlink, is used in approach 1, while approach 2 uses load control. Load control for terminals in macrodiversity is always done at the RNC based on the information of only one BS: this means that load control is virtually done by one BS, although the procedures are executed at the RNC. The BS can be different for the uplink and downlink. The other BSs in the active set regard the call as a noncontrollable variable rate call, and use the same SIR thresholds as they would be using in the nonmacrodiversity case.

In the uplink, the BS with the highest measured SIR value for the call provides the load control information for the call to the RNC. The RNC monitors the quality of the call, and if the quality is unacceptable, the SIR threshold value (used in the load control algorithm) for the call is adjusted with the appropriate factor. In the downlink, the BS with the smallest path loss to the terminal provides the load control for the RNC. The index of the BS with the smallest path loss to the terminal is broadcast at the same time the SIR threshold is broadcast from the mobile to the BS. In the load adjustment procedure the (transmitted) energy per bit of the call is multiplied by the amount of BSs in the active set to obtain a more realistic picture of the capacity usage of the call.

7.3.8.2   Admission Control Algorithms

In this section, the two different admission and congestion control approaches are proposed.

Received Power Admission Control (RPAC)

$$\text{Admit bearer service } i \text{ if } Z_k < Z_i^t$$

Z_k is the total received power (before a new user is admitted) at base station k, and Z_i^t is the service-vector-dependent threshold for bearer service i. The admission decision depends on bearer service i's bit rate, delay requirement, session duration, and so forth. The RPAC algorithm is based on local measurements at one BS. However, it might be wise to consult surrounding BS before admitting a very capacity-demanding bearer service.

For the case where the downlink is code limited, we propose an algorithm where one simply calculates the number of used channels at the BS. The busy-channel-based-admission control (BCAC) is defined as

BCAC

Admit bearer service if $N_k < N^t$

N_k is the number of used channels at BS k and N^t is a predefined threshold. Note that N_k includes the bearer service that seeks access and that different types of bearer services contribute with a different amount to N_k.

For the interference-limited case, we propose a downlink algorithm, which uses the transmitted power from several BSs. The reason for having a global algorithm is due to the orthogonality in the downlink. We call this algorithm transmitted-power-based-admission-control (TPAC). In a system that offers different types of bearer services, the algorithm would look like this:

TPAC

Admit bearer service i if $P_k < P_i^t$ for $\forall (k \in B)$

P_k is the transmitted power from BS k (before a new user is admitted), B is a set of BSs, and P_i^t is the service-vector-dependent threshold for bearer service i. Note that a new bearer service should be blocked if P_k exceeds the threshold at any BS in set B.

Even though admission control is used, unstable states might occur. Congestion control is needed to deal with these states. A simple congestion control algorithm for the uplink can look like this:

Stabilize the power vector (n is set to 0.25 in the fast-power-control algorithm). This creates bearers that will not achieve their SIR target. However, it will slow down the party effect process and give the CC algorithm some time to stabilize the system (see step 2).

Stabilize the system. This can be done by lowering bit rates (delay an unconstrained NRT bearer), perform an interfrequency HO, or remove a bearer.

Stabilize the power vector. If $n \neq 0$, go back to step 2 or else quit.

The fast PC algorithm will achieve a stable system in the order of 10 to 100 ms after an unstable event. However, the bearers experience a SIR slightly below the SIR target. The "active" CC phase has to be short (maximum about 0.2s). Otherwise, the slower outer loop PC tries to change the SIR target for the bearers due to bad quality and will counteract the CC and the fast PC algorithm (with n set to 0.25). The downlink algorithm is almost the same. It is activated when a BS transmits with its maximum output power, $P_{\max}$.

Stabilize the power vector. If $P_k = P_{max}$, go back to step 1 or else quit.

Most of the algorithms defined in this section use SIR thresholds, which are obtained by measuring the received signal to interference ratio. The instantaneous SIR value for a bearer i is defined as follows:

$$SIR_i = PG_i \frac{P_{rx,i}}{P_{int,i}} \tag{7.5}$$

where i is the index of the call (as everywhere in this chapter);
$P_{rx,i}$, is the received signal power;
$P_{int,i}$ is the received wideband interference power;
PG_i (the processing gain) is defined as:

$$PG_i = \frac{R_c}{R_i} \tag{7.6}$$

where R_c is the RF signal bandwidth and
R_i is the user bitrate for the call i.

The SIR defined in (7.5) can also be formulated as :

$$SIR_i = PG_i \frac{P_{rx,i}}{P_{int,i}} = \frac{E_{b,i}}{(P_{int,i} / R_c)} \tag{7.7}$$

where $E_{b,i}$ is the energy per user bit for user i.

The averaged[1] value of this measured SIR, denoted here as $SIR_{th,i}$, is presumed to be the *SIR threshold*, used by the fast-closed-loop power control with the effect of (optional) MUD and orthogonality removed. Due to the averaging, this quantity is stable and insensitive to PC errors. If the mobile is not able to attain its desired SIR value after MUD or orthogonality, the measured SIR must be corrected with the appropriate factor or set to the value that fast PC is actually using (i.e., with the effect of MUD or orthogonality included). For the downlink control, the SIR threshold values have to be signaled from MSs to BS at regular time intervals. These intervals could be between 100 ms to 1 second depending on the bit rate of the call, so that calls with higher bit rates would do the signaling more often. These signaling intervals were chosen as a compromise between the stability of the measured SIR (the SIR has to average over a period of time so that it can be thought of as the actual SIR threshold of the power control), and the changes in the SIR threshold value due to mobility of the terminal. In addition, the signaling overhead has been considered.

Some background motivation for the solutions presented is provided here. Mathematically, it can be presented as follows: In a CDMA system, the following always holds:

---

[1] Averaging period is a parameter; 200 ms is a typical value.

$$SIR_i = PG_i \frac{P_{rx,i}}{\sum_i P_{rx,i} + P_a + P_N} \tag{7.8}$$

where $\sum P_{rx,i}$ is the cell's total power

P_a is the interference power from other cells, and

P_N is the thermal noise power.

$$\Leftrightarrow \sum_i \frac{SIR_i}{PG_i} = \frac{\sum_i P_{rx,i}}{\sum_i P_{rx,i} + P_a + P_N} \to 1$$

when

$$\sum_i P_{rx,i} \to \infty \tag{7.9}$$

As we see from (7.9), the sum

$$\sum_i \frac{SIR_i}{PG_i} \tag{7.10}$$

is always less than one. When the SIRs are replaced with the SIR thresholds, (targets, defined earlier in this chapter) of the closed loop PC, we obtain the *load factor* of the system. It is desirable to keep this sum at a value of 0.4 to 0.7, depending on the situation.[2] This is because the location of the terminals affects the capacity of a CDMA system. The capacity is smallest when the mobiles are located at the border of the cell.

Admission control has to be done at several phases during the call setup procedure. Consider a call from subscriber A to subscriber B. Admission control must be done for A as well as for B. The connection between the terminal and the base station has to be built into both directions (down and uplink) in all call cases. This is the case because for a unidirectional datacall, a bidirectional signaling connection is also needed. In the actual call setup procedure, the specified control algorithms can be used in several ways, depending on the involved bearer services and parties. Negotiation of the bit rate limits for variable rate services are a part of the call setup procedure and are not specified here.

---

[2] If the system is functioning properly, the measured SIRs, sufficiently averaged, are equal to the SIR target of the call. Thus we can conclude that when the SIRs of (7.9) are replaced with the estimated (i.e., measured) SIR targets, we obtain a definition for the load factor of the system. This load factor is always between 0 and 1, and it approaches 1 when the load increases.

The basic principle is that admission control is always done when setting up an uplink or downlink connection for signaling or data, and during an active connection if a new bearer service not agreed upon in call setup is taken into use. For certain emergency signaling cases, admission control could be omitted. The admission control procedures are also executed during active set update and handover.

Uplink Admission Control Procedure

The uplink admission control operates in the following manner. The BS calculates the load factor, which is defined as:

$$\text{Load}_{\text{uplink}} = \sum_i \frac{SIR_i}{PG_i} \tag{7.11}$$

The BS then calculates the value of the upcoming load that is:

$$\text{Load}_{\text{upcoming}} = \text{Load}_{\text{uplink}} + \frac{SIR_{th,\text{upcoming}}}{PG_{\text{upcoming}}} \tag{7.12}$$

$SIR_{th,\text{upcoming}}$ is the (conservatively) predicted SIR threshold value that the call will have; PG_{upcoming} is the desired processing gain for the call.

The BS also estimates at regular intervals (adjustable, 20ms typical) the standard deviation of the changes caused by uncontrolled calls in the load factor. The value of this is defined to be std_{load}. If the equation

$$\text{Load}_{\text{upcoming}} < 1 - \eta - M \times std_{\text{load}} + margin_{handover} \tag{7.13}$$

η is a parameter controlled by uplink load control; $margin_{handover}$ is a parameter (0.05 is typical when the admission control procedure has been initiated by an existing call; 0 otherwise), whose purpose is to prioritize calls making an active set update (ASU); M is a selectable parameter ($M=5$ typical), holds, then admission control for the uplink is passed. The purpose of the last factor of (7.13) is to reserve enough capacity margin if the changes in the uncontrolled load are very big.

Downlink Admission Control

Downlink admission control is based on two criteria that both have to be true for admission to be passed if the call is new. If the call is initiated because of an ASU, then only the first criteria has to be satisfied. The downlink load factor is defined in a similar way as in the uplink:

$$\text{Load}_{\text{downlink}} = \sum_i \frac{SIR_{th,i}}{PG_i} \qquad (7.14)$$

Also, the load factor upcoming is defined as in the uplink case:

$$\text{Load}_{\text{upcoming}} = \text{Load}_{\text{downlink}} + \frac{SIR_{th,\text{upcoming}}}{PG_{\text{upcoming}}} \qquad (7.15)$$

In this formula, the SIR threshold is predicted (conservatively) by the BS. The first downlink criteria to be satisfied is as follows:

$$\text{Load}_{\text{upcoming}} < 1 - \varepsilon - M \times std_{\text{load}} + margin_{handover} \qquad (7.16)$$

Where ε is a parameter controlled by the downlink   load control. The two last factors are defined as in the uplink case. The second criteria that has to be satisfied is (only if the call is new):

$$P_{tot} < P_{threshold} \qquad (7.17)$$

$P_{threshold}$ is a maximum allowed total power (a parameter) and P_{tot} is the total transmitted power including pilot. If both of these criteria are satisfied, then admission control for the downlink is passed.

7.3.9   Load Control

The purpose of the load control algorithm is to adjust the bit rates of adjustable unconstrained delay data calls in such a way that an optimal load for the BS is achieved. The load control algorithm operates at specified time intervals (the length of the interval is a parameter).

Load Control for the Uplink Direction

The control for the uplink is based on the same load factor as was defined in the admission control, that is:

$$\text{Load}_{\text{uplink}} = \sum_i \frac{SIR_{th,i}}{PG_i} \qquad (7.18)$$

The value of the uplink load factor is at each time interval adjusted as close to the desired load $(1-\eta)$ as possible, by adjusting the bit rates of the adjustable calls. The load factor is adjusted as close to the desired value as possible, keeping the load under the desired limit $(1-\eta)$. The accuracy of this adjustment is limited by the available bit rates for the current adjustable calls. The adjustment is done by signaling the desired processing gain (bit rate) to the mobiles. The value of η is adjusted between certain time intervals (for example 100 ms) in the following manner. The received total power is stored in vectors of 100 ms. In the following, this vector is called P_{vect}. As well, the load factor for the last 100 ms is stored in a vector. This vector is called $Load_{vect}$. Then, the following quantity (η_{adj}) is calculated:

$$\eta_{adj} = \frac{\text{cov}(\log(P_{vect}), \log(Load_{vect}))}{\text{var}(\log(Load_{vect}))} \qquad (7.19)$$

The abbreviation var stands for variance and cov for covariance. A threshold is set by the operator for η. This threshold is called η_{th}. This threshold defines the load level as the operator wishes the system to operate (for example, approximately 70% of maximum capacity is obtained by using a value of 1.5 for η_{th}). Between the specified 100 ms time intervals, the following adjustment of η is performed:

$$\eta(t + 100 \text{ ms}) = \eta(t) + (\eta_{adj} - \eta_{th})k1 \qquad (7.20)$$

where $k1$ is a parameter (0.005 typical), and

   η_{th} is a parameter (1.5 typical).

This operation essentially scales the received power level relative to the constant interference level, which is produced by thermal noise and adjacent cells. The adjustment of η could also be done based on a received power threshold, but this would not utilize the capacity of the system maximally.

Load Control for the Downlink

Load control for the downlink is based on the same load factor as in the uplink direction, that is:

$$\text{Load}_{downlink} = \sum_i \frac{SIR_{th,i}}{PG_i} \qquad (7.21)$$

In the downlink, load control can either be done periodically or continuously (as fast as possible). If the load control is performed continuously, the load (which can be adjusted with a delay only slightly higher than the 10 ms frame length) is at all times kept as close as possible to the desired load $(1-\varepsilon)$ by adjusting the bit rates of the adjustable calls. As in the uplink case, the load is at all times kept below the desired level. If the load control is performed periodically (with intervals of 20-40 ms), then if the load factor is adjusted downwards, the bit rates for the calls with the lowest energy per bit (E_b) are first increased. If the load factor is adjusted upwards then the bit rates of the calls with the highest E_b are decreased. This results in lower transmitted total power, and thus an increase in system capacity.

There are also other methods for the selection of the subset of calls that are not discussed here.

The value of ε is adjusted (in periods of 60-200 ms) in the following manner. The total transmitted power including pilot is averaged over the adjustment period (this value is called P_{tot} in what follows). Then the following adjustment to ε is done:

$$\varepsilon(t + \Delta t) = \varepsilon(t) + \frac{(P_{tot} - P_{threshold})}{P_{threshold}} k2 \qquad (7.22)$$

$P_{threshold}$ is a parameter (same as in admission control), P_{tot} is the averaged (averaging period is Δt) total transmitted power including pilot, and $k2$ is a parameter (0.01 typical).

7.3.9.1 Measuring and Signaling Requirements

The measurement needed for the AC in approach 1 are given in Table 7.7

Table 7.7

AC Measurement Requirements

Measurement	Measuring entity	Range	Measurement frequency
Uplink_interference	BS	Performed at all BSs	1 or 2 kHz
Downlink_bs_power	BS	Performed at all BSs	0.5 or 1 kHz
downlink_busy_chan	BS	Performed at all BSs	-

The parameters of the AC algorithms in approach 1 are given in Table 7.8.

Signaling for AC in approach 2 is as follows:

- No information is broadcast from the BS to the MS;

- The MS has to signal its desired processing gain in the uplink (a function of bit rate and BER requirement).

Table 7.8

AC Algorithm Parameters

Parameter	Explanation
AC_THRESH_INTERF_UL	This parameter (one for each service) describes the maximum interference allowed at the BS. Used by uplink admission control.
AC_THRESH_PWR_DL	This parameter (one for each service) describes the maximum allowed BS (total) output power. Used by downlink admission control.
AC_THRESH_CHAN_DL	This threshold sets the maximum number of already reserved channels in the cell. If we have high mobility, we must reserve more (downlink) channels for HO, however, this is only needed in a code-limited environment. Used by downlink admission control.

7.3.9.2   Interaction With Other Algorithms or Entities

The admission control and congestion control will interact with the slow power control. Further, the performance of admission control and congestion control algorithms depends on the soft handover algorithm. In order to make a good local congestion decision in the uplink, an uplink power control that cancels the party effects, during a short amount of time is needed.

7.3.10   Algorithms Specific for Support of HCS

HCS is needed to give coverage to a huge area where traffic concentrations might occur locally (i.e., hotspots), with a moderate number of BS. Other obvious benefits with HCS are that the number of handovers for fast-moving users can be rather small, as well as the possibility to share the load between different frequency layers. This possibility to handle overloaded situations, by sharing load between frequency bands, improves the system's capacity. To achieve the benefits mentioned above, interfrequency handover is needed. The interfrequency handover is seamless. This is achieved by preparing the network for the change of BSs (i.e., buffer data at the new BS before the interfrequency decision is executed).

7.3.10.1   Requirements to Algorithms

Hard seamless handover is required when the user changes cell layer due to a change in speed (or cell coverage) or interference situation. The AS_MAX_SIZE is set to one when performing a hard interfrequency handover.

The mobile station reports to the network when above/below the HCS-related speed thresholds. The network then adds the "candidate" pilot channels transmitted on other frequencies (other cell layers) to the candidate set. The mobile does measurements in a frequency band different from the band where it currently is transmitting and receiving information. Further, it also reports the strength of the pilot channels, similar to the soft/softer handover algorithm. However, the candidate set may include pilot channels transmitted on other frequencies.

Algorithms

A sector (belonging to another frequency) is only added to the active set if

- The speed of the mobile station or the network indicates that we should change cell layer **and** the pilot strength is above HCS_LAYER_REPLACE **and** the mobile station speed is above HCS_LAYER_UP.

or

- The congestion control initiates an interfrequency handover, due to an overloaded situation **and** the pilot strength is above HCS_LAYER_REPLACE.

7.3.10.2   Measuring and Signaling Requirements

The measuring and signaling requirements are found in the soft handover section 3.1.7.

7.3.10.3   Interaction With Other Algorithms or Entities

Congestion control might use the possibility of moving users from one frequency band to another in the case of an overloaded situation (Figure 7.33).

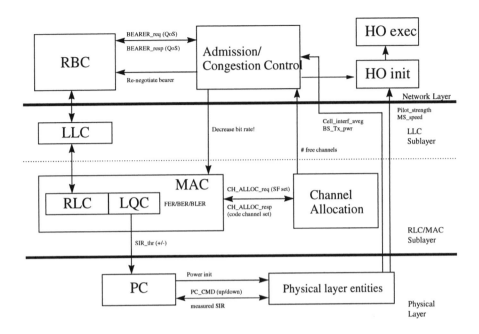

Figure 7.33 The interactions between RRM algorithms in UTRA FDD.

7.3.11 Conclusions

A robust framework of radio resource management algorithms for UTRA FDD that supports both RT and NRT bearers has been proposed. Modifications are likely to occur in order to improve performance and flexibility.

Several individual algorithms concerning power control, admission control, load control, congestion control, and handover have been presented. When constructing these algorithms, a considerable effort is made regarding the interaction between them to achieve a good overall system performance. Fast SIR-based power control is used in both the uplink and downlink. Further, the power control as well as the pilot symbols are chosen in such a way that adaptive antennas are supported. Admission control algorithms, for both the uplink and the downlink, are proposed to avoid an overloaded system. Although good admission control algorithm is employed, overloaded situations may still occur because of changing traffic concentrations, mainly due to mobility. Hence, uplink and downlink congestion control algorithms are proposed to handle these unstable situations. The proposed soft and hard handover algorithms can be service and load-dependent, to meet the different requirements from a variety of services. Further, an interfrequency handover algorithm for HCS support is also considered.

7.4   SCHEDULING

One of the key features of third generation cellular mobile systems is the availability of NRT packet data services. Different bit rates are offered to the subscribers in DL direction as well as in the UL direction. The sum of all the data of the different users belonging to the same cell that are to transmit at the same time exceeds the resources available for NRT users each frame. Therefore an authority (scheduler) located in a BS or a RNC has to divide up the resources and to assign them to the different users.

In this section, we consider the differences and similarities between FDD and TDD mode regarding scheduling possibilities. Two different scheduler algorithms are described and simulations are shown.

7.4.1   Layer 1 Aspects

The number of resources available for NRT users need not be constant during a long time period. This is an interesting point for the scheduler, since it has to assign the resources to the different users. If the number of available resources varies during packet sessions, the scheduler has an additional degree of freedom to assign resources very fast, if this is possible from the layer 1 point of view. Another issue is the capability of changing the number of resources allocated to the different users even when the entire resources are constant.

In TDD mode, a physical RU consists of a code/TS pair. Each physical code/TS pair can be allocated on a frame-by-frame basis. However, the allocation of RUs can also be valid for a longer time period. Since the RUs are assigned by using an FACH (possible in every frame), it would be possible to vary the number of RUs allocated to a certain user during the packet session. This gives freedom to take different channel conditions (retransmissions) of each user into account. Additionally, it is possible to use temporarily free resources since they can be released very fast.

In FDD mode, a physical resource that is allocated is a channelization code with a certain spreading factor (SF), or a branch of the code tree, including several SFs. The degrees of freedom for assigning physical resources are quite different in the UL and DL directions. In UL, an arbitrary channelization code is selected out of the entire code tree, because every MS has its own scrambling code. The MS has only to agree with the network about the maximum bit rate. In DL, there is just one scrambling code available for all of the connections of the BS. Therefore, the code tree has to be shared by all NRT users in a cell. Both the DPCCH part and the DPDCH part are transmitted together in the DPCH. Therefore, only one channelization code is needed. The usual procedure of allocation is to select a certain channelization code (with a certain SF). This selection is usually valid for the entire connection duration. The reason is that the TFCI can only be decoded if the channelization code is already known (DPCCH and DPDCH use the same one). Consequently, in DL a dynamic change of the channelization is not possible.

7.4.2 Scheduling Algorithm for TDD and FDD (Quasi-Round Robin)

The main goal of this algorithm is to achieve as many "satisfied" users as possible. To fulfill this requirement, it has to be agreed how a satisfied user is defined. Different definitions can have a severe impact on the way a good algorithm looks. It is a big difference if the satisfied user criterion focuses on an average data rate, or maximum data rate, or both. The criterion agreed upon in ETSI defines a user to be satisfied if the average data rate during the entire connection duration is higher or equal to 10% of the nominal data rate. Several different attempts to achieve the mentioned main goal are possible.

Let us consider a couple of NRT users with the same nominal data rate. The satisfied user criterion should be based on the ETSI decision. The amount of data to be transmitted may be very different for each of the NRT users. The lower the amount of data of a certain user, the shorter the allowed connection duration to fulfill the satisfied user criterion. Let all of the resources be allocated to a few users having a high amount of data to transmit. These users will likely be satisfied but they occupy the resources for certain time duration. After all of their data are transmitted, resources are allocated to the users with less data. The problem is that these users could be unsatisfied as they are not able to achieve the 10% data rate limit due to too little time left. To avoid this, the available resources are divided up to all of the users (or at least to a number of users worth mentioning) that have data to transmit. The data for every user added to a unique queue in the order they are "generated." The data that are to be retransmitted are also appended to the queue. This queue is read out by using a strategy similar to the "round robin" principle. To every user a certain amount of resources is allocated as long as resources are available. Figure 7.34 shows the principle of such a simple scheduler.

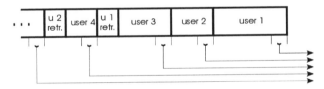

Figure 7.34 Scheduler principle.

The same amount of resources is allocated to every user as long as some resources are available. Such a scheduling scheme can be applied to a TDD and FDD system, respectively. To avoid differences in signaling aspects, it is assumed that the amount of resources allocated to each user per frame as well as the entire resources available for NRT users per frame are constant.

The study environment is a Manhattan grid with 384 Kbps UDD bearers in DL direction. The used retransmission scheme is ARQ I.

A user is unsatisfied if its average data rate during the connection duration is less than 38.4 Kbps according to the ETSI criterion. A PDU is the minimum number of RUs that are allocated to one user.

In this simulation, a PDU size of 3 is used. Since the number of bits per frame that can be assigned to a certain user is different in TDD and FDD (PDU size and spreading factor), it was decided to allocate resources in terms of number of PDUs.

A second axis in Figure 7.35 and Figure 7.36 shows the corresponding SFs for FDD.

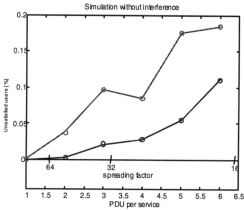

Figure 7.35 Scheduling performance (upper curve: high load, lower curve: 90% of high load).

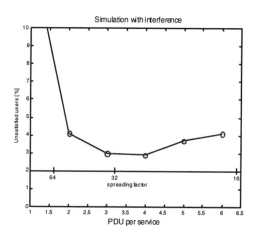

Figure 7.36 Performance of the scheduling algorithm if interference and retransmissions are considered.

Figure 7.35 points out the "performance" of the system without any interference (no retransmissions). The larger the number of resources allocated to a single user, the larger the number of unsatisfied users. These are rather users with a small amount of data to transmit. They stay in the queue too long and so they do not fulfil the 38.4 Kbps average data rate criterion. The lowest number of unsatisfied users is given at the lowest number of PDUs per service (an allocation of 1 PDU per frame to one user offers a data rate of about 80 Kbps). This is no longer true if interference and therefore retransmissions are taken into account. Data that are to be retransmitted are put into the queue again. Therefore, a small amount of resources (PDU or SF) do not satisfy most of the users any longer.

By increasing the allocated resources to each of the users, the number of unsatisfied users decreases. The reason is that each user has to make sure that all the data (including the retransmitted ones) are transmitted within the same time as it was necessary in the case without retransmissions. So for fulfilling the average data rate criterion, the allocated resources per user per frame have to be increased. By going on with increasing the assigned resources, the situation becomes the same as in the case without interference, and the ratio of unsatisfied users will increase. In Figure 7.36 the discussed behavior can be seen as a curve. Since this simulation has been performed with a TDD simulator the interference situation in an FDD system is likely to be different. But the used algorithm can be implemented in an FDD simulator as well because it uses an allocation scheme that can be used in both modes.

7.4.3   Scheduling Algorithm for FDD

Here, we show the expected gain by using a certain scheduling algorithm compared to when no scheduling is used (i.e., the users begin to transmit with a certain rate as soon as they have something to transmit). The impact of errors in the initial power setting is also investigated. Further, whether the scheduling algorithm should rely on local or global information is investigated in the downlink, while a local algorithm is investigated in the uplink. The reason for only considering a local approach in the uplink is because it has been argued in [23] that the interference levels, which are a good estimate of the current load, are quite similar in neighboring BSs.

7.4.3.1   Basic Methods

The basic strategy behind the scheduling algorithm in this section is to assure that the network operates at a certain predefined load, as long as there are NRT users in the system with something to transmit. In a real system, the total load consists of load generated by the RT users and the load generated by the NRT users. The scheduling algorithm allows a certain number of NRT users to transmit, so that the gap between the load generated by RT users and the target load is filled. The idea is shown in Figure 7.37. The scheduling algorithm cannot control the RT users. However, in the simulations, only NRT users are modeled. In order to avoid the queues becoming infinitely large, a simple admission control is used.

To control the system in such a way that the predefined load is reached, load measurements are required; one for the uplink and one for the downlink. In this book, the load measures are uplink interference and downlink output power. We choose uplink interference as the uplink measure because the cellular network is likely to be planned in such a way that it provides coverage to a certain area, given that the uplink interference remains below a certain value. The chosen uplink target load, corresponding to certain uplink interference, should be a little lower than the maximum interference for which the system is planned. In the downlink, the limiting factor is the output power of the BS. Thus, the downlink-scheduling criterion will be the downlink transmission power. The transmission power target that the scheduling algorithm aims to reach will typically be a little below the maximum allowed output power from the BS.

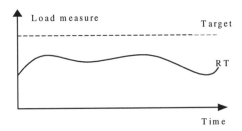

Figure 7.37 Gap between the load generated by RT users and the target load. The gap will be filled by NRT users scheduled in a certain way.

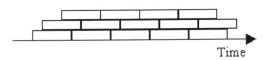

Figure 7.38 The figure shows three users. Note that the three users' frames do not start at the same time.

Based on how much interference is generated or on how much power is required, the BS assigns different connections a certain transport format (i.e., the scheduling decision is made in the network). The uplink decision is made in the BS while the downlink decision may be made in a BS or in an RNC. The transport format indicates the rate that the connection should be using. In this study, only the direction carrying data is considered (i.e., the bit rate of the return channel, where ACKs and NACKs are transmitted, is assumed to be much lower than for the data channel).

In both the uplink and the downlink, there are time offsets between the frame structures of the users (i.e., user one's and user two's frames do not begin at the same time). This is shown in Figure 7.38. This means that there will be a time gap when no one is transmitting in the short periods that occur when one user has finished its transmission, and a new user is scheduled to begin its transmission. This time gap is considered in the simulations, since it affects the effectiveness of different scheduling schemes.

7.4.3.2   Uplink Scheduling

In the uplink, scheduling is performed based on uplink interference measurements.

Analysis

Consider the uplink in a single cell WCDMA network. The received quality is then defined as $\gamma_i = C_i/(I_{tot} - C_i)$ γ_i is the SIR that user i generates at the BS. C_i and I_{tot} are the received signal strength and the total received interference at the BS. Further, the total interference at the BS is defined as

$$I_{tot} = \sum_{i=1}^{M} C_i + N$$

M is the total number of MSs in the cell and N is the background noise. From the relation for γ_i (SIR) and the equation above the total interference is expressed as follows

$$I_{tot} = \frac{N}{1 - \sum_{i=1}^{M} \dfrac{\gamma_i}{1 + \gamma_i}}$$

Further, assume that the E_b/N_0 is the same for all bit rates (i.e., γ is proportional to the bit rate). In reality this is not true, since the E_b/N_0 value depends on how much overhead the pilot bits add, as well as on the interleaving depth. Now, consider a system with M_1 RT users creating a certain uplink interference. In order to reach the interference target, some NRT users are allowed to transmit. We now investigate how much is gained by letting just one NRT user transmit compared to letting $M_2 = M - M_1$ NRT users transmit in parallel. The interference situations for these two scenarios are

$$I_{target} = \frac{N}{1 - \sum_{i=1}^{M_1} \dfrac{\gamma_i}{1 + \gamma_i} - \dfrac{\gamma_{max}}{1 + \gamma_{max}}} = \frac{N}{1 - \sum_{i=1}^{M_1} \dfrac{\gamma_i}{1 + \gamma_i} - \sum_{j=M_1+1}^{M} \dfrac{\gamma_j}{1 + \gamma_j}}$$

This means that

$$\frac{\gamma_{max}}{1 + \gamma_{max}} = \sum_{j=M_1+1}^{M} \frac{\gamma_j}{1 + \gamma_j}$$

and by using the assumption that γ_j is the same for all M_2 we see that

$$\gamma_i = \frac{\gamma_{max}}{M_2 + (M_2 - 1)\gamma_{max}} \tag{7.23}$$

from where one can calculate how much is lost in throughput by having several users sending in parallel compared to keeping the number of transmitting users low. The maximum solution from an interference point of view would be to simply let one sender at the time transmit.

Algorithms

The algorithm under investigation measures the uplink interference, I_m, and investigates how much it can increase the rate without exceeding the uplink interference threshold, I_t. The analytic expressions above are valid for a single cell scenario. To capture the effects of a multicell scenario, an expansion factor, f, is used. The factor is chosen to be 1.67, since it is commonly argued that 40% of the total interference is intercell interference in a macrocell environment. A bit rate increase is accepted as long as

$$f \cdot \left(\frac{\gamma_{rate}}{1 + \gamma_{rate}} \right) < \frac{N}{I_m} - \frac{N}{I_t} \qquad (7.24)$$

This is an algorithm that alters the bit rate a certain amount, depending on the current uplink interference level. The rates that are used in the simulations are 120, 240, and 480 Kbps. It is more favorable to have few users with high bit rates compared to having many users with low bit rates. Thus, the possible bit rate increase based on (7.24), is always used to give an active user a higher bit rate, unless all active users are using a 480 bearer, or if no active user can reach the next higher possible bit rate with the indicated bit rate increase. When a new user starts transmitting, it will always be the user with less time left until it becomes unsatisfied. It is of course also possible to decrease bit rates. A reduction in bit rate is ordered when

$$\frac{N}{I_m} - \frac{N}{I_t} < X$$

where X is a chosen value.

During macrodiversity, the mobile station may receive different commands from different BS. The MS then follows the command from the BS that requires the smallest amount of output power. Another scheme to use could be one where the MS follows the command that results in the lowest bit rate.

Simulations

The studied environment consists of 4×4 BSs. The cell radius is chosen to be 500m and a wraparound technique is used. The propagation model follows the model defined in [1]. The parameters for shadow fading are different. The standard deviation is chosen to be 10 dB and the correlation distance is 110m. At each location, the shadow fading is divided into two equally large components. One is the same for all BS, while the second one is BS-dependent.

The packet arrival process follows a Poisson distribution and is modeled so that all the information (on average 96 Kbit) that a certain user wants to transmit arrives at the same time. By using such an approach, the simulation time is reduced and it is also possible to calculate when a user will be unsatisfied. Whether a user is satisfied or unsatisfied is decided based on the same requirements as in the ETSI simulations (i.e., the 10% threshold and the admission decision).

Fast SIR-based power control is used. Further, gain based soft handover with a handover margin of 3 dB and a maximum set size of 2 is modeled. Admission control is used. A new user is blocked if the number of users queued in the cell, where the user seeks access, exceeds 20 MSs (soft handover legs are not included, during soft handover the MS is only counted at the BS that requires the smallest amount of output power). Further, to assure coverage, a second admission threshold is used. New users are blocked if the uplink interference is more than 1 dB above I_t in the scheduling algorithm. I_t is chosen to be 10 times larger than the background noise N. The interference level where the scheduling algorithm starts to decrease bit rates equals the admission threshold. This means that it is the 20-MSs threshold that blocks new incoming users when scheduling is applied.

A user will be unsatisfied if the average bit rate over the air interface is less than 38.4 Kbps. This corresponds to a 384-Kbps service in the ETSI simulations. With the chosen admission thresholds, unsatisfied users are avoided. However, there will be blocking in the system.

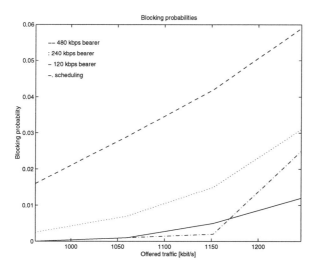

Figure 7.39 Blocking probability for three unscheduled single bit rate bearers as well as for when scheduling is used.

In Figure 7.39, the blocking probability for three different single bit rate bearers is compared to the scheduling algorithm earlier presented. From the figure, we see that it is better to use a low bit rate bearer if we are only allowed to use one single bit rate bearer. The reason for that is better statistical averaging. Earlier in this section, it was argued that it is more advantageous to minimize the number of users transmitting in parallel. However, for a single bit rate bearer it seems that the gain from minimizing the number of users is not as large as the gain from statistical averaging. We also see from Figure 7.39 that the proposed scheduling works quite well. However, it seems that the blocking probability is increasing rapidly as the offered traffic gets close to what the system can handle. One possible explanation could be that the number of times that the user is ordered to cease transmitting is increasing, and this leads to a less efficient use of the spectrum, based on the discussion related to Figure 7.39. The less efficient use of the spectrum leads to more users in the queue, and finally results in a higher blocking probability.

Another interesting characteristic of the studied approaches is the delay. It is plotted for a 480 and 120 Kbps bearer and for the scheduling approach in the following Figures 7.40-7.42. All simulations are run with an offered traffic of 1,246 Kbps.

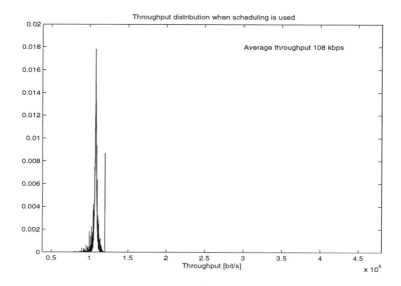

Figure 7.40 Delay distribution for the case where a 120-Kbps bearer is used.

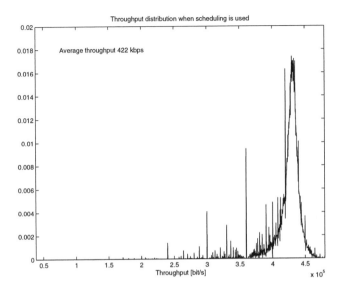

Figure 7.41 Delay distribution for the case where a 480-Kbps bearer is used.

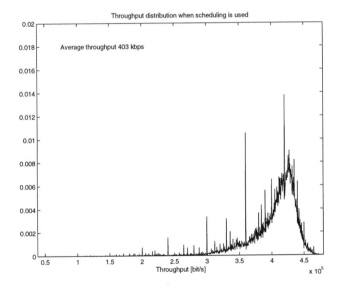

Figure 7.42 Delay distribution for the case where scheduling is used.

Evidently (Figure 7.40 and Figure 7.41), the average bit rate is much higher for the 480-Kbps bearer case compared to the 120-Kbps cases. However, as showed previously, the problem with the 480-Kbps bearer was its high blocking probability. The scheduling approach (Figure 7.42) has a lower blocking probability than the 480-Kbps case and almost the same average throughput.

7.4.3.3   Downlink Scheduling

In the downlink, the scheduling algorithm uses information about the total transmitted power from one BS or several BSs. Commercial systems are likely to have maximum output power limitations on a channel basis as well as on a BS basis. However, in this study, only the maximum output power limitation on a BS basis is considered.

Analysis

Consider the downlink in a single cell WCDMA network. The received quality is then defined as

$$\gamma_i = \frac{p_i g_i}{\alpha(P_{tot} - P_{sch} - p_i)g_i + P_{sch} g_i + N}$$

where p_i is the output power from the BS to user i, g_i is the gain between the BS and user i, and α is the orthogonality factor. Further, P_{tot}, P_{sch}, and N correspond to the total output power from the BS, the power allocated for the downlink syncronization channels and the background noise. By using this expression we see that

$$p_i = \frac{\gamma_i}{1 + \alpha\gamma_i}(\alpha P_{tot} + P_{sch} + N/g_i - \alpha P_{sch})$$

The total output power from the BS is defined as

$$P_{tot} = \sum_{i=1}^{M} p_i + P_{bc}$$

where M is the total number of MSs connected to the BS and P_{bc} is the total power of the downlink broadcast channels. By using the last three equations we see that

$$P_{tot} = \frac{P_{sch}(1-\alpha)\sum_{i=1}^{M}\frac{\gamma_i}{1+\alpha\gamma_i} + N\sum_{i=1}^{M}\frac{\gamma_i}{g_i(1+\alpha\gamma_i)} + P_{bc}}{1 - \alpha\sum_{i=1}^{M}\frac{\gamma_i}{1+\alpha\gamma_i}}$$

Further, assume that the E_b/N_0 is the same for all bit rates, (i.e., γ is proportional to the bit rate). In reality this is not true, since the E_b/N_0 value depends on how much overhead the pilot bits add, as well as on the interleaving depth. Now, consider a system with M_1 RT users require a certain output power from the BS. If there still are resources available, some NRT downlink connections will be established. We now investigate how much is gained by letting just one NRT user transmit compared to letting $M_2 = M - M_1$ NRT users transmit in parallel. The total output power for these two cases will be

$$P_{tot} = \frac{P_{sch}(1-\alpha)\left(\sum_{i=1}^{M_1}\frac{\gamma_i}{1+\alpha\gamma_i}+\frac{\gamma_{max}}{1+\alpha\gamma_{max}}\right)+N\left(\sum_{i=1}^{M_1}\frac{\gamma_i}{g_i(1+\alpha\gamma_i)}+\frac{\gamma_{max}}{g_{max}(1+\alpha\gamma_{max})}\right)+P_{bc}}{1-\alpha\left(\sum_{i=1}^{M_1}\frac{\gamma_i}{1+\alpha\gamma_i}+\frac{\gamma_{max}}{1+\alpha\gamma_{max}}\right)}$$

$$(7.25)$$

and

$$P_{tot} = \frac{P_{sch}(1-\alpha)\left(\sum_{i=1}^{M_1}\frac{\gamma_i}{1+\alpha\gamma_i}+\sum_{j=M_1+1}^{M}\frac{\gamma_j}{1+\alpha\gamma_j}\right)+N\left(\sum_{i=1}^{M_1}\frac{\gamma_i}{g_i(1+\alpha\gamma_i)}+\sum_{j=M_1+1}^{M}\frac{\gamma_j}{g_j(1+\alpha\gamma_j)}\right)+P_{bc}}{1-\alpha\left(\sum_{i=1}^{M_1}\frac{\gamma_i}{1+\alpha\gamma_i}+\sum_{j=M_1+1}^{M}\frac{\gamma_j}{1+\alpha\gamma_j}\right)}$$

$$(7.26)$$

In order to get an easy analytic expression, assume for a moment that g_{max} in (7.25) equals g_j in (7.26). Then, to have (7.25) equal to (7.26), the following is required

$$\frac{\gamma_{max}}{1+\alpha\gamma_{max}} = \sum_{j=M_1+1}^{M}\frac{\gamma_j}{1+\alpha\gamma_j} \qquad (7.27)$$

From (7.27), we see that

$$\gamma_j = \frac{\gamma_{max}}{M_2+(M_2-1)\alpha\gamma_{max}} \qquad (7.28)$$

Equation (7.28) is very similar to the corresponding uplink expression in (7.23). The only difference is the orthogonality factor. This means that one loses less in throughput in the downlink compared to the uplink, when several NRT users are transmitting in parallel instead of letting them transmit one at the time. However, the assumption that g_{max} equals g_j is mostly not valid. This means that there is a gain dependence, which most likely should be considered in the downlink scheduling decision.

7.4.4 Summary

Some aspects of scheduling that are common in FDD and TDD have been considered. Restrictions to both systems from the layer 1 point of view were compared. A scheduling algorithm that can be implemented in TDD as well as in FDD was investigated and simulations have been performed. An optimum for the algorithm regarding the number of resources allocated to every user per frame could be found.

Some WCDMA-specific features that may have an impact on the effectiveness of a scheduling algorithm were identified. Further, some uplink and downlink characteristics that should be considered when designing scheduling algorithms were discussed. An uplink scheduling approach was compared to unscheduled transmissions. We see that the scheduling algorithm can be used to make a tradeoff between blocking and delay.

References

[1]   ETSI Technical Report 101 112, "Selection Procedures for the Choice of Radio Transmission Technologies of the UMTS (UMTS 30.03)," Version 3.2.0, April 1998. Available at http:/www.itu.int/imt/

[2]   M. Berg, S. Pettersson and J. Zander, "A Radio Resource Management Concept for 'Bunched' Personal Communication Systems," *Proc. Multiaccess, Mobility and Teletraffic for Personal Communications Workshop, MMT'97*, Melbourne, Dec. 1997.

[3]   U. Dropmann, X. Lagrange and P. Godlewski, "Architecture of a Multi-Cell Centralized Packet Access System," *6th IEEE International Symposium on Personal, Indoor and Mobile Radio Communications, PIMRC'95*, Toronto, Sept. 1995, pp. 279-283.

[4]   Kronestedt, Frodigh, Wallstedt, "Radio Network Performance for Indoor Cellular Radio Systems," ICUPC'96

[5]   Broddner, Lilliestråhle, Wallstedt, "Evolution of Cellular Technology for Indoor Coverage," ISSLS'96, Melbourne, Feb 1996

[6]   S.A. Grandhi, J. Zander and R. Yates, "Constrained Power Control," *Wireless Personal Communications*, (Kluwer) Vol. 2, No. 3, Aug 1995.

[7]   J. Zander, "Radio Resource Management - An Overview," *Proc. IEEE Vehicular Technology Conference, VTC 96*, Atlanta, Georgia, May 1996, pp. 661-665.

[8]   M. Berg, "A Concept for Hybrid Random/Dynamic Radio Resource Management," *Proc. PIMRC'98*, Boston, USA, 1998.

[9]   H. Lou and A.S. Cheung, "Performance of Punctured Channel Codes with ARQ for Multimedia Transmission in Rayleigh Fading Channels," *46th IEEE Vehicle Technology Conf.*, 1996.

[10] G.J.R. Povey, "Time Division Duplex – Code Division Multiple Access for Mobile Multimedia Services," *Proc. PIMRC'97*, pp. 1034-1037.

[11] L. Chen et al, "A Dynamic Channel Assignment Algorithm for Asymmetric Traffic in Voice/Data Integrated TDMA/TDD Mobile Radio," *International Conference on Information, Communications and Signal Processing ICICS'97*, pp. 215-219.

[12] H.C. Papadopoulos and C-E.W. Sundberg, "Shared Time Division Duplexing (STDD) with Fast Speech Activity Detection," *ICC'96*, pp. 1745-1749.

[13] ETSI TR 101 112, "Selection Procedures for the Choice of Radio Transmission Technologies of the Universal Mobile Telecommunications System UMTS (UMTS 30.03)," Version 3.2.0, April 1998

[14] U. Dropmann, C. Lamare, X. Lagrange and P. Godlewski, "Dynamic Channel Allocation for Micro- and Pico-Cellular System with Overlapping Coverage," *ITG-Fachtbericht Mobile Kommunikation*, Neu Ulm, Germany, 1995.

[15] C. Mihailescu, X. Lagrange and Ph. Godlewski, "Dynamic Resource Allocation in Locally Centralized Cellular Systems," *Proceedings of IEEE Vehicular Technology Conference*, Ottawa, Canada, 1998, pp. 1695-1700

[16] C. Mihailescu, X. Lagrange and Ph. Godlewski, "Locally Centralized Dynamic Resource Allocation Algorithm for the UMTS in Manhattan Environment," *Proceedings of IEEE International Symposium on Personal, Indoor and Mobile Radio Communications*, Boston, USA, 1998.

[17] J. Blanz, et al., "Performance of a Cellular Hybrid C/TDMA Mobile Radio System Applying Joint Detection and Coherent Receiver Antenna Diversity," *IEEE Journal on Selected Areas in Communications*, Vol. 12, No. 4, May 1994, pp. 568-578.

[18] J-E. Berg, "A Recursive Method for Street Microcell Path Loss Calculations," *Proceedings of IEEE International Symposium on Personal, Indoor and Mobile Radio Communications*, Toronto, Canada, 1995, pp. 140-143.

[19] C.Y Huang and R.D Yates, "Call Admission in Power Controlled CDMA Systems," VTC '96.

[20] Z. Liu and M.E Zarki, "SIR Based Call Admission Control for DS-CDMA Cellular Systems," *IEEE Journal on Selected Areas in Communication*, Vol. 12, No. 4, 1994.

[21] F. Mario, G. Paolo, "Dynamic Channel Allocation for ATDMA," *RACE Summit*, Cascais, November 1995, pp. 299-303.

[22] Katzela, M. Naghshineh, "Channel Assignment Schemes for Cellular Mobile Telecommunication Systems: comprehensive Survey," *IEEE Personal Communications*, Vol. 3, No.3, June 1996.

[23] J. Knutsson, P. Butovisch, M Persson and R.D. Yates, "Evaluation of Admission Control Algorithms for CDMA Systems in a Manhattan Environment," *Proc 2nd CDMA International Conference*, Seoul, 1997.

[24] Zander, J, "Performance of Optimal Power Control in Cellular Radio Systems," *IEEE Trans Veh Tech*, No. 1, Feb 1992.

[25] Zander, J, "Radio Resource Management in 3rd Generation Personal Communication Systems," *IEEE Comm. Mag*, No 8, Aug 1998.

[26] R. Yates. "A Framework for Uplink Power Control in Cellular Radio Systems," *IEEE J. Sel. Areas Commun*. Vol. 13, No. 7, September 1995, pp. 1341-1348.

[27] M. Berg, S. Petterson and J. Zander, "A Radio Resource Management Concept for Bunched Personal Communication Systems," *MMT '97 Conference*, Melbourne, Australia, December 1997.

[28] M. Pizarroso, J. Jimenez, ed., "Common Basis for Evaluation of ATDMA and CODIT System Concepts," CEC deliverable MPLA/TDE/SIG5/DS/P/001/b1, September 1995.

[29] Zander, J, "Performance Bounds for Joint Power Control & Link Adaption for NRT bearers in Centralized (Bunched) Wireless Network," *Proc PIMRC99*, Sept. 1999.

[30] M. Andersin, "Power Control and Admission Control in Cellular Radio Systems," Ph.D. Thesis, Royal Institute of Technology, May 1996.

[31] J. Knutsson, P. Butovitsch, M. Persson and R. D. Yates, "Downlink Admission Control Strategies for CDMA Systems in a Manhattan Environment," *VTC*, May 1998.

Chapter 8

UTRA Interworking

The spectrum allocation for UTRA in Europe is shown in Figure 8.1. The close proximity of frequency allocations sets the demands for UTRA interworking. In this chapter, the interworking aspects are studied and methods that allow such interworking are presented.

Frequency [MHz]

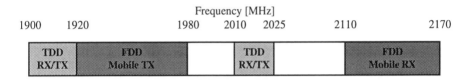

Figure 8.1 Reserved frequencies for UTRA operations.

This chapter gives an overview of the UTRA and GSM interworking. Technology and physical premises of a multimode terminal are discussed in the beginning, but this is extended into special UTRA FDD system issues such as compressed mode and the impact to the UTRA FDD link performance. The interference between uplink and downlink in UTRA TDD is analyzed by simulations, and recommendations for UTRA TDD operation are given. The GSM-UTRA handover and required synchronization measurements are also discussed so that UTRA system synchronization performance is better understood. Adjacent channel interference in the UTRA FDD system is also discussed.

8.1   COEXISTENCE AND COMPATIBILITY

This section first gives the technology baseline and then expands to analyze the link performance of the compressed mode, a special method to provide the compatibility between UTRA FDD and GSM or UTRA TDD. The last part of this section is dedicated to UTRA TDD uplink and downlink interference evaluation to understand the nature of a TDD system.

8.1.1 UTRA/GSM Multimode Terminal Considerations and Interworking Issues

The multimode terminal must be capable of operating in all of the three modes (UTRA FDD, UTRA TDD, and both bands of the GSM). This does not only require the capability to synchronize, receive, transmit, and perform the required monitoring in all systems, but also to handle the intersystem monitoring requirements to seamlessly switch the operation mode when needed. This is the most demanding task of a multimode terminal, since duties of different modes are carried out simultaneously.

8.1.1.1 Interworking Requirements

There are two different approaches to perform interworking functions. The handover measurements, either radio signal strength indication (RSSI) or synchronization measurement, require transmission on the current resource, and reception on the current and monitored resource. This can be accomplished either with:

- A dual receiver approach;

- Scheduling for different operations.

The dual receiver approach is the most direct, straightforward, and trivial way of monitoring. However, the requirement of simultaneous transmission and reception on adjacent bands sets extensive filtering requirements. Also, the duplicated receiver hardware is not desirable. There is no point of having a receiver that is inoperative for at least 50% of the time. This duplicated hardware will increase the terminal (receiver) cost at least by 100%. The circuits required by the alternative signal path would also degrade the receiver performance. Figure 8.2 exemplifies a block diagram of such a transceiver.

The scheduling-based approach has the advantage of a reduced receiver hardware complexity if scheduling is done correctly. This approach might also allow an integration of GSM and UTRA receivers into the same signal path. The monitoring times are critical and they determine the complexity of the synthesizer and the signal processing requirements.

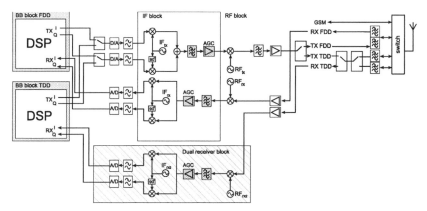

Figure 8.2 An exemplar block diagram of a multimode UTRA/GSM transceiver with a dual receiver.

The scheduling-based approach suffers from the same problems (i.e., interference, intermodulation, and impedance matching problems), as the dual receiver approach if the transmission and reception for the different modes or systems are performed simultaneously. It is therefore highly favorable if reception and transmission are done alternately. This is achieved with:

- Idle period monitoring;

- Frame stealing for monitoring (not allowed by 3GPP specifications);

- Compressed mode.

For a system not receiving and transmitting continuously, the idle periods can be used for measurements. This is applicable for both GSM and UTRA TDD. The only drawback is the increased power consumption due to the measurement when compared to two single-mode phones.

For a system with continuous transmission and reception, the frame stealing option forcefully generates a period for monitoring. On the other hand, frame stealing causes a degradation of QoS when reception and transmission are skipped in order to perform monitoring. This results in a trade-off between the errors introduced due to measurements, and the ability to prevent disrupted connections by enabling handover.

The compressed mode is an arrangement for UTRA FDD to generate gaps for monitoring. The compressed mode approach is a more elegant way to arrange the measurement period for resource monitoring than frame stealing. In this procedure, the terminal receives data in a time-compressed manner, so that the remaining data transmission time in a frame can be used for monitoring. For further details on compressed mode, see [1].

The scheduling approach has a clear need for strict time specifications that guarantee undisturbed operation. The scheduling-based approach is considered to be the best approach, since it provides low cost and harmonized implementation possibilities for the multimode terminal.

8.1.1.2   Technical Framework

Major concerns for the RF section of a UTRA mobile station are the filtering and synthesizer requirements. If simultaneous reception and transmission are allowed, the interference at the receiver will naturally increase. This applies especially when monitoring UTRA TDD or GSM 1800 from UTRA FDD. Figure 8.3 shows the cellular spectrum allocation around 2 GHz.

frequency [MHz]

1710	1785 1805	1880 1900 1920	1980	2010 2025	2110	2180
GSM 1800 UPLINK		GSM 1800 DOWNLINK	DECT RX/TX	UTRA TDD RX/TX	UTRA FDD UPLINK	UTRA TDD RX/TX	UTRA FDD DOWNLINK

Figure 8.3   Reserved frequency bands for cellular operations around 2 GHz.

The close location of transmission and reception frequency bands makes the filter bank design complicated for simultaneous UMTS TDD or GSM 1800 reception and UMTS FDD transmission. The impedance matching of the different filters over several frequency bands becomes difficult. This would require a highly integrated and thoroughly tested filter bank. In a practical multimode UTRA terminal, the filters are integrated into a single package. Therefore, the physical limitations of the receiver filter leads to limitations of how much the adjacent channel signal is attenuated.

The transmission of UTRA FDD will not be attenuated enough when monitoring another system/mode, and the belonging transmitter will be apparent as an additional high power interfering source. The best way to avoid transmitter interference at the receiver is to interrupt the transmission. For intermode measurements, the uplink compressed mode in conjunction with downlink compressed mode is needed, especially when monitoring UTRA TDD or GSM 1800 from UTRA FDD. The transmission is suspended while the monitoring is performed. This is the only method to perform seamless handover monitoring in the mobile station.

8.1.1.3   Impact of Uplink Compressed Mode for UTRA FDD Link Performance

Using the compressed frames generally causes certain degradation, so its use should be reduced to a minimum. The introduction of uplink compressed mode has certain impacts to the uplink range, and those impacts are studied with the link level simulations in [2]. In this section, the results of those simulations are presented in order to better understand the effect of compressed mode.

The main impact to the link performance comes from the power control of the link. Also, the selected coding method has impact (see [2] for simulation details). In typical link-level study, the power control dynamics are assumed to be infinite, which corresponds to the situation elsewhere but not at the cell edge. As we see from Figure 8.4, the impact of the uplink compressed mode to the BER curves in terms of the average required E_b/N_0 is not significant.

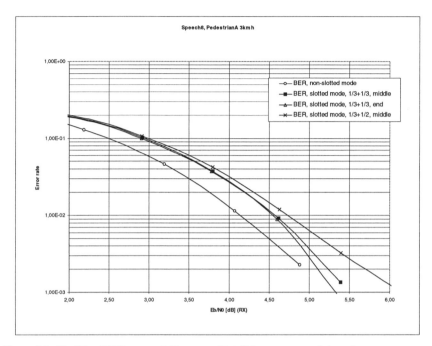

Figure 8.4   Traditional BER versus E_b/N_0 curves with infinite power control dynamics.

The situation in Figure 8.4 corresponds to a typical case where most terminals in the cell are not in the power-limited situation. This is an even more valid assumption when considering microcell environment. In Figure 8.5, the impact of the uplink-compressed mode to the UTRA FDD coverage is presented with the so-called headroom versus received E_b/N_0 curves. The figure shows the effect of compressed mode from a coverage point of view. The case is interesting when mobiles are in the power-limited situation at the cell edge. This corresponds typically to the mobiles in a macrocellular environment at the edge of the UTRA coverage area where soft handover is not possible. The headroom is the difference between maximum transmitter power over the whole simulation and average receiver power during the noncompressed frames (in decibels). The E_b/N_0 (RX) is average received E_b/N_0 during the noncompressed frames (decibels).

The headroom value is needed for the link budget calculations. If the mobile is at the cell edge, it cannot use the full dynamic range for the power control anymore, but instead it has to increase the target E_b/N_0 for the power control, because the allowed headroom over the average transmit power is decreased.

From the Figure 8.5 it we see that especially with 1/3-rate coding, the impact of compressed mode to the headroom is less than 1 dB, while the impact is higher with the 1/2-rate coding. From the studies for intersystem handover and especially uplink compressed mode, the following can be concluded:

- The effect to the link performance is marginal when the terminal is not in the transmission power-limited case, such as in the system coverage area edge.

- The performance degradation increases as the terminal approaches the coverage edge.

- From the overall system capacity point of view, the impact is not significant, also as in the normal operation with continuous coverage, soft handover alleviates the problems related to power limitations.

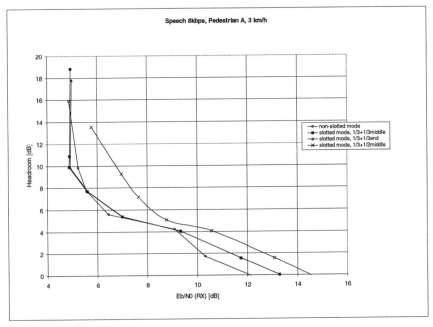

Figure 8.5 Speech service, headroom versus E_b/N_0. This gives the effect of compressed mode for the network coverage planning.

Note that the channel model used in [2] is the most sensitive to impacts due to the compressed mode. The channel offers very little diversity gain, which highlights the power limitation impacts.

8.1.1.4   Multimode Terminal Implementation Considerations

The ETSI consensus decision in January 1998 raised the terminal cost issue as one of the key factors along with harmonization of UTRA with GSM. The low-cost aspect and the harmonization strongly favor multimode terminals. The best engineering solution is a common antenna and common switches and separate filters [3]. It outperforms independent and separate antennas and receiver terminals, which are expensive and bulky.

Figure 8.6 shows a block diagram of a multimode superheterodyne transceiver. The GSM (900/1800) receiver part is not depicted in the picture. The antenna switch provides the signal path for the GSM transceiver. The actual implementation depends on the duplexer and other filter properties.

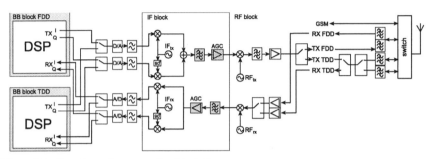

Figure 8.6 Superheterodyne transceiver block diagram.

Terminal Receiver Impact

The time-continuous transmission in UTRA FDD requires steeper filters than UTRA TDD. The analogy to GSM is much higher in the UTRA TDD mode than in UTRA FDD, which has RF section requirements similar to the first generation analog phones (e.g., NMT).

The current phone designs (e.g., GSM 900/1800, PDC, AMPS, and IS-95) have a noise figure between 8 to 12 dB [3]. The higher noise figure relates directly to higher frequency band operation. Therefore, it is reasonable to assume the UTRA receiver noise figure to be typically 7 dB and the extreme value to be 9 dB. If a value of 5 dB was assumed for the system simulations in light of the current technology, the increase of 2 dB is justified due to an extra attenuation in the signal path due to the multimode terminals. The implementation margin of 2 dB is required for mass production. There should be no difference between UTRA single mode and multimode terminal receiver noise figures.

Terminal Transmitter Impact

The extra loss due to the switches has to be balanced. The only way to do this is to increase the power amplifier output in order to have the required power at the antenna connector. This affects the power consumption and reduces the active time compared to single-mode terminals. The thermal problems have to be encountered.

8.1.1.5   Conclusions

The need for interworking between UTRA and GSM is evident and UTRA/GSM multimode terminals must be competitive from the first day. An evolution from GSM is better suited for the engineering perspective, but there will be a revolution on the service and feature side. The effect of implementing GSM in multimode UTRA terminals is negligible. The specifications for UTRA multimode and single mode terminals must be the same. The two different UTRA modes have a potential commonality in the RF section of the terminal, but this tie must be kept all the way throughout the standardization phase to allow efficient terminal implementation.

For interworking of different modes, methods to perform a different task with the same hardware should be developed. Two examples of such system features are the scheduling of different tasks to allow for an optimized implementation, and the compressed mode to perform the intersystem/mode measurements. The synthesizer and filter requirements should not be too stringent when compared to the existing systems.

8.1.2   Evaluation of Interference Between Uplink and Downlink in UTRA TDD

An ideal TDD system offers high capacity, dynamic asymmetry between uplink and downlink, and no coordination requirements between operators and base stations. This subsection studies the performance of UTRA TDD and how close to the ideal TDD system UTRA TDD could be. Intermode and intersystem handovers between GSM and UTRA are covered in Section 8.2. This section is based on the work presented in detail in [4].

8.1.2.1   Interference Between Uplink and Downlink in TDD

Since both uplink and downlink share the same frequency in TDD, those two transmission directions can interfere with each other. This kind of interference occurs if the base stations are not synchronized. The interference is present also if a different asymmetry is used between up- and downlink in adjacent cells, even if the base stations are frame-synchronized. Frame synchronization requires an accuracy of a few symbols, not an accuracy of chips.

The interference between uplink and downlink can also occur between adjacent frequencies. Therefore, the interference between uplink and downlink can take place within one operator's band, and also between two operators.

The interference between uplink and downlink can occur between two mobile stations or between two base stations. In FDD operation, the duplex separation prevents the interference between uplink and downlink. The interference between a mobile station and a base station is the same both in TDD and in FDD operation, but this interference scenario is not considered in this section.

Mobile-to-Mobile Interference

Mobile-to-mobile interference occurs (Figure 8.7) if the mobile M2 transmits and the mobile M1 receives on the same (or on an adjacent) frequency in adjacent cells. Mobile-to-mobile interference is statistical because the locations of the mobiles cannot be controlled. Therefore, mobile-to-mobile interference cannot be avoided completely by the network planning.

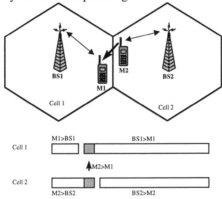

Figure 8.7   Mobile-to-mobile interference at the cell border.

Base Station-to-Base Station Interference

Base station-to-base station interference occurs in Figure 8.8 if base station BS1 transmits and base station BS2 receives on the same (or on an adjacent) frequency in adjacent cells. Base station-to-base station interference depends heavily on the path loss between the two base stations, and, therefore, the network planning greatly affects this interference scenario.

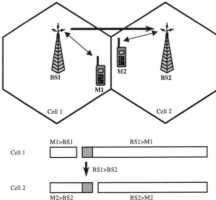

Figure 8.8   Base station-to-base station interference.

8.1.2.2   Intraoperator Interference

A simulation platform was generated to evaluate the interference scenarios in TDD operation [4]. The simulation first generates two cells; the cell causing the interference, and the one experiencing it. Hexagonal cells were used so that the two cells could be made adjacent without overlaps. A random distribution of the mobiles within the interfering cell was used. The simulation scenario is shown in Figure 8.9. Here, the left cell is the cell experiencing the interference, and the right cell includes a few random test locations where the test interference is calculated. A path loss exponent of 3.7 and slow fading with a standard deviation of 12 dB has been assumed both between the mobile stations and between the base stations. The outage probabilities are evaluated for those timeslots where uplink and downlink transmission occurs at the same time in adjacent cells.

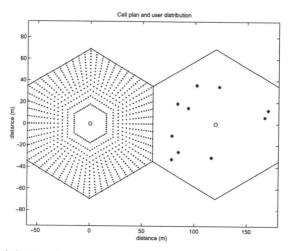

Figure 8.9 Simulation scenario.

Mobile-to-Mobile Interference

Mobile-to-mobile interference within one operator's band occurs at the cell borders. Figure 8.10 shows the percentage of the cell area where the outage probability exceeds the required maximum levels (1%, 2%, 5%, or 10%) as a function of the base station separation. In case of continuous coverage, the base station separation is equal to twice the cell radius. Using that base station separation, an outage level of 2% is exceeded in about 8% of the cell area.

One solution for this problem is to use intelligent DCA techniques. In DCA, such timeslots are selected for the connection where the interference levels are low enough. DCA can be applied because the UTRA TDD mode includes a TDMA component. DCA is used also, for example, in the DECT system. Note that DCA is effective if only a few time slots are used for the connection (i.e., for low and moderate bit rate connections). DCA does not work properly for those connections where almost all timeslots are used. The maximum bit rate per timeslot in UTRA TDD is 144 Kbps.

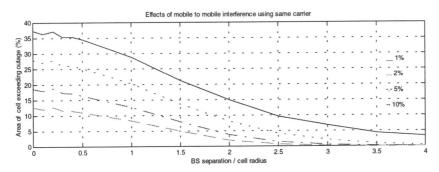

Figure 8.10 One operator, MS-to-MS interference, percentage of the cell area where the given outage probabilities are exceeded.

If an operator does not use UTRA TDD to provide a continuous coverage, the interference problems can be greatly reduced be keeping the base station separation large.

Base Station-to-Base Station Interference

The interference between base stations depends on the locations of the base stations. In Figure 8.11, the outage probability is shown as a function of the base station separation. It is assumed that the path loss formula between two base stations is the same as between a base station and a mobile station. For continuous coverage (base station separation = 2 × cell radius), the outage probability is about 28%. Such a high outage probability indicates that the adjacent cells need to be synchronized, and they need to have the same asymmetry if the system is fully loaded. Alternatively, the BS-to-BS interference can be reduced significantly by using space-selective transmission via smart antennas at the considered base stations. Note that the interference reduction that can be obtained via smart antennas (on the uplink as well as the downlink) is not studied in this section.

By introducing an additional 12-dB protection between the adjacent base stations in Figure 8.11, the outage probability is reduced from 28% to 8%. The 12-dB protection is still not enough to bring the outage probability to reasonably small figures, but more protection is desirable. This additional protection was assumed to be obtained by antenna direction, or antenna tilting, or by careful network planning.

The interference between base stations can be particularly strong if the path loss is low between the base stations. Such cases could occur (e.g., in macrocells), if the base stations are located in masts above the rooftops.

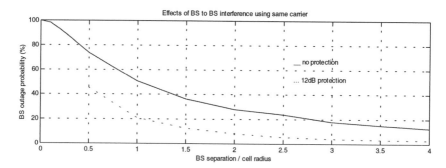

Figure 8.11 One operator, BS-to-BS interference, outage probability. A 12 dB protection could be obtained by antenna techniques or by network planning.

8.1.2.3   Interoperator Interference

Mobile-to-Mobile Interference

The interoperator interference between mobiles can occur anywhere where two operators' mobiles are close to each other and transmitting with a fairly high power. An adjacent channel protection of 35 dB is assumed.

The percentage of the cell area where problems exist is only less than 0.5% if the required outage probability is 2%. The mobile-to-mobile interference between operators is not a problem if the mobiles are equally distributed over the cell area, since the adjacent channel protection mitigates the problem. Mobile-to-mobile interference is more difficult within one carrier and interference; however, it can be more difficult than shown here if the mobiles are not equally distributed but are close to each other (e.g., in meeting rooms).

Base Station-to-Base Station Interference

The outage probability increases to very high values if the different operators' base stations are located close to each other. If the base stations are colocated, the interference levels will be intolerable. These outage probabilities show that cooperation between TDD operators' network planning is required, or the networks need to be synchronized and the same asymmetry applied if the systems are fully loaded and smart antennas are not used. Otherwise, the overall capacity for each user is, at best, effectively halved. Note that in FDD operation the inter-operator interference can be avoided best by colocating the base stations.

One solution to avoid interoperator interference is to use very high requirements for the adjacent channel filtering. However, with the technology level after a few years, it is not possible to considerably improve the filtering characteristics to obtain higher adjacent channel protection values more than about

45 dB. For the second adjacent carriers, the channel protection value should be about 55 dB.

From the synchronization and coordination point of view, the higher the transmission power levels and the larger the intended coverage area, the more difficult the coordination for the interference management. In particular, macrocell-type of antenna locations easily result in a line-of-sight connection between base stations causing strong interference.

8.1.2.4   TDD Operation in FDD Uplink Band

If there is less traffic in the uplink direction in UTRA FDD than in downlink direction, part of the uplink spectrum could be utilized for TDD operation. According to [5], TDD operation should not be prohibited in the FDD uplink band.

If TDD is used in FDD uplink, the TDD base station transmission can interfere with the reception in FDD base station. The interference analysis is taken from Section 8.1.2.2. Therefore, we notice that without using special methods such as smart antennas, the co-location of operator's TDD and FDD base stations is difficult if TDD is operated in the FDD uplink band. Note that DCA techniques are not effective against FDD interference, since FDD transmission is continuous.

8.1.2.5   Unlicensed TDD Operation

Unlicensed operation with UTRA TDD is possible if DCA techniques are applied together with a TDMA component. DCA techniques cannot be applied for very high bit rates, since almost all timeslots are needed. Therefore, unlicensed operation is restricted to low to moderate bit rates if there are several operators in the same area that do not use special measures (e.g., smart antennas).

8.1.2.6   Operational Differences Between UTRA FDD and TDD

UTRA FDD and TDD are harmonized to a large degree on the physical layer. There are, however, a few differences in the operation of FDD and TDD networks. In the TDD radio network-planning more coordination between the operators is desirable. Also, the TDD uplink range is shorter than in FDD and that needs to be considered in the planning. The real-time radio resource management algorithms are partly different between FDD and TDD: power control, handovers, dynamic channel allocation, resource allocation, and load control are different. The synchronization requirements in TDD between operators and between adjacent cells have an effect on the network operation.

8.1.2.7   Conclusions

The interference between uplink and downlink in UTRA TDD has been evaluated in this section. This interference can occur between two terminals or between two base stations. The results show that within one carrier (one operator), both mobile-to-mobile interference and base-station-to-base-stations can cause problems. The solutions to tackle the problem include base station synchronization, careful network planning, smart antenna techniques, and dynamic channel allocation.

The results also show that between different operator's TDD networks, the most difficult interference occurs between base stations when the base stations are colocated or are in close proximity. That problem occurs also if TDD is used in the FDD uplink band. This problem can be solved by avoiding colocation and by coordination between operators including network planning, base station locations, and common synchronization, or by using special means (e.g., smart antennas) at the base stations.

For UTRA TDD operation dynamic channel allocation, synchronization of base stations and cooperation between the operators is preferred to obtain the maximum performance. The different interference scenarios for single antenna systems are summarized in Table 8.1.

Table 8.1

UTRA TDD Interference Summary

	Outage without synchronization or with different asymmetry	Solutions
Intraoperator: MS-to-MS	High outage	DCA
Intraoperator: BS-to-BS	High outage	Network planning, smart antennas, DCA
Interoperator: MS-to-MS	Low outage	
Interoperator: BS-to-BS	High outage if base stations close to each other	Coordination between operators, no colocation of base stations, DCA, smart antennas

BS=Base station, MS=Mobile station

8.2   GSM – UMTS HANDOVER MEASUREMENTS

After the deployment of UMTS, the two modes of UTRA (FDD and TDD) and the existing GSM system will be operating in parallel in different frequency bands. There is a need for seamless handovers between these two systems for coverage reasons, especially in the early phases of the UMTS deployment, and also for capacity reasons at a later stage.

The different phases of a handover are divided into three steps. The first step, *initialization*, is where the need of a handover is indicated either by the mobile or by the network. At the second step, the mobile performs *measurements* to find the best candidates for handover and reports the results to the network. In the third step, the *actual handover* is performed.

Further, the second step can be divided into two more steps. First, the synchronization information of the target cell must be found and then the different cells are sorted in order of received quality to ensure that the most appropriate candidate is chosen for the handover. To acquire the synchronization is generally more difficult, since a synchronization channel that contains information of the target cell needs to be found and decoded. This section evaluates the time needed to find the synchronization information of the target cell when using a schedule-based approach.

To perform the measurements, it must be possible for the mobile station to monitor channels on another frequency. In case of UTRA TDD and GSM, this can be achieved during the time when nothing is received or transmitted (i.e., the mobile station is not using all slots for transmission or reception). UTRA FDD, however, uses continuous transmission and reception. As explained in the prior section, UTRA FDD should use the compressed mode technique to perform measurements on the candidate frequencies.

8.2.1 Synchronization Channels

This subsection presents the synchronization channel structure of the different modes/systems. The synchronization channels provide the information needed for the mobile station to synchronize to the desired base station. This information might include, for example, scrambling code and frame timing.

8.2.1.1 UTRA FDD

In case of UTRA FDD, the SCH is divided into one primary and one secondary channel, as shown in Figure 8.12. The primary synchronization channel consists of a primary synchronization code c_p, which is the same for all UTRA FDD base stations. By detecting the primary synchronization channel, the chip- and slot synchronization is found. The secondary channel is built up of a 17-ary alphabet of short codes c_s. The secondary synchronization channel indicates the group of the scrambling code used in the base station and also facilitates the frame synchronization.

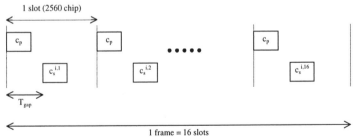

Figure 8.12 Primary and secondary synchronization channel for UTRA FDD.

If the synchronization channels are detected, the scrambling code can be determined by correlating the codes of the group with the CCPCH on a symbol-by-symbol basis. After this, it is possible to decode the BCCH information.

Since the synchronization channels are both transmitted with relatively low power, they have to accumulate over several slots to be detected. Also, at least two secondary synchronization channel slots need to be detected correctly to decide which secondary SCH code was used.

8.2.1.2   UTRA TDD

In UTRA TDD, each slot is assigned to uplink or downlink transmission. The proposed scheme is very flexible and allows many different uplink and downlink configurations. Three examples are shown in Figure 8.13. The theoretical downlink/uplink asymmetry ranges from 1:7 or 1:15 (this figure depends on the different number of downlink slots used for the SCH) to 15:1 [6].

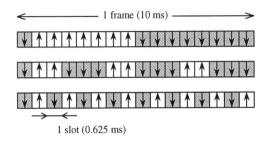

Figure 8.13 Three examples of possible up- and downlink configuration of UTRA TDD slots.

One or two slots of each frame are always allocated for downlink transmission, and the synchronization channel (SCH) and the common control physical channel (CCPCH) are placed in these two slots [6]. Since the secondary synchronization channel is a sequence of code words, it is assumed that the secondary synchronization channel is repeated after eight time slots as shown in Figure 8.14.

To avoid interference from mobile station to mobile station in UTRA TDD, adjacent base stations are frame-synchronized. Therefore, an additional offset, T_{offset}, is added to make it possible to distinguish between primary synchronization channels of different base stations.

Similar to the synchronization concept used in UTRA FDD, the primary synchronization channel facilitates the slot and chip synchronization, and the secondary synchronization channel indicates the frame synchronization and the code group of the use by the base station.

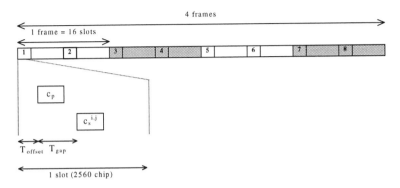

Figure 8.14 Primary and secondary synchronization channel for UTRA TDD.

In UTRA TDD, the location of the CCPCH is very flexible. It can be put in one or both of the downlink slots and it could be using one or more spreading codes [6]. Therefore, the synchronization channels must point where the CCPCH is located. As in UTRA FDD, the synchronization channels have to accumulate over several slots to be detected with greater certainty.

8.2.1.3   GSM

In the GSM system, the channels are organized in hyper-, super-, and multiframes. There are two types of multiframes, as shown in Figure 8.15. The first one contains 51 frames and is used by the common channel. The traffic channels use a multiframe structure of 26 frames.

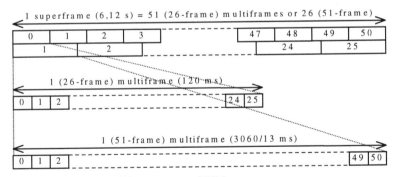

Figure 8.15 The super- and multiframe structure of GSM.

The information needed for the mobile station to synchronize to the base station is on the FCCH and the SCH. The FCCH is used to synchronize the frequency of the mobile station and also facilitates slot synchronization. If an FCCH is detected, this also shows that the mobile station is listening to a beacon carrier. The FCCH and SCH are mapped onto the common channel and then mapped onto timeslot zero of the beacon carrier, as shown in Figure 8.16.

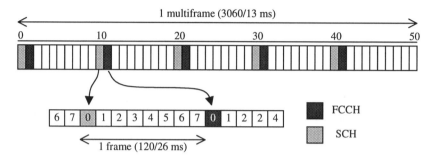

Figure 8.16 Mapping of FCCH and SCH.

8.2.2 Measurement Possibilities

Since the other modes/systems are operating on other frequencies, the single receiver of the mobile station must change to another frequency to perform the measurements. This can only be achieved if the terminal does not receive anything at that moment. In this subsection, we discuss the measurement possibilities in UTRA FDD, UTRA TDD, and GSM.

8.2.2.1 UTRA FDD

A special technique called compressed mode is used on the downlink of UTRA FDD to create idle slots and, thereby, enable measurements on other frequencies/modes/systems [1]. The basic principle of this technique is shown in Figure 8.17. The information is concentrated to a smaller part of the frame, resulting in an idle part that can be used for measurements. This concentration of information is achieved by decreasing the spreading factor or by changing the channel-coding scheme.

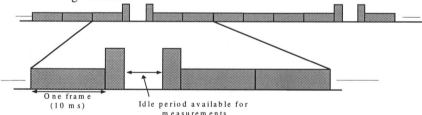

Figure 8.17 Compressed mode.

In Figure 8.17, one method, where the idle period is placed in the middle, is shown. Figure 8.18 shows other possible ways to place the idle period.

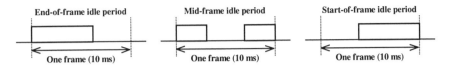

Figure 8.18 Three possible locations of idle period during the compressed mode.

There are some drawbacks with the compressed mode. By using a smaller spreading factor, the number of available codes for the other users served by the same base station decreases. Also, a higher signal-to-interference ratio might be required to keep the desired quality of service when changing the coding scheme.

8.2.2.2   UTRA TDD

In UTRA TDD, the slots that are not used for transmission or reception (i.e., the idle slots), can be used to perform measurements. If, for example, a low-rate service is used and the mobile station only uses two slots per frame, one for reception and one for transmission, the other slots are available for measurements.

For cases with higher rate services, other schemes must be considered. This might include rescheduling of the resources to create idle slots for measurements.

8.2.2.3   GSM

In GSM, there are three possibilities for the mobile station to perform measurements:

- *Between reception and transmission.* Here, approximately 2 bursts per frame can be used for measurements.

- *Between transmission and reception.* Here, approximately 4 bursts per frame can be used for measurements.

- *In the idle frame of each multiframe.* Here, approximately 12 bursts per multiframe can be used for measurements.

Due to the adjustment time when switching frequencies, only the last two options are sufficiently long to be used for measurements.[1]

8.2.3   Measurement Time

The evaluation was done by simulating the time needed to detect the necessary synchronization information. In this subsection, the simulation tool and the assumptions are described first. Then the results for the possible handover cases are presented.

---

[1] This is only valid for a connection that uses one timeslot per frame.

8.2.3.1   Simulation Tool and Assumptions

The simulation tool compares the measurement slots with the location of the synchronization information, and measures how much time it takes until enough information is detected. This is done for different offsets between the source and target system/mode and, thereby, an average and a maximum time are derived.

The measurement time when measuring on UTRA is highly dependent on the power of the synchronization channels. In the following subsections, it is assumed that the total power of the synchronization channels is the same for the both UTRA modes. Since UTRA TDD transmits the synchronization channels in only two slots per frame, the power of the synchronization channel is eight times larger in UTRA TDD than in UTRA FDD. Each synchronization channel is detected after an accumulation over four frames (i.e., 64 SCH slots for UTRA FDD and 8 SCH slots for UTRA TDD). It is also assumed that the serving base station provides no additional information about the adjacent cells.

The time it takes to switch frequencies is assumed to be 500 μs due to the synthesizer settling time [3]. After this time, there is no remaining frequency offset. The offset between the primary and secondary synchronization channel in UTRA, T_{gap}, is assumed to be to zero.

8.2.3.2   Measurements From UTRA FDD

When measuring from the UTRA FDD system, the compressed mode operation is used. The results from the simulations are shown in Table 8.2. The performance is compared for several different rates of the frames operating in compressed mode.

In UTRA TDD, eight slots per frame are used for measurements, yielding a 4-ms long measurement slot. If the measurement slot is always placed the same way in each slot when performing handover from UTRA FDD to UTRA TDD, the measurement period must be long enough to guarantee that at least one SCH will be caught. Since the slots are not long enough, the location of the measurement slots for two consecutive frames is shifted according to Figure 8.19.

Table 8.2

Measurement Time From UTRA FDD to Different Modes/Systems

Frames in compressed mode	UTRA FDD		UTRA TDD		GSM	
	Average [ms]	Max [ms]	Average [ms]	Max [ms]	Average [ms]	Max [ms]
10%	1983	2101	2003	2816	802	2359
11%	1785	1891	1803	2536	825	2934
13%	1586	1681	1604	2256	558	1575
14%	1388	1471	1404	1976	990	2433
17%	1190	1261	1204	1696	-	-
20%	992	1051	1004	1416	474	1554
25%	794	841	804	1136	346	1013
33%	596	631	605	856	289	654
50%	398	421	405	576	205	515
100%	199	211	205	296	139	325

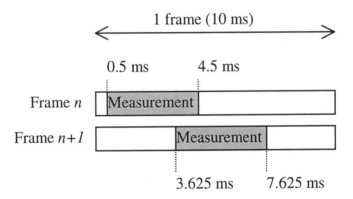

Figure 8.19 Measurement slots for handover from UTRA FDD to UTRA TDD.

8.2.3.3   Measurements From UTRA TDD

For UTRA TDD, the same performance as for UTRA FDD is achieved by using the same slots for the measurements. However, UTRA TDD has greater freedom in choosing the slots, since it can use any of the slots not already used for reception or transmission. The results of using a different number of consecutive slots per frame are shown in Table 8.3.

Table 8.3

Measurement time from UTRA TDD to different modes/systems when the number of consecutive slots per frame is varied.

Nbr of slots used	UTRA FDD Average [ms]	Max [ms]	UTRA TDD Average [ms]	Max [ms]	GSM Average [ms]	Max [ms]
4	557	631	-	-	265	652
5	382	421	-	-	212	468
6	290	312	-	-	172	468
7	236	252	-	-	148	371
8	199	211	-	-	139	325
9	172	181	-	-	121	184
10	149	155	140	146	113	184
11	135	141	131	146	103	184
12	121	125	122	146	92	184
13	110	114	114	146	81	137
14	102	105	105	146	72	137
15	94	97	96	146	63	137

8.2.3.4   Measurements From GSM

In Table 8.4, the simulation results of performing measurements from GSM are presented. This simulation is performed with different numbers of multiframes used for the measurements, ranging from using only one tenth of the multiframes up to using every multiframe. Furthermore, it is assumed that it is possible to measure between the transmission and reception of a slot in GSM.

Table 8.4

Measurement Time From GSM to Different Modes/Systems

Multiframes used for measurements	UTRA FDD Average [ms]	Max [ms]	UTRA TDD Average [ms]	Max [ms]	GSM Average [ms]	Max [ms]
10%	2406	2416	2147	2416	18009	50456
11%	2166	2176	1935	2176	6752	17336
13%	1926	1936	1722	1936	5373	16376
14%	1686	1696	1510	1696	3709	10141
17%	1446	1456	1297	1456	3235	11576
20%	1206	1216	1085	1216	6177	20456
25%	966	976	872	976	4048	14456
33%	726	736	659	736	1975	5816
50%	486	496	447	496	982	2456
100%	254	269	242	297	536	1256

8.2.4   Network-Aided Monitoring

The introduction of network-aided monitoring would reduce the measurement times even further. The commonality between the GSM and UTRAN can make it possible to relay timing information and guide terminals to reduce the monitoring needed to achieve synchronization.

Figure 8.20 shows a simplified example of GSM monitoring, where the SCH is captured directly with one monitoring step. The location of the SCH is transmitted with accuracy information (including the time required to make the frequency hop) to the mobile station and the mobile station performs the synchronization according to the network instructions. The network-aided monitoring allows considerably faster synchronization.

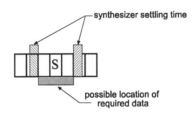

Figure 8.20 An example of network-aided synchronization measurement. The timing information with accuracy is transmitted to the base station that performs the direct GSM synchronization from received SCH with one monitoring. S denotes the SCH.

8.2.5   Conclusions

We have demonstrated that it is possible to perform handover measurements between the different modes and systems. The required measurement times are depicted in Table 8.5, where the average times for the measurements are summarized. In this table, it is assumed that eight slots per frame can be used for the measurements when measuring from UTRA, and that all of the multiframes can be used when measuring from GSM.

Table 8.5

Average Measurement Time When Doing a Handover Between Different Modes/Systems

	Measurement time [ms]	To		
		GSM	UTRA TDD	UTRA FDD
From	GSM	536	242	254
	UTRA	139	205	199

The results show a shorter measurement time for UTRA compared with GSM. This improved performance might be needed to enable very fast handovers in UTRA, but can also be used to reduce the measurement activity in both UTRA and GSM, and still perform better than the average GSM-to-GSM handover. There are *several monitoring combinations* that provide seamless handovers between the modes/systems and they vary depending on the need.

8.3 ADJACENT CHANNEL INTERFERENCE IN AN UTRA FDD SYSTEM

UTRA FDD allocates separate frequency bands for different cell hierarchies (macro – microcell deployment) [7,8] or different operators (macro – macrocell deployment) [9,10]. The network performance of the UTRA FDD cellular systems is interference-limited, and thus adjacent band interference can reduce the coverage or the capacity of the different layers or networks.

There is a trade-off between carrier spacing and adjacent channel interference requirements. The effects of adjacent channel interference on capacity with two adjacent-frequency operators or layers can be evaluated through simulations to optimize the system performance.

8.3.1 Adjacent Channel Interference

Adjacent frequency bands interfere with each other due to the imperfection of power amplifiers and receiver filters.

- Imperfect transceivers induce out-of-band power in adjacent bands. The ACLR (adjacent carrier leakage ratio) quantifies the transceiver out-of-band transmitted power.

- The selectivity of the receiver filter is not infinite in the adjacent band. The power of adjacent bands that cannot be suppressed becomes interference. The ACS (adjacent carrier selectivity) represents the capability to separate adjacent channels.

The adjacent channel power ratio (ACP) quantifies the level of interference experienced by a receiver due to the transmission in adjacent channels. Figure 8.21 illustrates the ACP. The practical limitations come mainly from the mobile implementation. In the downlink, ACP = $ACLR_{mobile}$ and in the uplink ACP = ACS_{mobile}. The in-band transmitted signal power after filtering is given by the following equation:

$$P_{TX} = \int_{fc1-W/2}^{fc1+W/2} S_{xx}(f) \, |H_{TX}(f)|^2 \, |H_{PA}(f)|^2 df \qquad (8.1)$$

where Sxx is the signal power spectrum at the input of the transmitter filter, $|H_{TX}(f)|^2$ is the transmitter filter response, and $|H_{PA}(f)|^2$ represents the nonlinearity of the power amplifier in the frequency domain. The out-of-band received power after the receiver filtering is given by the following equation:

$$P_{RX,a} = \int_{fc2-W/2}^{fc2+W/2} S_{xx}(f)\left|H_{TX}(f)\right|^2\left|H_{PA}(f)\right|^2\left|H_{RX,a}(f)\right|^2 df \qquad (8.2)$$

where $\left|H_{RX,a}(f)\right|^2$ is the receiver filter response.

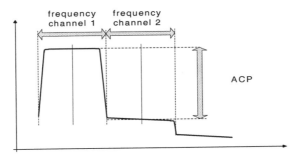

Figure 8.21 Adjacent channel power (ACP).

Tight spectrum mask requirements for a transmitter and high selectivity requirements for a receiver would guarantee low adjacent channel interference. However, these unrealistic requirements would have a detrimental impact on the UTRA mobile stations. The source of imperfections in ACLR is mainly the non-linear properties of the power amplifier [11]. Strict ACP requirements have a direct impact on the efficiency of the power amplifier and, therefore, on the active time and price of the mobile station. Practical values of the adjacent channel power attenuation are based on technology limits of power amplifiers. As seen in Figure 8.22, increasing the carrier spacing more than 5 MHz has no significant impact on the ACP. This is due to the 4.096MHz amplifier harmonic distortion bandwidth.

The adjacent channel interference (ACI) is the long-term average of the interference caused by the operation in the adjacent frequency channel. The total ACI in the network can be summed over the mobile stations and integrated over time,

$$ACI = \int \left(\sum_{MS_i} \left(\frac{P_{TX}}{ACP} \cdot fading/L_p \right)_i \right) dt \qquad (8.3)$$

where MS_i is a single mobile station operating in the adjacent frequency channel, *fading* is the multipath fast-fading with the Rayleigh amplitude distribution, and L_p is the path loss.

8.3.1.1   The Spatial Near-Far Problem

The ACI is high if the transmitted power is maximized by the spatial near-far problem that occurs if two CDMA operators do not colocate their base stations. A mobile station connected to one operator can be far from its serving base stations but close to a base station of another operator. If microbase stations are located far from the location of the macrobase station in HCS, the same situation occurs for the mobiles connected to the macrobase station.

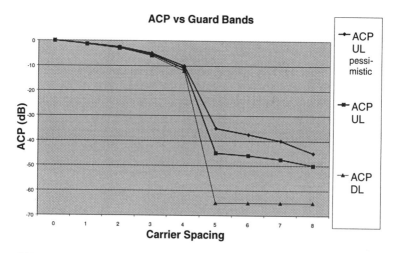

Figure 8.22 Adjacent channel power (ACP) as a function of carrier spacing.

On the uplink, the worst situation is when a terminal is transmitting at high power nearby an adjacent band base station, and thereby causes much ACI to the other base station. An unacceptable interference level occurs at the base station due to spectrum spillover to the adjacent frequency if the terminal has infinite transmitter power. In the practical cases, the call will be dropped due to the interference in the uplink caused by the interfered base station ACP properties.

8.3.2 Capacity Loss Evaluation

By comparing network situations with and without ACI for the downlink, we see that additional outage occurs in a few locations. Around the base station is a dead spot due to insufficient mobile transmitter power in the adjacent channel to overcome the ACI, and the mobile cannot be served. In other locations, the number of terminals that can be served by the base station decreases with increasing ACI.

On the uplink, from the base station point of view, the ACI randomly varies in time according to the spatial interference distribution. Higher interference together with C/I-based power control induce higher target power at the base station receiver. This causes the terminals to transmit at higher power and leads to additional outage.

In the interoperator case, the two networks mutually impact each other in both the uplink and the downlink. In the interlayer case, high transmit power terminals of the macrolayer induce a loss in the uplink of the microlayer and microbase station transmissions reduces the macrolayer downlink coverage and capacity.

8.3.2.1 Simulations

The interlayer and interoperator cases were studied in [7] with a static snapshot simulator. The simulations apply a large number of cells in rural macro or urban HCS environments. Propagation models describe the signal loss in the environments. Mobile stations are added one by one to the network to increase the load toward the operating point. The mobile stations are static during the simulation. The effect of mobility is emulated with fast-fading channels.

The base stations and mobile stations are deployed in such a way so that situations with high ACI occur. A mobile station located close to a base station of another layer or an operator inflicts coverage loss on the uplink. This is the most serious problem. Furthermore, the mobile station will inflict an uneven downlink coverage.

8.3.2.2 Interoperator Case

The simulations are performed with different ACP values separately for the uplink and downlink. A macrocell propagation model with high mobile station speed is used, which takes into account the path loss, shadowing, and fading. The base stations of the adjacent frequencies belong to different operators. The base stations have various deployments. In the simulations, the effects of ACI are studied with different network loads. The network is loaded with the number of mobile stations with an outage probability less than 5%.

The vehicular environment used with the interoperator ACI experiments applies two overlapping base station grids of two operators as shown by Figure 8.23. The cell diameter is either 1,800 or 5,400 m.

Figure 8.23 The cell deployment of the interoperator study.

In the uplink case, the average interference power as a function of load with various ACP values is shown in Figure 8.24.

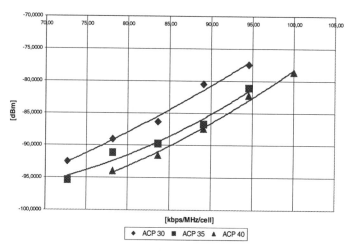

Figure 8.24 The average interference power as a function of load in the uplink.

Increasing the load, the interference increases in a similar way for the various ACP values. The difference in the interference levels is less than 5 dB. Table 8.6 and Table 8.7 show the capacity loss due to ACI for the uplink and the downlink, respectively. The capacity reduction due to the ACP is expressed as percentages of the full capacity that could be obtained in a single-layer or single-operator case.

Table 8.6

Capacity Loss Due to the Adjacent Channel Interference in the Uplink

ACP	Capacity			
	1,800-m cells		5,400-m cells	
	Kbps/MHz/cell	%	Kbps/MHz/cell	%
single operator	99.8	100	71.5	100
50 dB	92.8	93	55.1	77
45 dB	92.8	93	55.1	77
40 dB	90.8	91	53.6	75
35 dB	88.8	89	52.1	73
30 dB	87.8	88	50.1	70

Table 8.7

Capacity Loss Due to the Adjacent Channel Interference in the Downlink

ACP	Capacity			
	1,800-m cells		5,400-m cells	
	Kbps/MHz/cell	%	Kbps/MHz/cell	%
single operator	57.8	100	55.1	100
60 dB	57.6	99	54.4	98
50 dB	57.4	99	54.5	98
40 dB	55.8	96	53.8	97
30 dB	53.2	92	50.9	72

8.3.2.3   Interlayer Case

On the macrolayer downlink, the transmission power is insufficient to overcome the adjacent carrier interference. The call is dropped and this defines an outage area around the adjacent carrier base station. Figure 8.25 illustrates this.

Figure 8.25 Downlink adjacent carrier layer or operator outage area or dead zone.

ACI reduces downlink capacity. Compared to the single layer scenario, ACI increases the transmitted power of the base stations, and the base station may reach its maximum output power more often. On the uplink, the focus is on the interference induced on the microbase station from mobiles connected to the macrobase station.

From the microbase station point of view, ACI randomly varies in time according to interference space distribution around the base stations. Higher interference together with C/I-based power control inflict a performance loss. This forces the terminals to increase the transmit power. As long as the mobiles don't reach maximum power limit, services can be provided. High interference levels may propagate in the network, leading to unstable operation.

In the uplink, coverage and capacity results are strongly related to the time and space distribution of the fast-moving mobiles. The position of the micro-BTS antennas should be as high as possible but still below rooftops (20m is the average height of buildings in major cities), and the user density on the macrolayer should be minimized. To master the macrolayer connected mobiles transmitting power near the microbase station, it is recommended to deploy microbase stations in locations where fast-moving users are well covered by the macrolayer. Use of adaptive antennas in the macrolayer may play a role in reducing the interference.

The simulated loss of capacity [7] is shown in Figure 8.26. The capacity reduction is shown as a function of ACP. The reference is the full capacity in a single-layer case.

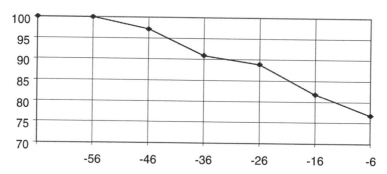

Figure 8.26 Capacity loss in the microlayer uplink as a function of ACP. The reference is a single-layer case.

8.3.3 Conclusions

The practical value of mobile terminal ACLR and ACS value is 32 dBm. Inter-operator results show a capacity loss of around 10 percentage units, with an ACP of 32 dB and with 1,800m cell radius. However, the results strongly depend on the used models. In the interlayer case, the base station location shift plays a major role in capacity loss. Interlayer results show a capacity loss of around 10 percent with an ACP of 32 dB in the uplink with a 1,500m base station shift. But a 32dB ACS leads to dead zones around microbase stations. If there is an escape mechanism that stops UL transmission when in the dead zone, the microlayer uplink performance will not be affected. However, it makes the macrolayer coverage look like a colander. Uncoordinated networks using adjacent frequencies may be the usual spectrum allocation case between operators, but whether HCS will be used as a deployment strategy is not taken for granted.

8.4 CONCLUSIONS

The need for interworking between UTRA and GSM is evident and UTRA/GSM multimode terminals must be competitive from the first day. The specifications for UTRA multimode and single-mode terminals must be the same. The two different UTRA modes have a potential commonality in the RF sections of the terminal, but this tie must be kept all the way throughout the standardization phase to allow efficient terminal implementations.

For the interworking of different modes, methods to perform the different tasks with the same hardware are preferred. In UTRA FDD, both uplink and downlink compressed mode is needed to perform the intersystem/mode measurements.

For UTRA – GSM handover, the use of the compressed mode can be reduced with the availability of timing information between UTRA and GSM. For the handover from UTRA FDD to GSM 1800 and to UTRA TDD, the compressed mode is needed in the uplink direction, due to the close proximity of the frequency bands concluded in this chapter. The impact of such an operation with up to every eighth frame in compressed mode was studied in [2], and it was found that with a proper coding scheme, the impact is marginal for practical ranges (i.e., less than 1dB).

The interference between uplink and downlink in UTRA TDD can occur between two terminals or between two base stations. The results show that within one carrier (one operator) both mobile-to-mobile interference and base station-to-base station can cause problems. The solutions to the problem include base station synchronization, careful network planning, smart antenna techniques, and dynamic channel allocation. Between different operator's TDD networks, the most difficult interference occurs between base stations when the base stations are colocated or located in close proximity. The same problem occurs if TDD is used in the FDD uplink band. These problems are avoided by colocation and coordination between operators including network planning, base station locations, and common synchronization, or by using smart antennas at the base stations.

In this chapter, the physical procedures for handovers between UTRA FDD, UTRA TDD, and GSM are discussed. The simulations show, when UTRA is compared to GSM, a shorter measurement time. This improved performance might be needed to enable very fast handovers in UTRA, but can also be used to reduce the measurement activity in both UTRA and GSM and still perform better than the average GSM-to-GSM handover. There are *several monitoring combinations* that provide seamless handovers between the modes/systems, and different needs at the network management level should set the actual monitoring parameters for the intersystem handover.

In UTRA FDD, the interoperator interference results greatly depend on the used models. The capacity loss is less than 10% in every case. In the interlayer case, the base station location shift plays a major role in capacity loss, and network planning becomes important. Uncoordinated networks using adjacent frequencies may in the future be an usual sight between operators, but an HCS deployment strategy is not self-evident and may require some planning as in GSM.

References

[1]   Gustafsson, M. et al, "Compressed Mode Techniques for Inter-Frequency Measurements in a Wide-band DS-CDMA System," *PIMRC –97 proceedings*, Helsinki Finland, September 1997.

[2]   A.Toskala, O-A.Lehtinen and P.Kinnunen, "UTRA GSM Handover From Physical Layer Perspective," *4th ACTS Mobile Communications Summit Proceedings*, Sorrento, Italy, June 8-11, 1999, pp 57-62.

[3]   O.A. Lehtinen, "UTRA/GSM Multimode Terminal Considerations and Interworking Issues," *FRAMES Workshop Proceedings*, Delft, The Netherlands, Jan. 18-19, 1999, pp 137- 144.

[4]   H. Holma, G. Povey and A. Toskala, "Evaluation of Interference Between Uplink and Downlink in UTRA/TDD," *VTC'99*, Fall, Amsterdam.

[5]   ERC/DEC/(99)HH. http://www.ero.dk

[6]   M. Haardt, A. Klein, W. Mohr, and J. Schindler, "The Physical Layer of the TD-CDMA Based UTRA TDD Mode," *Proc. ACTS Mobile Summit*, Sorrento, Italy, June 1999, pp. 171-176.

[7]   N. Guérin, "Adjacent Carrier Interference in HCS with WCDMA," *FRAMES Workshop Proceedings*, Delft, The Netherlands, Jan. 18-19, 1999.

[8]   S. Hämäläinen, H. Lilja, J. Lokio and M. Leinonen, "Performance of a CDMA Based Hierarchical Cell Structure Network," *IEEE PIMRC'97* Vol.3, Sept. 1997, pp. 863-866.

[9]   S-J.Park, H-B.Ha, J-T. Chung, Y-S. Shim and D-Y.Lee, "Frequency Coordination Between Adjacent Carriers of Two CDMA Operators," *IEEE VTC'96,* Vol.3, April 1996, pp. 1458-1461.

[10] M.Rinne, S. Hämäläinen and H. Lilja, "Effects of Adjacent Channel Interference on WCDMA Capacity," *IEEE ICT'99*, Cheju, Korea, June 15-18, 1999.

[11] S-W.Chen, "Nonlinear Distortion in Digital Wireless Communications," *Wireless Communications Conference*, Aug. 1996, pp. 20-26.

Chapter 9

Mobile Station Positioning

Mobile station (MS) positioning refers to the determination of the coordinates of a desired mobile user. Recently, there has been an increased interest in MS positioning. The primary reason is that the FCC has mandated that in the United States, as of October 2001, 67% of all MS-originated emergency calls must be located with an accuracy of 125m. There are also many other possible applications for MS positioning. The most important criteria, which has to be taken into account is when developing MS positioning systems, include accuracy, coverage, capacity, capacity reduction, and cost. In CDMA, there is an inherent hearability problem to overcome, which is closely related to the well-known, near-far problem. One possible solution to this problem is to include idle slots in the serving BS output, when downlink signals are used for positioning measurements. If uplink signals are used, powering up the MS transmitter power might solve the problem. Both these solutions reduce capacity of other services. This chapter mainly considers UTRA FDD. Note that UTRA TDD can adopt the methods presented herein, with some minor modifications. Due to the harmonized parameters, it can be expected that the performance would be similar. The fact that the nodes are synchronized in UTRA TDD simplifies the problem of obtaining timing, assuming that the synchronization is accurate enough.

The outline of this chapter is as follows. In Section 9.1, a general introduction to the subject of MS positioning is given. Sections 9.2 and 9.3 describe, respectively, some uplink and downlink methods. Section 9.4 evaluates the performance of these methods, and finally, Section 9.5 gives concluding remarks.

9.1 BACKGROUND

A general introduction to the subject of MS positioning is presented in this section.

9.1.1 MS Positioning Applications

Different applications of MS positioning services set different requirements on positioning systems. Some applications require more MSs to be located simultaneously than others, some applications are related to calls and others are not, and of course different applications require different accuracy. Some of the possible applications are briefly described in the following [1].

Government Applications

In some countries, there are plans to legislate certain accuracy requirements for MS positioning. For example, the FCC has decided that in the United States, as of October 1, 2000, 67% of the MS-originated emergency calls must be located with an accuracy of 125m [2]. Positioning of the MS can also be used for law enforcement purposes, for example to locate stolen property, or criminals could be tracked using MS positioning. These applications typically require that only a few MSs are located simultaneously.

Operator Applications

Operators can use MS positioning services to provide new services, but also for network planning purposes and to use the network more efficiently. For example, operators might want to offer calls with a lower price from a certain home area to compete with fixed-line operators. Another possible application is to use MS location data in handover algorithms. Operator applications typically require that many MSs are located simultaneously.

Commercial Applications

These applications consist of pure positioning services offered by the operator, as well as value-added services that are based on the positioning service and offered by some, possibly independent, service provider. The subscriber can use the position provided by the system, for example, in car navigation or to locate himself or herself on the map. Another application is fleet management, which can be utilized by transportation companies. One important class of commercial applications are those who provide information about nearest accommodation, transportation, restaurants, or any other services. Commercial applications may also be related to other telecommunication services. The subscriber could, for example, order a taxi without needing to know its position.

9.1.2   Criteria to Evaluate MS Location Methods

The most obvious criteria to evaluate location methods is accuracy, but it is not the only one. Many different criteria have to be taken into account when developing MS positioning systems. Some of the most important criteria are described in the following.

Accuracy. Different applications have different accuracy requirements. A 500m accuracy could be enough for fleet management applications, but of course the better the accuracy, the better services and more applications can be provided.

There are many statistical measures of accuracy of positioning systems (e.g., root-mean-square error and average error). Descriptions of different accuracy measures and their relationships are found in [3] and [4]. Overall accuracy of cellular-based location system is hard to determine, because the accuracy of the system differs depending on environment and network geometry.

Positioning coverage. The ideal case would be that the coverage area of the MS location system is the same as the coverage area of other cellular services. Near-far effects limit coverage of location methods. Also, in network border areas or in rural networks, it might be impossible to use a sufficient number of BSs.

Capacity reduction of other services. Cellular-based MS positioning system reduces capacity available for other telecommunication services. If specific positioning channels are used or transmitting power of existing channels is increased due to positioning, the capacity of other services decreases. Signaling needed in MS positioning also has to be taken into account.

Response time. Some applications require that position estimates can be calculated very quickly when requested. The response time of the location system is especially important in emergency call positioning.

Capacity of the location service. Determines how many MSs can be located during a certain time interval within some specified area, for example within one cell. A high capacity is needed, for example, for some network planning algorithms.

Cost. Cost of the positioning system for operators, manufacturers, and the subscriber is an important factor in practice.

Power consumption. Extra MS power consumption caused by MS positioning services has to be taken into account in developing the positioning method.

User privacy. The subscriber should have the opportunity to prevent the positioning of his MS. Authorities would also most likely want to have the possibility of locating the MSs, for which positioning is not allowed by the subscriber.

9.1.3 MS Position Estimation Methods

The most promising cellular MS position estimation methods are based on measurements of the propagation delays between the MS to be located and several BSs. Other cellular-based methods to estimate MS position are also briefly described. Cellular-based MS positioning methods and also other positioning systems are discussed in [5]. A more theoretical approach to positioning methods can be found in [5]. Sections 9.2 and 9.3 describe some particular methods in more detail.

9.1.3.1   Time of Arrival (TOA) and Time Difference of Arrival (TDOA) Methods

If there is LOS between the MS and BS_i, the range d_i between them is

$$d_i = c\Delta t_i \tag{9.1}$$

where c is the speed of light, and Δt is propagation delay of the radio signal between MS and BS_i. When the propagation delays between the MS and at least three BSs are known, the position of the MS can be calculated (see Figure 9.1). Measurement and non-LOS (NLOS) errors cause errors to the position estimate.

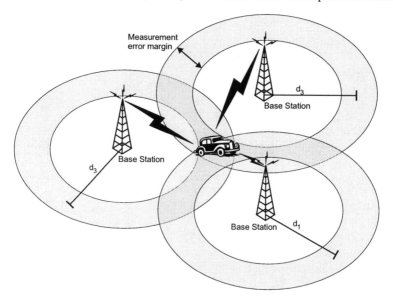

Figure 9.1   The position estimate of the MS based on TOA.

Due to errors, the circles do not generally intersect in one point. Because of this, the position estimate is calculated, for example, by minimizing the distance between the MS and TOA circles in a least-squares sense. With more than three TOA measurements, the accuracy is improved. With two measurements, there is an ambiguity in position of the MS; it is not known in which intersection of the circles the MS is. Different algorithms for calculating the position estimate using TOA or TDOA measurements are described for example in [6] and [7]. TDOA methods are based on the time difference in the reception of the signals from different BSs. If this difference between the MS and base stations BS_1 and BS_2 is

Δt, and there is LOS between the MS and the BSs then MS is located on a hyperbola

$$(d_1\text{-}d_2) = c\Delta t \qquad (9.2)$$

where d_1 is range from MS to BS_1, and d_2 is the range from MS to BS_2 and c is the speed of light. Two TDOAs can be measured with three different base stations, and the MS is located in the intersection of the hyperbolae defined by the TDOAs (see Figure 9.2).

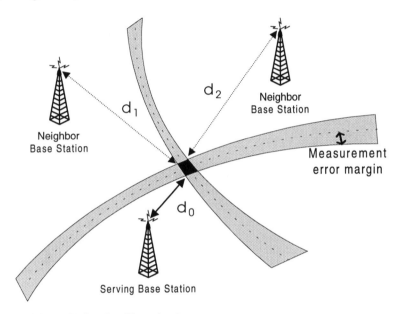

Figure 9.2   TDOA-based position estimation.

Similar to the TOA method, the position estimate can be calculated by minimizing the distance of the MS to the hyperbolae in a least-squares sense, if there are more than two TDOA values available. One advantage of the TDOA method with respect to TOA method is that errors that are common to all BSs sum up to zero, when TDOA values are estimated. Such errors include common multipath delay and synchronization errors.

According to simulations in [8, 9] the accuracy of the TDOA and TOA methods can be improved using more than three measurements. However, in those simulations it was assumed that all measurement and NLOS errors were similarly distributed. In practice, error distribution of different measurements can be different. For example, if measurements with NLOS error could be somehow detected, the accuracy of the location estimate can in some cases improve, if such measurements are not used in location estimation [9]. Knowledge of the distribution of measurement errors, for example variances of different measurements, can be utilized in position estimation to improve accuracy.

The accuracy of TOA and TDOA-based location systems depends highly on the geometric relationship between the MS to be located and the BSs involved in location measurements. This dependence is verified by simulations in [7]. The best results are achieved when the BSs are located as symmetrically as possible with respect to the MS. One measure of the accuracy based on the geometry of the system is geometric dilution of precision (GDOP). GDOP is defined as the ratio of root-mean-square position error to the root-mean-square ranging error [4]. Influences of the network geometry should be taken into account when developing cellular- based positioning systems.

9.1.3.2   Angle of Arrival (AOA) Method

AOA methods are based on the assumption that BSs can measure angles of arrival of signals transmitted by the MS. If there were LOS between the MS and two BSs and AOA measurements were available, the MS would be in the intersection of the lines defined by angles of arrival (see Figure 9.3). In practice, the AOA estimates are erroneous and similar to TOA and TDOA-based methods, error margins of the AOA estimates define error margins for the MS position estimates. As TOA and TDOA methods, more than two measurements can be used to improve accuracy. Multipath causes severe errors, especially when AOA methods are used in urban environments. This method requires that BSs are equipped with antennas that can measure AOA values. Methods to estimate AOA values using antenna arrays are described, for example, in [10].

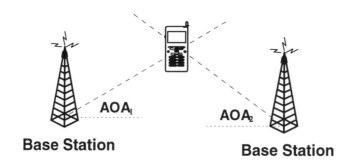

Figure 9.3   AOA-based position estimation.

9.1.3.3   Combination of AOA and TOA Methods

If the AOA and TOA values are both measured with respect to one BS, intersection of the line determined by the AOA value, and the circle determined by TOA value can be used as an estimate of the MS position (see Figure 9.4). The main advantage of this method is that measurements are needed only between the MS and one BS. The main disadvantages of the method are the same as in the AOA method; multipath causes severe errors, especially in urban environments.

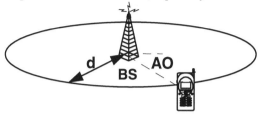

Figure 9.4 TOA- and AOA-based location estimation.

9.1.3.4   Cell Information

Identity of the serving cell gives some information about the position of the MS. Of course, position estimates based on cell identity would not usually be accurate enough, but in the future, urban pico- and nanocells' knowledge of the cell identity might give accuracy that is sufficient for some applications.

9.1.3.5   Received Signal Level

Received signal level is used to estimate the distance between the MS and BSs. However, rapidly changing propagation conditions and differences of the channels in different environments will cause so many problems, that the received signal-level-based method would probably not be a feasible solution in practice.

9.1.3.6   GPS

Although GPS is not actually a cellular-based positioning system, it is presented here because it is the most widely used existing positioning system and it can be integrated with mobile phones. GPS is a satellite-based system whose basic principle is similar to the TOA method. When a GPS receiver hears the spread spectrum signals sent from at least four satellites, it can calculate its position by determining distances to the satellites. The accuracy of GPS is very good. Mean position error of a normal GPS receiver is about 50m and down to 2m with differential GPS. Differential GPS uses information received from fixed GPS receivers to correct position estimation errors [11].

There are several problems with installing a conventional GPS receiver into a MS. A GPS receiver needs a special antenna, which is typically relatively large. Furthermore, the GPS receiver increases the manufacturing price of the MS and the power consumption. There are many locations where GPS doesn't work, because the receiver cannot hear signals from four different GPS satellites (e.g., indoor environments and so-called urban canyons in city centers). One aspect to take into account is that GPS is a US military-controlled system.

9.1.4   Problems in MS Positioning

Hearability

Measurements needed for MS position estimation are done by the MS or the BSs or even by some separate receiver. In order to take the measurements, the measuring unit has to receive the signal with SIR high enough to last a long time. Because of the near-far problem, CDMA systems utilize power control. This means that an MS close to a BS will have a reduced transmitter power. Consequently, other BSs cannot hear the MS. Similarly, an MS cannot hear other BSs when it is close to the serving BS. When TOA or TDOA values are measured by the BSs, one possible solution to the hearability problem is to increase the power of the MS to be located for a certain period of time. This kind of power up function (PUF) has been simulated in [12]. More detailed descriptions of two PUF based methods are presented in Section 9.2. When the MS performs the measurements, periodic idling of the BS can solve the hearability problem. See Section 9.3 for further details. In Section 9.4, the performances of these methods are evaluated and compared.

NLOS

In order for TOA and TDOA methods to work properly, there should in general be LOS between the MS and the BSs involved in the positioning measurements. In practice this is not true, especially in urban environments. Due to multipath, the signals may travel excess path lengths when there is NLOS. This NLOS error could be several hundred meters and can cause errors of the same magnitude to the position estimate [9, 13]. One proposed method to detect and reduce NLOS errors is presented in a paper by M. Wylie and J. Holzman [14]. Their method is based on the assumption that variance of the TOA measurements from or to a BS is higher in an NLOS situation than in a LOS situation. The variance of the LOS measurements has to be known in advance in order to use the method. One assumption of the method is the range between a BS and an MS can be approximated by a low-degree polynomial of time for a certain time interval. It should be verified with real data whether this is true or not. The method of Wylie and Holzman can be used only when the MS is moving. Another limitation of the method is that it can detect and correct the NLOS problem only when the magnitude of the NLOS error is different in different measurements, or when there is both LOS and NLOS situations between the MS and a specific BS at different times. There are some models based on measurements that model NLOS errors statistically (for example the CoDiT model [13]). However, existing channel models do not model the magnitude of NLOS error related to the position of the

MS. Without such models or measurement data, new methods to detect and correct NLOS errors cannot be developed.

Inter-BS Timing

TDOA and TOA methods require that the relative timing of the BSs is known. Usage of round-trip delays, from MS to every BS involved in positioning measurements and back, could be used to avoid this problem. Such a solution requires both uplink and downlink measurements, which means that the hearability problem has to be solved for both uplink and downlink transmissions. This would reduce capacity of other cellular services. Reference MSs with known positions can be used to measure the relative timing of BSs. If the hearability problem is solved by idling the downlink, the receivers can be located in the BSs. GPS is another alternative to get the timing of BSs. In this case, all BSs must have a GPS receiver.

9.1.5 GSM

As a reference, we mention that positioning is currently being standardized for GSM. A TDOA-based uplink method is currently being standardized in T1P1. The method forces the MS into a fake handover. The adjacent BSs then listen for the access bursts transmitted by the MS. After the uplink method is standardized a downlink method, called observed time difference (OTD), will also be standardized.

9.2 UPLINK METHODS

Uplink methods are based on the principle that the MS transmits a high power signal and that the BSs in the vicinity measure the TOA of that signal. With the TOAs available, the position can readily be computed. In the two following sections methods that use known and unknown transmitted data are presented.

9.2.1 Uplink With Known Data

The most straightforward way of designing a positioning system for UTRA FDD is to let the MS transmit a special positioning pulse, or training sequence, at an increased power. In the uplink, the near-far problem can be combated by increasing the transmitter power of the MS and by transmitting the known signal for a long period of time. This has some serious effects on system performance. If the MS transmits a known sequence, it has to replace the normal speech channel with this signal, which will potentially cause a speech interruption. The problems described above will of course be worse if the signal has to be transmitted for a long period of time. Typically, the power increment is quite large (e.g., 25 dB

[12]). Such a large power increment will have some serious effects on the system capacity. See Section 9.4.2 for details.

In the receiver, the received signal is correlated with the known sequence to get the TOA of the transmitted signal. The correlation is performed coherently for a time that is in the order of the coherence time of the channel. Several such correlation outputs are combined noncoherently to avoid the effects of fast-channel fading.

We conclude this section by listing some pros(+) and cons(-) with the known-UL method.

- (+) Positioning is fast since the positioning pulse is relatively short;

- (+) Easy to implement in the BSs;

- (-) Simultaneous positioning of closely located MSs cannot be performed;

- (-) Cannot locate all MSs easily;

- (-) Power increase reduces system capacity;

- (-) Increased MS complexity;

- (-) Possible speech interrupt, since the pulse is much stronger.

9.2.2   Uplink With Unknown Data

In the unknown UL method, the MS transmits its normal data at an increased power. Since the data is useful, the receiver can correlate with the received signal with the code used by the MS for a long time, without causing any speech interrupt. Since the transmitted symbols are unknown, the correlation has to be performed noncoherently at every symbol interval. This will cause a loss compared to coherent combining; on the other hand, the correlation time can be much longer. Because of the increased correlation length, the power increment does not have to be as large.

In the unknown UL method, the MS to be positioned can optionally be ordered to use a *shorter scrambling code* than the LC ordinarily used for scrambling. This shorter scrambling code (SSC) should typically have the time-length of one symbol period. The case when the symbol period is shorter than the SSC is discussed separately at the end of this section. An alternative to this solution is to order the MS not to use any scrambling code at all, but the system may then be susceptible to the possibly bad auto and cross-correlation properties of the short code (SC). Thus, if no scrambling code at all is to be used, the MS may have to be ordered to change it's SC to an SC that is not used in normal operation, so as to make the MS signal identifiable by a nonactive[1] BS.

---

[1] Nonactive BSs are those BSs to which the MS under consideration is not connected. All other BSs are active BSs.

By switching to short codes, the nonactive BSs can find the timing of the MS signal relatively easy by using a simple matched filter matched to SC+SSC (or only to SC if the scrambling code is omitted), noncoherent detection, and combining over a relatively long period of time. Note that the MS is still transmitting its normal traffic data, so there will be no information loss or capacity loss due to a new noninformation-carrying signal being transmitted. Thus, there is not such a need for a radical increase of the MS power, which would cause an increase in interference, and thus a capacity loss and possibly even RF blocking of the active BS. The advantage of using short codes is that it becomes much easier for the nonactive BSs to correlate with the MS signal. Typically, a matched filter with length equal to the length of the code is needed. Additionally, the use of interference cancellation is simplified when short codes are used.

A potential problem with the SC and SSC approach described above is that the obtained timing is only unambiguous up to one symbol interval. However, as we describe in detail below, this is not a major drawback, since the nonactive BSs do not necessarily need to decode the datastream, they merely have to find the timing. When using the obtained timing for positioning purposes, the timing ambiguity causes an ambiguity in the distance. For example, if the chip time is 0.24 µs and the length of the SC is 256 chips, the time ambiguity will correspond to a distance ambiguity of about 18 km. Hence, the ambiguity will only be a problem for large cells, and in that case it can be resolved with, for example, power measurements and sector information. The problem of resolving ambiguities is further discussed at the end of this section.

Variable symbol length: As mentioned above, special care has to be taken in the case when the symbol length is allowed to vary. The problem is that a solution with an (SSC+SC)-matched filter in the nonactive BSs will not readily work, since the SSC is longer than one data symbol (longer than the SC). This case occurs, for example, when the spreading factor is variable. The spreading factor may be varied to allow different data rates. For example, if the slowest data rate is B symbols per second and the corresponding SC has length N, the length of the SC will be $N2^{-k}$ for a data rate equal to $B2^k$. Below are a number of ways to deal with this problem:

1. Order the MS to alter the rate so that the length of the SC is N.
2. Order the MS to use an SSC of length N, without any rate alterations. The nonactive BSs listen to only a part of the MS signal that uses an SC with length N, for instance a physical control channel that may be transmitted along with a data-carrying physical data channel.
3. Order the MS to use an SSC with length equal to $N2^{-k}$ (equal to the length of the SC and the symbol length).

Method 1 may be the most suitable alternative for channels that have a rate close to the slowest rate (e.g., speech), by using rate matching (e.g., code puncturing) and a power increase to make up for the lost coding gain. This is quite convenient since it is likely that a power increment will be used anyway, as mentioned above.

Method 2 can always be used for systems that have a physical channel structure consisting of a variable spreading factor; variable-rate physical data-channel carrying information (physical data channel); and a parallel, fixed spreading factor, fixed rate, parallel-channel-carrying layer 1 control information (physical control channel). The physical control channel rate could be chosen to have a data rate of B symbols per second and thus a corresponding SC of length N. Nonactive BSs can then listen to the physical control channel with a simple matched filter solution. The power of the physical control channel can be increased if needed to secure hearability in nonactive BSs.

Method 3 works well as long as the SC length (symbol-length), and thus also the SSC length, is long enough not to make the time-ambiguity in the nonactive BSs unmanageable. This is strongly related to the cell size and structure. Apart from the power measurements mentioned previously, there are a few other factors that will help resolving the ambiguity, for example

- If the cell planning is sectored, a crude direction estimate will be obtained.

- Measuring the received MS power in several different sectors at the same site can refine the direction estimate obtained from the sector information. The direction estimate is obtained from the antenna angular response and direction. If one or several of the BSs have an antenna array, the direction estimate will be even better.

- For BSs in the active set, it is possible to estimate the round-trip delay, and thus the distance, to the MS.

- For BSs with a reasonably good C/I value, when symbol synchronization is achieved, it might be possible to detect the data bits, and hence achieve slot and/or frame synchronization.

- The likelihood of choosing the correct location estimate can be improved also by a purely computational method. If we have a round-trip measurement for at least one link, all the TOAs can be converted to round-trip measurements. Thus, we have a set of possible radii from each BS. In the ideal case, the corresponding circles would intersect in one point, and that point is the MS position. The basic idea to resolve the ambiguity is that in the generic case, only the circles corresponding to the true radii will intersect in exactly one point, and hence the ambiguity would be resolved. In practice, we have measurement errors and even the circles with resolved ambiguity would not intersect exactly in the same point. However, in this case, it is possible to devise a cost-function, which is a function of one radius from each BS and the MS position. The minimum of this cost-function with a set of fixed radii would yield an estimated position of the MS. One of these minima would be obtained for each combination of the radii. It is likely that the combination yielding the smallest minimum would correspond to the radii with resolved ambiguities. If the SSC is very short, there would be a large number of combinations. The number of combinations can be reduced by first applying the ambiguity resolving methods described previously. The method described above can considerably reduce the effects of ambiguities.

We finally note that if the MS user is using a packet channel (i.e., it is not transmitting data all the time), the cellular system must fix a channel on which the MS transmits continuously during the positioning (e.g., a physical control channel).

We conclude this section by listing some pros and cons with unknown UL

- (+) Since the positioning signal is user data it can be transmitted during a longer period of time;

- (+) No special positioning pulse is needed;

- (+) Small MS complexity;

- (+) Simple to implement interference rejection;

- (-) Simultaneous positioning of closely located MSs cannot be performed;

- (-) Cannot locate all MS easily;

- (-) Power increase reduces system capacity.

9.3   DOWNLINK METHODS

Downlink methods are based on the principle that the MS estimates the TOAs of signals transmitted by adjacent BSs. To combat the near-far problem, it is possible to use DTX in the BSs. A method called idle slot downlink (IS-DL) positioning is presented below. In the method, the signal transmitted by the BSs is the normal synchronization channels (primary and secondary).

As mentioned above, in a normal DL positioning situation, there will be a severe hearability problem if the MS is close to the serving BS. The timing of other BS cannot be found unless data is collected during a very long time. It is likely that the required time would be longer than the time during which the MS can be considered stationary and the BS clocks can be considered stable. Simulations have shown that it would indeed be very difficult, if not impossible, to satisfy a 67%, 125m accuracy requirement (proposed by FCC for E911 services). By using idle slots in the DL (i.e., the BSs cease *all* transmission for a period of time), the hearability problem is drastically reduced. The idle time-intervals should preferably be made quite short and distributed in time, in order to minimize the effect on BER and FER. There is a slight capacity reduction due to the idle intervals. See Section 9.4.2 for details.

In order to reduce the complexity of the TOA estimation algorithm, it is essential that the MS knows the timing of the slotting in advance. This can be achieved in the following ways:

- **Regular intervals.** The slotting is performed regularly and continuously. Since the BSs are asynchronous, the slotting of adjacent BSs might overlap, thus affecting the performance. This can be avoided by coordinating the slotting times between adjacent BSs.

- **Pseudorandom slotting.** The slotting follows a pseudorandom slotting pattern, known in advance by the MS. This procedure has the advantage that it reduces the probability that the slotting of the BSs involved overlap.

- **On demand.** If location requests are infrequent it is unnecessary to perform slotting continuously.

Assuming that the MS knows the timing of the slotting, there are a few different ways the TOA estimation can be implemented. Basically, the choice of TOA estimation is a trade-off between MS complexity and positioning response time. The following examples illustrate different ways the TOA algorithm can be implemented.

- The MS records and saves the received data during the slotting for subsequent postprocessing. By doing this, the correlations and noncoherent combining can be performed during the time between the idle periods. The obvious disadvantage is of course the additional complexity (e.g., memory) required in the MS. Accurate estimates should be obtained from a modest number of idle slots.

- The MS performs correlation and noncoherent combining on-line. Due to the limited number of matched filters and sliding correlators available in the MS, it is only possible to correlate with one or a few BSs at a time during the slotting. This means that a large number of idle slots will have to be used, resulting in delayed estimates.

The time required for positioning an MS can be significantly reduced if the MS tries to find the TOA of other BS signals not only when being positioned. In this case, the MS simply delivers the already available TOAs when a location request is ordered. This would likely demand only a small extra computational complexity in the MS, since it searches for other BSs for soft handover purposes anyway.

In order to speed up the search for other BSs, the MS can be provided with a list of likely candidates. This list can be based on, for example, prior information such as a rough initial position and previous BSs successfully involved in positioning in that region.

In the discussions above it was assumed that the relative timing of the BSs is known. It can be obtained, for example, with GPS. However, IS-DL makes it possible to get inter-BS timing in a more attractive way. If every BS has a reference MS with known position, preferably at the BS site, which listens for the other BSs synchronization channels, inter-BS timing can be obtained. Hence, GPS is not required. This is an advantage that is unique to IS-DL; it is not feasible with UL methods. Furthermore, a reference MS does not have to be installed into every BS, since it is enough to get timing between two BSs with one reference MS. This is advantageous, since it might be undesirable to install a reference MS, or a GPS, into, for example, micro-BSs.

The following example illustrates some of the conceptual differences between UL and DL methods. An UL method that uses a 25-dB power-up during one frame will cause all frames transmitted by other mobiles at the same time to be lost. Since the frames are not aligned, some MSs will lose two frames. This is the effect of positioning only *one* MS. The power-up will cause an *increased* average interference level in the system. Similarly, for IS-DL methods that slot an entire frame at the time, all frames in the DL will be lost. However, during this slotting, *all* MSs belonging to the slotting BS can be positioned and the average interference level in the system is *decreased*. Additionally, an IS-DL method will typically use shorter slotting intervals and distribute them in time to reduce the FER and BER effects further.

We conclude this section by listing some pros and cons with IS-DL, as compared to UL methods:

- (+) All MSs can be positioned simultaneously. This is useful for location statistics for network planning, dynamic network control, and so forth. For UL methods, only one MS at a time can be located in an area spanning several sites.

- (+) The use of GPS or other external timing references is not needed.

- (+) The over-the-air capacity reduction is limited and constant (neglecting signaling). For UL methods, the capacity reduction increases with the number of positioning requests. For applications that require repeated updates (e.g. navigation and tracking), UL methods would cause a significant load on the system.

- (+) Performance is adjustable. By changing how often the slotting is performed the operator can make a trade-off between positioning performance and capacity reduction.

- (+) The slotting can possibly be used for interfrequency measurements.

- (-) MS complexity. The extent of additional MS complexity needs further investigation. It can however, be expected that the amount of additional MS complexity is modest, since the computations are similar to those performed in normal cell-search. It should also be noted that some UL methods (e.g., [12]), would cause increased MS complexity.

9.4 COMPARISON AND EVALUATION

In this section, the methods described in the previous sections are evaluated and compared. Many of the evaluation criteria discussed in Section 9.1.2 are discussed in the previous sections. In this section, we investigate the position accuracy and the capacity reduction.

9.4.1   Position Accuracy

When carrying out the simulations, much attention has been focused on the channel model. Important factors to consider when selecting a channel model for positioning include the following:

- The channel model should be based on physical, measurable parameters. Such parameters include delay spread, angle of arrival, fading statistics, and power delay profile shape.

- Delay spread is important, due to the fact that many positioning techniques use TOA estimates to position the mobile. The accuracy of these estimates depends on the delay spread of the channel impulse response. Therefore, the delay spreads generated by the model should conform to measurements.

- The model should represent the general channel behavior in a range of typical environments, corresponding to geographically diverse conditions.

The channel model used is a version of the CODIT model, slightly modified to conform to empirical channel measurements [14]. The CODIT model is a measurement-based wideband channel model, defined for different environments such as urban, suburban, rural, and so forth. A 6-dB lognormal slow fading is added.

The simulation setup used to simulate the various positioning methods is described in the following. The air interface follows [15]. The algorithms are tested in two environments, urban and suburban. The BSs have three sectors and are located in a hexagonal pattern and the closest inter-BS distance is 1.5 km for urban and 6 km for suburban. Mobiles are randomly placed until full system load is achieved. Each mobile has a speed of 50 km/h. The received signals (transmitted signal plus interference and noise) are filtered and sampled with a 4-bit A/D converter in the MS.

For the known UL method, the MS replaces one frame with a training sequence, transmitted at a power increased with 25 dB. Coherent correlation is used during one slot at a time. The results of these correlations are combined noncoherently (16 slots). This is to avoid effects of channel fast fading.

For the unknown UL method, the power is increased with 10 dB during 0.5s. The correlation is performed coherently during each symbol. These correlation results are combined noncoherently.

For the IS-DL method, two different numbers of measurements are used to position the MSs, 16 and 48, respectively. The measurements are performed during the idle slots. In the case with 48 measurements, the TOAs are estimated from 16 measurements and three of these TOA estimates are combined. The slotting times are distributed to get a total of one frame idle time during one second.

The results obtained when positioning 300 randomly selected MSs are presented in Figure 9.5. The curves are the circular probabilities as a function of the radius. In other words, the probability that the estimated position lies within a certain distance from the true position. In the figure, the 67%, 125m requirement is shown by straight lines.

We see in Figure 9.5 all the presented methods fulfill the 67%, 125m requirement. For IS-DL, the case with 16 idle periods is enough to get good position estimates. When the number of idle periods is increased to 48, the performance is improved.

The performance is better in the suburban environment for both UL and DL. This is due to the fact that there are fewer cases with NLOS than in the urban case. By comparing the results for the different methods, one can draw the conclusion that there is no major difference in performance between IS-DL and the different UL methods.

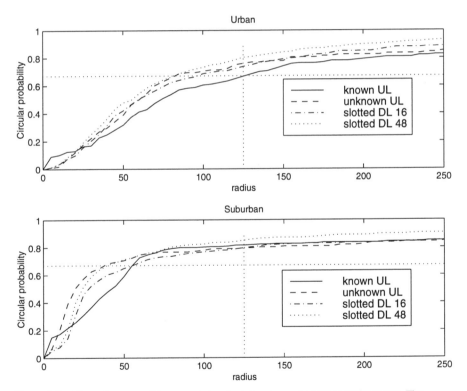

Figure 9.5 The performance of the presented methods in urban and suburban environments. The curves are the circular probabilities (i.e., the probability that the estimated MS position is a certain distance from the true position).

9.4.2   Capacity Reduction

Downlink

In order to ensure that the MS can hear the other BSs, it is necessary that the idle period covers at least one slot. The periods can, however, be divided so that the first idle period covers the first half of the slot, and the next one covers the second half. Of course, the length of the idle periods should be kept as small as possible in order to minimize the capacity reduction. There is however a trade-off, because each idle period contains a guard interval, and thus the more idle periods, the more total guard time. In this book we consider three idle period lengths. These are 2, 5, and 10-symbol intervals, plus a guard time of 1 symbol. The symbols are assumed to be 256 chips long.

Link simulations using the COSSAP chain have been carried out to investigate the impact of the idle periods. In Figure 9.6 below, the BER/FER is shown for different lengths of the idle periods. The idle period reoccurrence rate is 10 Hz. For the cases when the idle period is shorter than one slot, the reoccurrence denotes the frequency with which one entire slot is covered. For example, when the idle period length is 6 symbols, the actual idle period frequency is 20 Hz. In other words, for the idle length of 11, there will be one 11-symbols-long idle period every 10th frame, and for the idle length of 6, there will be one 6-symbols-long idle period every 5th frame. The environment is vehicular A (see e.g., [15]) and the service is speech, using 10-ms interleaving. Due to the large simulation times involved, no other environment or service was simulated. There is no reason to believe that there would be any major differences for other cases. Contrary, it can be expected that for larger interleaving lengths the impact will be smaller. In the simulation, the MS does not know that there is an idle period. By utilizing this knowledge, the performance could be slightly improved. Furthermore, the effect of the decrease of overall system interference, due to the idle slots, is not taken into account.

From Figure 9.6, we see that there is about a 0.1dB (about 2%) loss in BER and FER. The differences between the different idle period lengths are very small.

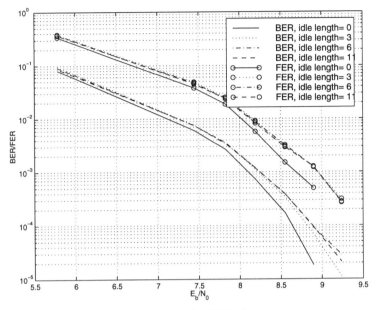

Figure 9.6   Example of BER/FER performance using IS-DL scheme.

Uplink

The simulations carried out in this section are primarily for the uplink method with unknown data (see Section 9.2.2). The results should, however, be quite similar for all uplink methods.

When performing uplink positioning, at least three BSs must receive transmission from the MS. But when the MS is close to the serving BS, its power is kept low, so as not to cause strong interference to the others users. In this situation, it's possible that only the closest BS hears the MS. One possible solution, as discussed above, is to increase the power of the MS to be located for a certain period of time. Of course, the price to pay for this power-up function, is capacity reduction. Interference cancellation can help to enhance the final performance.

The system-level simulator models a CDMA network, with several BSs. During the simulations, MSs are placed randomly in the system area, every 400 ms, using uniform distribution. After that, moving of the MSs and power control is simulated for 400 ms and after that, the MSs are again placed randomly. During each of 400-ms period, the same amount of MSs sends a location signal. The MSs sending location signals are chosen randomly before each 400-ms simulation period. The starting time of location signal transmission of each MS is also chosen randomly using even distribution. At the receiver, interference cancellation is simulated, so that the interference caused by the location signals is reduced by a certain factor.

The number of the base stations used in the simulations was 25. The simulator calculates the outage of the system based on a certain SIR threshold. Outage of the system is the percentage of MSs, which do not receive transmission from BSs with SIR above the defined threshold. Power control, soft handover, and small and large-scale fading are all modeled in the simulator. We took the link-level assumptions and large-scale fading model from ETSI simulations [20] (see Table 9.1). Large scale fading L_p is modeled by equation $L_p=239.33+36log_{10}(d)+\sigma$, where d is distance from BS and σ is a lognormally distributed random variable. Small-scale fading is modeled by Rayleigh distribution.

Table 9.1

Link-Level Parameters

Uplink target SIR	6.5 dB
Chip rate	4.096 Mcps
Distance between BSs	5 km
Max mobile speed	120 km/h
Noise power	-102 dBm
Large-scale fading variance	8 dB

Other parameters can be set and the most interesting ones are given in Table 9.2.

Table 9.2

More Link Level Parameters

Target outage level	5 %
Interference cancellation gain	0.2, 0.3, 0.5
Duration of location signal	400 ms

The number of MSs to be located and magnitude of power rise used for location signal have been changed in the different simulations. Table 9.3 contains the results and specifies these variables. In particular, we were interested to know what capacity can be achieved with IC when either 10 or 25 dB are used as power increase for the location signal. The IC gain depends on the power of the signal to be canceled, and expresses how much of the interfering signal strength remains after cancellation. Of course, IC gain depends also on the speed of the MS, which is connected to the fading phenomenon. Despite the different strengths of the interfering signals at the receiver, we used a fixed IC gain value that can be read as the mean reduction of the signal energy. The results presented were obtained with

IC gain equal to 0.2, 0.3, and 0.5. We have tested several IC gain values just to have an idea about how the capacity gain is sensitive to different possible cases. Since interference cancellation is based on the estimation of the estimated signal, it is clear that when these signals are powerful, detection and estimation are easier unless the number of interfering signals is too big. In our case, the signals that are canceled are the ones used for positioning purposes, and thus are transmitted with increased power. This should allow us to consider even a strong cancellation factor as 0.2.

As we would expect, raising the number of mobiles to be located, and consequently the interference caused by the location signals, caused capacity to get worse and worse. On the other hand, the use of interference cancellation helps to make the capacity loss less severe. Nevertheless, if the system is too loaded, even IC doesn't work so well. In our simulator we implemented an ideal interference canceller. The capacity loss is defined as the difference between the number of MSs in the whole system when power of the signal is increased for positioning purposes. Note that the temporary capacity loss in the serving cell can be worse. Table 9.3 shows some results with IC gain equal to 0.2.

Table 9.3

Percentage of capacity lost using high-power location signal and recovered after IC with gain equal to 0.2.

Number of MSs to be located in 400 ms	Location signal power increase	Capacity loss due to uplink location signals	Percentage of capacity recovered with IC gain = 0.2
1	10 dB	0.65%	0.46%
10	10 dB	4.7 %	3.43%
1	25 dB	7 %	5%
10	25 dB	22 %	11.8%

Other results for IC gain equal to 0.3 and 0.5 are shown in Table 9.4 and Table 9.5. The case of 25-dB power increase was not considered, since only half of capacity loss is recovered when IC gain is 0.2.

Table 9.4

Percentage of capacity lost using high-power location signal and recovered after IC with gain equal to 0.3.

Number of MSs to be located in 400 ms	Location signal power increase	Capacity loss due to uplink location signals	Percentage of capacity recovered with IC gain = 0.3
1	10 dB	0.65%	0.26%
10	10 dB	4.7%	3%

Table 9.5

Percentage of capacity lost using high-power location signal and recovered after IC with gain equal to 0.5.

Num. of MSs to be located in 400 ms	Location signal power increase	Capacity loss due to uplink location signals	Percentage of capacity recovered with IC gain = 0.5
1	10 dB	0.65%	0.001%
10	10 dB	4.7 %	1.92%

The capacity after IC in the case of a location signal with 25-dB power increase is too degraded when too many MSs ask for positioning. Therefore, the use of such a high power should be avoided in order to find a trade-off between capacity of location services and the capacity of other services.

When we use a 10-dB power increase, the performance seems to be quite good, especially if we have an IC gain lower than 0.5.

9.5   SUMMARY

Three TDOA-based positioning methods for the UTRA FDD mode were presented. The downlink method, IS-DL, has a better performance than the two uplink methods. The difference is, however, not major. Hence, the choice of method should be based on other considerations. Some key issues that influence the choice include:

- Expected number of location requests per cell and time unit. The more requests, the more favorable the DL methods become;

- Relative timing issues. The use of reference mobiles is greatly simplified in IS-DL, primarily since it can be located in the BS;

- MS complexity.

For the UTRA TDD mode, the presented methods can be used with some minor modifications. Due to the harmonized parameters, it can be expected that the performance is similar. If the synchronization is accurate enough, the problem of known timing is solved.

References

[1]   M. Silventoinen and T. Rantalainen, "Anytime, Anywhere...Big Brother is Watching You," *Mobile Europe*, Vol 5, pp. 43-50.

[2]   FCC Report and Order, CC Docket No. 94-102.

[3]   F. van Diggelen, "GPS Accuracy: Lies, Damn Lies and Statistics," *GPS World*, January 1998.

[4]   D. J. Torrieri, "Statistical Theory of Passive Location Systems," *IEEE Transactions on Aerospace and Electronic Systems*, Vol. AES-20, No. 2, March 1984.

[5]   T.S. Rappaport, J.H. Reed and B.D. Woerner, "Position Location Using Wireless Communication on Highways of the Future," *IEEE Commuinications Magazine*, October 1996.

[6]   H. Saarnisaari, "Mobile Positioning Algorithms," AC090/OUL/PT/PI/I/013/a1, FRAMES contribution for PT Workpackage, 1998.

[7]   G.A. Mizusawa, "Performance of Hyperbolic Position Location Techniques for Code Division Multiple Access," Master's Thesis, Virginia Polytechnic Institute and State University, Virginia, 1996. http://borg.lib.vt.edu/theses/public/etd-447221779662291/etd-title.htm

[8]   P. Lundqvist, "Evaluation Sheet for the Uplink TOA Positioning Method," T1P1.5/97.034.

[9]   J.J. Caffery and G. L. Stüber, "Radio Location in Urban CDMA Microcells," *Proceedings of IEEE PIMRC* 1995, Vol. 2.

[10]   H. Krim and M. Viberg, "Two Decades of Array Signal Processing Research," *IEEE Signal Processing Magazine*, July 1996.

[11]   E. D. Kaplan, ed., *Understanding GPS - Principles and Applications*, Norwood MA: Artech House, 1996.

[12]   Tdoc SMG2 UMTS A36/97, "Mobile Station Location Using Power Up Function," Motorola, September 1997.

[13]   J. Jimenez, et al., "Final Propagation Model" R2020/TDE/PS/DS/P/040/b1, June 1994.

[14]   M.P. Wylie and J. Holzman, "The Non-Line of Sight Problem in Mobile Location Estimate," 1996 *International Conference on Universal Personal Communications*.

[15]   Tdoc SMG2 359/97, "Concept Group Alpha - Wideband Direct Sequence CDMA: Evaluation Document," Draft 3.0.

T1P1 documents can be downloaded from WWW sites:
http://www.t1.org/index/0515.htm   (from 1997)
http://www.t1.org/index/0509.htm   (from 1998)

Chapter 10

TDD Demonstrator

ACTS FRAMES is a radio project centered on defining, specifying, and validating an air interface for UMTS. The project developed a unified multimode air interface which was designed to meet the needs of third generation mobile systems. The elements of the air interfaces were evaluated within the ETSI concept groups, and after extensive debate, an agreement was reached by consensus within ETSI on its radio interface for the third generation mobile system. The solution, called UTRA, draws on both WCDMA and TD-CDMA technologies.

Within the FRAMES project, a hardware demonstrator was developed to help in the validation and evaluation process of aspects of the UTRA radio interface [1]. As a result of the decision within ETSI:

- In the paired band (FDD) of UMTS, the system adopts the radio access technique formerly proposed by the WCDMA group.

- In the unpaired band (TDD), the UMTS system adopts the radio access technique proposed by the TD-CDMA group.

The demonstrator was adapted from its original concept of TDMA with and without spreading ([2, 3]). This chapter outlines the development of the TDD TD-CDMA demonstrator ([4, 5]), describes the key system concepts and technologies used in the demonstrator, and provides a brief history of the project to date. The main objective of the demonstrator was to allow the evaluation of the TD-CDMA techniques in the real world, and to allow joint application trials with other ACTS projects.

The demonstrator hardware provides a flexible and powerful signal processor platform for both bench trials and trials in the field. The TDD TD-CDMA test-bed development is twofold. First, it will implement a real-time hardware platform according to the system specification. The test-bed was adapted to the TDD TD-CDMA as proposed in the Delta Concept Group evaluation report [6] from ETSI SMG2, and then to demonstrate the feasibility of the proposed radio interface to gain the required credibility for the proposal. Furthermore, successful demonstrations and trials with well-selected applications will be a key factor in speeding up the acceptance of the UMTS as a standard.

10.1 FUNCTIONAL DESCRIPTION

The demonstrator consists of two mobile stations, one base station, and one control terminal (Figure 10.1, [7]). The air interface is TDD TD-CDMA and the external access interface to fixed networks is ATM/AAL5 for all three stations.

The Central Control Terminal (CCT) provides the man/machine interface to the demonstrator. From this graphical interface, both mobiles and the base stations are operated. The CCT is composed of BCT (bearer) and ACT (analysis), which are respectively, the control and analysis interfaces of the demonstrator. The BCT provides the demonstrator with the different operation mode configurations for both mobiles and base stations, together with monitoring of the system operations. The message flow resulting from the operation of the system is displayed to the user to analyze the performance and validate the system integrity. From ACT, the data stored in the system, to analyze the performance validity, is acquired.

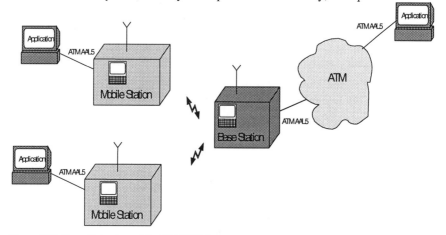

Figure 10.1 Overall structure of the FRAMES demonstrator.

The functionality of the mobiles and base stations was designed in a layered way and an overview of this structure is shown in Figure 10.2.

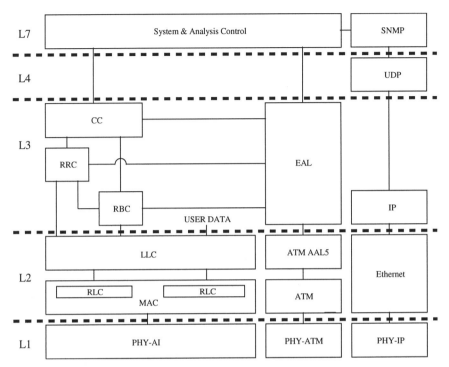

Figure 10.2 An overview of the layer structure.

10.2 LAYER 1 - PHYSICAL LAYER

The physical layer is responsible for the transmission of user data and signaling over the TD-CDMA air interface. It is also responsible for MS synchronization and performing air interface-related measurements.

TD-CDMA frame and slot structure is based on a fundamental TDMA structure with a spreading feature, with a time frame length of 4.615 ms (60/13 ms). The time frame is divided into 1/8 slots of length 577 ms. Up to 8 CDMA codes can be assigned within a single timeslot. The basic way of supporting multiple bit rates is to assign multiple slots and/or codes to a user. The bit rate for a specific service is fine tuned by selecting an appropriate combination of burst type, number of slots, and coding rate.

The data modulation used is QPSK, or (for the highest data rates) quaternary QAM (sometimes referred to as 16QAM) is chosen. This higher level modulation is a flexible way of extending the modulation to support higher bit rates without increasing the bandwidth requirement. The bandwidth efficiency is doubled compared to QPSK.

The spreading modulation is linearized GMSK. It is the spreading modulation that determines the spectral characteristics of the transmitted signals. The spreading modulation using linearized GMSK provides a near constant-envelope transmit signal when combined with QPSK data modulation.

TD-CDMA supports the joint detection of the user signals, which are active within the same timeslot and the same frequency band in the same cell. Furthermore, TD-CDMA supports joint detection of user signals that are active within different cells at the same time in the same frequency band. Joint detection can be used both in the base station and in the mobile to jointly detect the user signal with the worst interferes. It requires training sequences carefully selected to support simultaneous estimation of the channel impulse responses from several received bursts.

10.3 LAYER 2

10.3.1 Radio Link Control/Medium Access Control

The MAC entity is common to all bearers; however, the RLC functions of the RLC/MAC protocol are bearer-specific, and RLC entities for each bearer are created during bearer setup.

The MAC entity is responsible for handling the initial access procedure and resource allocation to bearers.

The purpose of the radio link control procedure is to assure good enough signal quality using FEC and ARQ techniques. In addition, there may be control tasks dedicated to the RLC. Many of these are the closed-loop type involving both MS and BS.

For each bearer connection there is one RLC entity. Control channels, which are maintained between active data transmissions, are also considered as separate bearer connections.

There are two MAC/RLC operating modes, real time (RT) and nonreal time (NRT). The former is used for radio bearers that have severe delay variation constraints and the quality is mainly fulfilled by forward error correction and power control. The latter is used for radio bearers with relaxed delay variations that allow the use of backward error correction. The main difference between the two is that for RT services, the resource allocation is made for the duration of the call (unless modified by the bearer modification procedure), whereas for NRT services the allocation is made for a relatively short period of time. The resource allocation and release mechanisms for the two operating modes are described below.

LLC Header

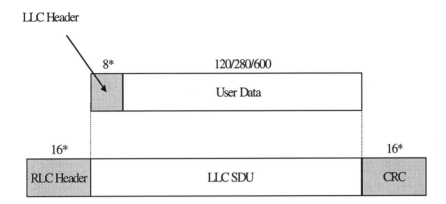

Figure 10.3 LLC and RLC PDU mapping.

The lengths of the LLC and RLC headers (Figure 10.3) and the RLC CRC field are set to 0 for RT bearer services.

RT Operating Mode

Capacity for RT services is allocated in a circuit-switched manner (i.e., the capacity is allocated for a bearer until a specific release procedure is executed). The MAC resource allocation function administers resources in the RT mode by allocating the bearer the maximum resources that the call requires, and the call retains exclusive use of those resources throughout its duration even if only a fraction are used for most of the time.

NRT Operating Mode

Capacity allocation for an NRT bearer is made for relatively short periods of time, and the resource is automatically released at the end of the allocation period [8] in the FRAMES demonstrator. This is referred to as immediate allocation and is used where type I selective retransmission ARQ is employed. It is the transmitting end that selects what is to be sent and the receiving end acknowledges, requesting retransmission of what has been missed.

The BS MAC makes allocations of NRT resources to MS for uplink or downlink data transmission. These allocations are made for relatively short periods of time, specified by the MAC resource allocation process at the time of allocation. Allocations take the form of sets of slots/codes. Each set, over an interleaving period, represents an ARQ retransmission unit (RLC PDU). The allocation period is equivalent to one or more interleaving periods.

10.3.2   Logical Link Layer (LLC)

Within CTSYS two LLC modes of operation have been defined: HDLC mode and minimum mode. In the FRAMES demonstrator only minimum mode is supported. In minimum mode the main function of the LLC sublayer is to relay the data to the lower or upper layer and provide flow control between the LLC and the RLC. The functions of the LLC are summarized as follows:

- Establishing, releasing, and maintaining data links between the MS and BS.

- The transfer of data between the two ends of the data link.

- The operation of flow control within the layer 2 data link and with layer 3. The LLC, as the upper segment of layer 2, acts as a gateway for the transfer of all information between layers 2 and 3.

- The segmentation of received user data (LLC PDUs) onto LLC SDUs, which are passed to the RLC/MAC, and the reassembly of LLC SDUs received from the RLC/MAC into LLC SDUs.

- Relay transfer of primitives between RLC/MAC and layer 3. There are some functions performed by the RLC/MAC that depend upon information passed from layer 3, for example, paging and broadcast channel message formatting and transmission. Similarly, measurements may be passed by RLC/MAC to layer 3. In the layer model adopted, the LLC acts as a relay between RLC/MAC and layer 3 for these messages.

In minimum mode, the LLC provides flow control between itself and the RLC/MAC sublayer. The LLC is restricted in the number of LLC-SDUs it may transfer to the RLC without receiving any acknowledgments. It must then wait for an acknowledgment prior to sending a further LLC-SDU. However, transfer of data to the LLC is unconstrained (i.e., PDUs can be transferred to LLC as soon as they become available from the RLC although they must be transferred in the correct order). There is no application of HDLC framing and CRC error detection, since it is assumed that the RLC provides the required QoS. The LLC also provides rudimentary flow control between itself and layer 3, insofar as it will discard all LLC PDUs received beyond its buffer limit.

The LLC behavior is dependent on whether a RT or NRT bearer service has been selected. For the RT bearer services the LLC does not perform its segmentation/reassemble function, and there must be a one-to-one mapping between a LLC PDU and RLC PDU. The LLC also does not add its LLC header in the case of RT bearer services. Therefore, the LLC must receive user data in blocks of 160, 640 or 1,440 bits depending on the bearer service selected. Nominally, one such block will be received every 10 ms, although some buffering will be provided.

For the NRT bearer services, the LLC must receive user data in blocks that are multiples of either 120, 280, or 600 bits.

10.4 LAYER 3

10.4.1 Radio Resource Control (RRC)

The RRC has the overall responsibility for the radio resources in the FRAMES demonstrator. The RRC is the protocol used mainly for conveying information for radio resource management algorithms that are located in MS and BS. This information includes initial parameter settings and measurement reports.

RRC-level paging is required when the system control requests the BS to set up a bearer to a MS, and the MS is in idle mode (i.e., where no radio bearer exists). The BS first sends the paging message to the MS, which responds with a paging response message.

Radio Bearer Control (RBC)

The RBC protocol is responsible for the setup, maintenance, reconfiguration, and release of radio bearers.

Call Control

Call Control controls the establishment and release of all calls over the radio interface. It provides the signaling protocol between MS and BS that is required to manage all connections.

External Adaptation Layer (EAL):

This is the interface between FRAMES and network demonstrators. The information to be passed over the interface is formatted into ATM cells and routed by allocation of VCI/VPI (Virtual Circuit Identifier/Virtual Path Identifier) addresses. Some messages are addressed to the FRAMES control processor while others are intended to pass over the air unaffected by FRAMES.

The higher layer primitives are classified into 3 categories:

- **Control**: these messages concern local commands between both parts of the interface that are not transferred across the air interface.

- **Signaling**: these messages concern "end-to-end" transport of higher layer signaling messages across the FRAMES air interface. They do not have to be interpreted by FRAMES.

- **User data**: these messages provide an "end-to-end" transport of user-related information across the FRAMES air interface.

Each primitive is given a VCI address. There are 3 kinds of VCI (one per primitive category), but all primitives are transmitted on a single VPI.

RAINBOW network demonstrators implement a FRAMES adaptation layer (FAL) to interface with FRAMES through ATM/AAL5 standard. On the other hand, FRAMES implements a common EAL to interface with network demonstrators. It is pointed out that the exchange of information between demonstrators is based on AAL5 CPCS SDUs standard, but this does not necessarily imply that ATM cells are transparently transmitted over the radio interface. See Figure 10.4.

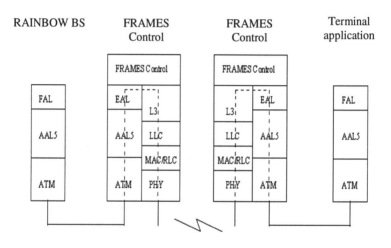

Figure 10.4 The logical interconnection of EAL and FAL interfaces.

In standalone demonstration, FRAMES runs its own call-control, bearer-control, and system-control software. These are made up by pulling together a number of low-level control modules, which provide the basic functionality of the FRAMES air interface. During joint demonstrations, most of the local control functionality is replaced by the joint projects, which use the EAL to access the air interface. A number of control messages have been defined for use by the joint demonstration projects. These messages provide functions such as call initiation and clear-down. Figure 10.5 shows all the messages and states supported by this interface.

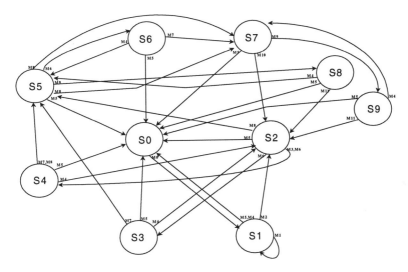

Figure 10.5 EAL state diagram.

10.5 LAYER 7 - SYSTEM CONTROL AND ANALYSIS

This module is responsible for configuring the MS/BS during system startup, processing commands received from the CCT, and generally tracking the status of the MS/BS.

The CCT provides the man/machine interface to the demonstrator. The CTT consist of a graphical user interface (GUI) and simple network management protocol (SNMP) manager that are written in Java. The manager interacts with the agent, compiled in C code running on the control unit, by means of the SNMP protocol. The method routines defined for the actions performed when a leaf in the MIB is accessed, send messages to other tasks in the system to indicate that a relevant data structure has been changed by the CCT. The logical loop is therefore completed, and the operation requested from the CCT is performed by the system.

10.6 ARCHITECTURE

The FRAMES demonstrator consists of two mobile stations and one base station, and one or more control terminals (Figure 10.6). The connections between stations and control equipment are done through Ethernet networks.

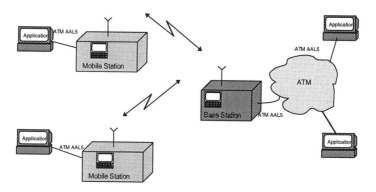

Figure 10.6 Overall structure of the FRAMES demonstrator.

The FRAMES demonstrator system provides a test-bed for evaluation and validation of aspects of the UTRA concept by analysis, lab tests and field trials, and also provides joint demonstrations with other ACTS projects [9], [10].

The hardware platform has been used to validate the basic functionality of TDD TD-CDMA. In addition to technical validation, the platform will be used to demonstrate the proposed concept with a selected set of applications. These demonstrations consist of the FRAMES internal demonstration campaign with user trials, and joint trials with the ACTS project RAINBOW. The main validation goals for the FRAMES demonstrator of TDD TD-CDMA are:

- High bit rate data;

- Efficient packet access services;

- Flexible services;

- Joint detection.

To support feasible validations and demonstrations of the specified radio interface, the FRAMES demonstrator implements the lower-layer components for the TDD TD-CDMA demonstrator. In addition, the relevant higher layer components are also being implemented.

The BS and MS platforms each consist of three units (control, baseband, and RF). The control unit is connected to a CCT. The physical medium used for the connection will be Ethernet with TCP/IP communication. The user terminal or network will be connected directly to the control unit via an ATM AAL5 interface. The control unit software is based on a layered architecture. Drivers for the hardware devices sit at the lowest layer. These include drivers for the 7110, VME bus, and the ATM NIC card. The real-time operating system (RTOS) sits above the drivers and provides abstraction from the hardware. The operating system also includes the socket interface to TCP/IP and UDP (including Ethernet drivers) and SNMP support. The FRAMES demonstrator application software is implemented as a set of tasks.

The control unit is also responsible for system boot including Digital Signal Processor (DSP) software download and configuring and controlling the baseband and RF units (Figure 10.7, Figure 10.8).

Figure 10.7 Demonstrator RF unit.

The baseband transmitter is responsible for generating all the necessary downlink physical and logical control channels for TDD TD-CDMA. MS and BS transmitters are similar.

Similarly, the baseband receiver is responsible for the processing of received bursts from the baseband receiver front-end.

Figure 10.8 Demonstrator baseband and control unit.

The implementation work proceeded as follows:

In the first phase of work, the selected system functionalities were simulated. Link-level concepts were simulated using the COSSAP environment. Higher layer functionalities are tested using SDL (specification and description language, ITU-T Z.100) descriptions. The outcome of the phase was a COSSAP simulation model presenting the essential parts of the baseband processing.

In the second phase of work, the hardware platform was built and software modules were developed, including the generation of DSP code from the COSSAP simulation model. Integration of the demonstrator system proceeded as follows:

- Integration and testing of a single MS-BS link supporting bearer services up to 144 Kpbs;

- Integration and testing of two simultaneous MS-BS links to demonstrate joint detection;

- Integration and testing of a single MS-BS high-data-rate link supporting bearer services up to 512 Kbps.

10.7 TEST-BED SPECIFICATION OVERVIEW

The FRAMES project has aligned the initial test-bed specifications to adapt to the ETSI decision regarding UTRA selection. Therefore, the FRAMES demonstrator was used to validate the functionality of the TDD TD-CDMA radio interface as originally specified within FRAMES, and further elaborated within the delta concept group in ETSI/SMG2. However, due to the requirement to freeze the specifications, not all parameters are in line with the current ETSI SMG working assumptions.

The frame structure is presented in Figure 10.9. Both MSs and the BS have 2 timeslots active in the downlink (TS 0 and 1) and the uplink (TS 5 and 6). The two active timeslots are needed to support high bit rates. The demonstrator can process up to 8 codes per slot. The switching point between uplink and downlink is fixed midway through the frame.

Power control is only provided in the uplink. Its range is about 60 dB and its rate is 217 Hz. It counteracts the slow propagation variation due to shadowing effects. A downlink pilot channel is implemented to perform downlink synchronization. It is transmitted on timeslot 0.

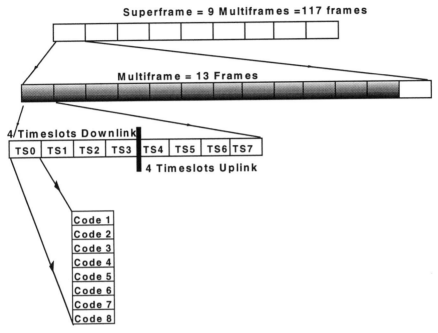

Figure 10.9 Frame structure.

Several different bearer services are provided by the FRAMES demonstrator, which are summarized in Table 10.1. High-bit-rate data services to be supported include both circuit and packet-switched bearers. The demonstrator was adapted through the purchase of additional hardware to support the processing of two timeslots in both the uplink and downlink direction. The so-called circuit-switched mode uses the real time RLC protocol, whereas the packet-switched mode uses the nonreal time RLC protocol.

Convolutional encoding is used for error protection. Services needing a BER equal to 10^{-3} use a ½ coding rate, while the service needing a BER equal to 10^{-6} uses a 1/3 coding rate. Both convolutional encoders have a constraint length equal to 7.

There are some limitations on the bearer services supported when the two mobiles are used simultaneously, because the BS can only process 8 codes per slot. Table 10.2 presents the bearer services available when two mobiles are connected to the BS. For instance, the 144 Kbps bearer service needs 9 codes per slot, hence, the BS cannot support two simultaneous 144 Kbps bearers.

Table 10.1

Bearer Services for One Mobile

Switching mode	User Bit rate	Max BER	Max delay	Mod.	No. of TS and codes	Service no.
Circuit	16 Kbps	10^{-3}	30 ms	QPSK	1 TS 1 code	1
Circuit	64 Kbps	10^{-3}	30 ms	QPSK	1 TS 4 codes	2
Circuit	64 Kbps	10^{-6}	30 ms	QPSK	1 TS 6 codes	3
Circuit	144 Kbps	10^{-3}	100 ms	QPSK	2 TS 9 codes	4
Circuit	512 Kbps	10^{-3}	100 ms	16 QAM	2 TS 16 codes	5
Packet	144 Kbps	10^{-6}	300 ms	QPSK	2 TS 12 codes	6
Packet	PS max*	10^{-6}	300 ms	QPSK	2 TS 17 codes	7

* The data rate for PS max (packet switched maximum) is dependent on the signaling channel configuration, which has not yet been defined, and the efficiency of the packet access protocol, which is still to be defined. The demonstrator will support two timeslots and hence a combined maximum data rate for signaling and traffic of 512 Kbps can be achieved. The QoS (delay, ARQ overhead) provided for this bearer service is still under evaluation.

Table 10.2

Bearer Services for the Two Mobile Tests

Mobile	Service number					
MS1	1	1	1	2	2	3
MS2	1	2	3	2	3	3

10.8 IMPLEMENTATION

The implementation work for the FRAMES demonstrator was divided into three phases. The division as separate phases was chosen to ensure suitable monitoring during the project.

Phase 0 was the first of the three implementation phases. The outcome of the phase was a COSSAP simulation model presenting the essential parts of the baseband processing.

Phase 1 was the second phase. The outcome of the phase is a reduced functionality demonstrator. That is a real system with all custom hardware designed and in use (Figure 10.10). The system was still not the final one (i.e. bit rates were lower than in the final demonstrator).

Phase 2 was the last phase. The outcome of the phase was the full functionality demonstrator.

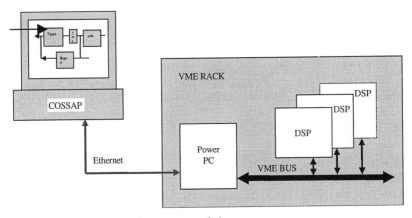

Figure 10.10 Phase 0 to Phase 1 demonstrator evolution.

After Phase 0 of the demonstrator was completed, the transition from Phase 0 to Phase 1 was implemented, involving partitioning, optimizing, and compiling the COSSAP modules to run on the DSP hardware as described in Figure 10.10. The COSSAP environment allowed the phase 0 modules that were written in C to be compiled for C40 DSP processors, and downloaded to the hardware. Unfortunately, the interactive functionality that allows COSSAP to capture data back from the DSPs was not currently operative so this was done via the lower level tools. This enabled us to test phase 1 modules and verify their operation against the phase 0 simulation [11].

In order to ease this transition, the interface that COSSAP normally uses between primitives has been modified for the phase 0 simulation. The modified primitive interface passes pointers to memory containing the data transferred from one primitive to the next. This significantly reduces the time that the resulting code spends simply copying data from one memory area to another.

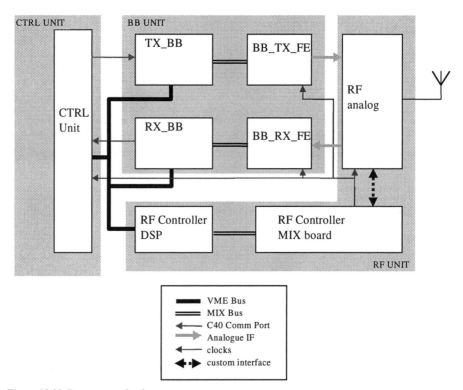

Figure 10.11 Demonstrator hardware structure.

The hardware structure of the demonstrator is outlined in Figure 10.11. The main hardware areas within each unit are indicated. The CTRL unit is implemented on the Radstone PPC604 board. The main part of the TX_BB runs on the Pentek 4270 Quad C40 board. The main part of the RX_BB runs on the Pentek 4285 Octal C40 board. The RF controller DSP is a Pentek 4284 single C40 baseboard. The remaining boards are custom MIX boards. Finally, the RF analog is constructed from custom and off-the-shelf modules and housed in a separate box.

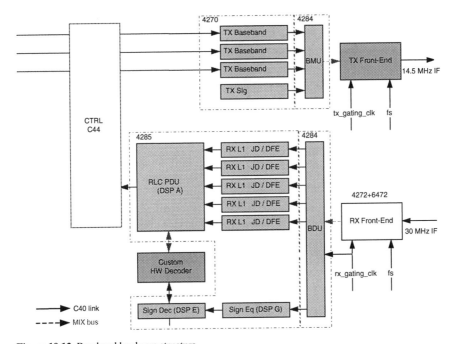

Figure 10.12 Baseband hardware structure.

Figure 10.12 illustrates the mapping of software modules onto hardware for the baseband unit.

For the transmit functions, the RLC PDU processing and the L1 PDU burst processing are both mapped onto a single DSP, and the processing is duplicated for three processors. These processors handle the transmission for one or more slots in the TDMA frame, depending on the frame size and active bearer service. A separate DSP is used for the signaling channels.

The receive functions are split into two parts. The RLC PDU processing is handled by a single DSP and a custom hardware decoder. The L1 PDU burst processing is spread over the remaining DSPs. This L1 PDU burst processing involves equalization (DFE) for the nonspreading option, and joint detection for the spreading option. The two options are not available simultaneously.

10.9 MEASUREMENT CAMPAIGN

The measurement campaign is focused on quantitative testing and performance evaluation of the TD-CDMA technology, with an evaluation of the test-bed performances based on lab tests. These quantitative tests have to be performed in a laboratory with a propagation channel emulator (both at the RF and baseband level), which allows researchers to control, reproduce, and characterize the test conditions. The bit error rate for each service will be evaluated as a function of the signal-to-noise ratio, in various propagation environments characteristic of the actual environment, including indoor, pedestrian, and vehicular for various delay profiles and mobility conditions.

Some extensive validation of joint detection features is being carried out in the so-called emulator mode configuration. It involves the baseband channel emulator [12], as developed within the project (BBCE), which performs the emulation of a mobile radio channel based on a set of predefined channel models (ITU). Only one demonstrator device is needed to carry through tests, which acts as transmitter and receiver simultaneously (Figure 10.13). The emulator is configured using a personal computer (PC).

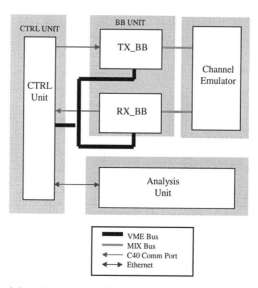

Figure 10.13 Baseband channel emulator configuration.

The BBCE offers the possibility to perform rather exhaustive testing of the joint detection algorithm in the uplink. The intracellular cochannel interference, introduced by the CDMA component, is emulated in the following way. The implementation is based on a preproduced interfering signal that consists of several different user signals, each spread by a different CDMA code, transmitted over a different emulated channel, and shifted in time to simulate the different delays of the channels. The resulting signals are added and stored in the channel emulator. The desired transmission is then emulated convoluting the desired user signal, which is spread by a remaining CDMA code, with a time-varying channel impulse response, and adding the preproduced interfering signal of the other preprocessed users. Therefore, this mode offers the possibility of testing joint detection under realistic uplink conditions. In addition, the baseband channel emulator is also capable of using measured channels characteristic of actual environments.

Figure 10.14 Measurement trials at France Telecom (CNET), Paris.

The joint detection algorithm is used to simultaneously equalize and despread the received signals from all active users within the same timeslot in a given cell. From the a priori knowledge of the codes used, this technique allows simultaneous estimation of the different channel experienced by the various users in order to further remove the interfering signals from the desired one. This novel technique is required to compensate for the moderate processing gain (PG=16) achieved, but allows a relaxation of the requirements set on the fast-power control by counteracting near-far effects on the uplink. This technique is therefore a basic feature of the proposed TD-CDMA radio interface, the performance of which is being carefully evaluated and validated.

First, the influence of the number of active codes within one slot is evaluated for various propagation conditions (indoor, pedestrian, and vehicular), and different service scenarios compatible with (Table 10.2). Then, the near-far resistance of the joint detection algorithm will be determined, depending on the received interferer power range, by varying the level of the interfering signals. Preliminary simulation results indicate that joint detection would still work properly with about a 20 dB signal amplitude. This permits recovery from an imperfect power control on a frame-by-frame basis. The exhaustive measurement set obtained from the test-bed in the emulator mode will be used for validation of the simulation results already provided within the project, using the same channel models as defined by ITU.

In addition, outdoor measurements can be undertaken to check that the demonstrator performs well in a real environment, and to roughly determine the coverage range around CNET premises under various mobility conditions (see Figure 10.14). Again, the performance of the joint detection algorithm can be evaluated using two transmitting mobile stations, and one receiving base station supporting two different bearer services under real propagation conditions. In this configuration, the effect of the actual RF frontends will be taken into account.

Some additional key features to be supported by UMTS, such as the ARQ mechanism implemented for the packet-switched mode, could be further evaluated in terms of the number of retransmision and corresponding delays in various propagation conditions.

10.10 SERVICE DEMONSTRATION CAMPAIGN

The demonstration campaign [13, 14] shows the ability of the FRAMES test-bed to provide high-bit-rate services supported by both circuit and packet-switched bearer services, and to demonstrate some enhanced UMTS functionalities such as ARQ mechanisms and bearer flexibility. These demonstrations will involve application terminals, which will provide multimedia applications through ATM/AAL5 links. Thus, a proprietary high-quality video conference service will be demonstrated that requires high-bit-rate bearer services (around 400 Kbps) with a high quality of service. Other typical Internet applications (video-conference, web browsing, white board, data file transfer, email, etc.) are also supported both locally and through remote ATM connections to the Internet network.

Figure 10.15 Demonstrator at the ACTS Mobile Summit 1999, Sorrento, Italy.

The terminals are formed by a PC (labtest PC or fixed PC, depending on mobile or fixed terminal) with Windows NT and an ATM AAL5 card. Classical IP protocol is implemented on the ATM card. Moreover, it will be possible to provide ATM video-conference service on AAL5 with the same terminals, by developing the interface to connect a video codec on the ATM AAL5 card. Figure 10.15 shows the FRAMES demonstrator at the ACTS Mobile Summit in 1999, and Figure 10.16 shows the videotelephony terminals attached to the base station during demonstrations at CNET in Paris in 1999.

The demonstration runs in PVC mode, and the application terminals are used only as a data source. No signaling exchange has been defined between the FRAMES demonstrator and the internal terminal, because these terminals cannot be easily modified to support the signaling exchange.

Bearer establishment/release consequently has to be performed manually, by internal functionality of the FRAMES demonstrator. The protocol stack used in the traffic plan for ATM videophony and IP application over ATM are presented in Figure 10.17.

Joint demonstrations were also performed with the RAINBOW project to demonstrate advanced call and bearer functionality as controlled by the RAINBOW access network. These joint demonstrations with collocated FRAMES and RAINBOW demonstrators took place in CSEM premises because no remote connections through national hosts were available.

Figure 10.16 Demonstrator with video application test terminal.

The interface sits between the FRAMES radio interface and RAINBOW network. The information to be passed over the interface is formatted into ATM cells and routed by allocation of VCI/VPI addresses. Some messages are addressed to the FRAMES control processor, while others are intended to pass over the air unaffected by FRAMES. It is pointed out that the exchange of information between demonstrators is based on ATM cells standard, but this does not necessarily imply that ATM cells are transparently transmitted over the radio interface.

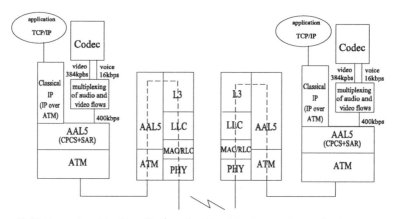

Figure 10.17 Protocol stack in the traffic plane for ATM videophony and IP applications.

Figure 10.18 shows the overall structure of interconnection between FRAMES and external applications/network. The RAINBOW network demonstrator implemented a FRAMES adaptation layer (FAL) to interface with FRAMES through the ATM/AAL5 standard. On the other side, the FRAMES demonstrator implement a common EAL to interface with the RAINBOW demonstrator.

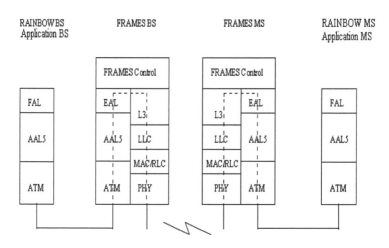

Figure 10.18 Overall structure of the interface for the joint trials.

References

[1] "Universal Mobile Telecommunications System (UMTS); Selection Procedures for the Choice of Radio Transmission Technologies of the UMTS," *UMTS Technical Report* Mar. 30, 1997.

[2] A. Klein, R. Pirhonen, J. Sköld and R. Suoranta, "FRAMES Multiple Access Mode 1 – Wideband TDMA With and Without Spreading," *Proc. International Symposium on Personal, Indoor and Mobile Radio Communications*, Helsinki, Sept. 1997.

[3] T. Makelainen et al., "FRAMES Demonstrator Implementation Roadmap," *ACTS Mobile Telecommunications Summit*, Granada, Spain, November 27-29, 1996.

[4] K. Richardson et al., "The FRAMES Demonstrator First Implementation Steps," *ACTS Mobile Telecommunications Summit*, Granada, Spain, November 27-29, 1996.

[5] P. Croft et al., "The Development of the FRAMES TDD TD-CDMA Demonstrator," *ISWC98*, Montreal, Canada, May 22-23, 1998.

[6] Concept Group Delta Wideband TDMA/CDMA, Evaluation Report, *Tdoc SMG2 368/97*, Cork, Ireland, Dec. 1997.

[7] M. Urios et al., "Overview of the FRAMES Demonstrator," *FRAMES 1999 Workshop*, Delft University of Technology, Delft, The Netherlands, Jan. 18 – 19, 1999.

[8] G. Noubir and B. Perrin, "FRAMES Demonstrator Packet Access Mode," *FRAMES 1999 Workshop*, Delft University of Technology, Delft, the Netherlands, Jan. 18 – 19, 1999.

[9] P. Anderson et al., "Investigations in the FRAMES Project on the UMTS Radio Interface," *ACTS Mobile Summit*, Sorrento, Italy, June 8 – 11, 1999.

[10] P. Croft, L. Girard, K. Richardson, H. Erben and C. Boisseau, "The FRAMES TDD TD-CDMA Demonstrator - Development, Evaluation and Trials," *FRAMES Workshop, ICCT'98*, Beijing, China, Oct. 21, 1998.

[11] P. Demain and D. Johnson, "Simulation Results from the TD-CDMA FRAMES Demonstrator," *FRAMES 1999 Workshop*, Delft University of Technology, Delft, The Netherlands, Jan. 18 – 19, 1999.

[12] Neill Whillans, "The Development of a Baseband Channel Emulator for use in the Evaluation of the FRAMES Demonstrator," *FRAMES 1999 Workshop*, Delft University of Technology, Delft, The Netherlands, Jan. 18 – 19, 1999.

[13] P. Croft, L. Girard, K. Richardson, H. Erben and C. Boisseau, "The Development, Evaluation & Trials Programme of the FRAMES TDD TD-CDMA Demonstrator," *WPMC'98*, Japan, Oct. 4 and Nov. 6, 1998.

[14] L. Girard et al., "The Evaluation & Trials Programme of the FRAMES TDD TD-CDMA Demonstrator," *ACTS Mobile Summit*, Sorrento, Italy, June 8 – 11, 1999.

Chapter 11

FDD Demonstrator

The ETSI decision in January 1998, with WCDMA in the paired bands based on FMA2, and TD/CDMA in the unpaired bands based on FMA1 with spreading, was a big success for FRAMES. In addition, FMA1 without spreading was adopted in the United States for the high-speed mode in UWC-136. From that point of view, FRAMES had a big impact on the UMTS and IMT-2000 standardization. The previous chapter described the TDD TD/CDMA demonstrator developed in FRAMES after the ETSI decision. In this chapter, we describe the FMA1 without spreading demonstrator, which was part of the FRAMES project until the ETSI decision. To complete the picture of demonstrators, this chapter also provides some background information on a WCDMA evaluation system; this material is an excerpt from [1] (and can also be found at http://www.ericsson.com/review).

11.1 WTDMA DEMONSTRATOR

During the third year of FRAMES projects, evaluation of the candidates for the next generation cellular wireless standards were proceeding. One critical aspect in the standardization process is the definition of terrestrial physical layer interface. The physical layer principles, along with resource allocation mechanism on MAC, together define how well the physical layer resources can be utilized. The prototyping environment presented in this chapter was built to analyze and validate effectiveness of the different design choices. As an example, we consider WTDMA-based air interface developed in the EU ACTS AC090 FRAMES project.

Validation of a candidate wireless communication system is a time-consuming task. Due to complexity of the task, it is important to design the system so that design parameters both on physical and MAC layers can be modified without complete redesign of the validator. Chapter 11 describes a wideband radio demonstration environment, where design parameters can be modified easily, and the effects can be analyzed and validated both off-line and on-line with the real-time prototype system. As an example, the case considered is WTDMA-based air interface developed in the EU ACTS FRAMES. The architecture and main capabilities of the WTDMA system are shown.

11.1.1 Development Approach

Keys for Flexible Prototyping

There are a number of methods that can be used to ensure reasonable design iteration times. The ones used in our prototyping system are

- Simulator-based design and implementation;

- Use of standard system components and interconnections;

- Configuration flexibility, centralized system control;

- Testing of analysis functions in simulator environment.

Development Environment

Simulator-based design and implementation in this case, means the use of software simulators to first implement and functionally simulate major parts of the system before implementation on the final hardware. Figure 11.1 illustrates the main components of the design platform.

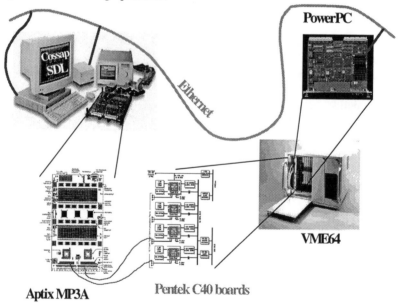

Figure 11.1 WTDMA demonstrator design environment.

Protocol and system control software has been implemented using SDL using Telelogic Inc. TAU-toolset, with some lower-control functionality and protocol data unit processing implemented in C. Most of the baseband algorithms are written using generic C in Synopsys COSSAP environment, while in addition some custom coding with assembler/C, and VHSIC hardware description language (VHDL) has been used. Both the simulation environments support generation of C-language implementation down to target platform. The selected approach offers simulation and testing of the functionality in a workstation environment, and a direct route down to actual hardware to iterate the design.

To efficiently utilize flexibility provided by the simulator-based design (Figure 11.2), the platform must be based on standard system components and interconnections. Use of standard components facilitates use of code generation both for hardware and software. The standard interconnections allows easy modification of the target. Our system uses C40 communication links, and MIX and VME buses. User and control/signaling data flows are internally separated to ensure flexible modifiability and deterministic timing. This separation combined with the idea of centralized system control is well suited for the selected design methodology, where user data processing and system control and signaling development is done on COSSAP and SDT, respectively.

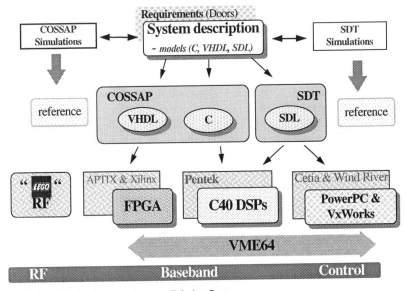

Figure 11.2 WTDMA demonstrator overall design flow.

WTDMA Test Scenario

The test case used is WTDMA-based air interface developed in the EU ACTS AC090 FRAMES. It used GSM frame period with 1/16 and 1/64 slots. Air interface supports both RT and NRT connections. A detailed description of the WTDMA parameters is found in [2] and [3].

The platform consists of an MS and BS. Functionally, both the MS and BS have been built to resemble each other. As the target of prototyping is to validate the concept, no product-related design aspects, such as power consumption or cost, have been considered. A HP UNIX workstation, connected through Ethernet to both MS and BS, is used for centralized control and for off-line and on-line analysis displays.

Internally, an MS or BS is divided to three separate logical units, namely RF, baseband, and service/control (CTRL).

11.1.2 Demonstrator Architecture

RF Unit

The RF unit is split into three subunits, RF front-end, RF MIX board, and RF DSP board. RF front-end module upconverts the Tx intermediate frequency (IF) signal (generated by the baseband unit) into the final RF channel, amplifies it to the required level, and performs the transmission. RF front-end receives the RF signal located on a specific RF channel, amplifies and filters it, and downconverts to a fixed low intermediate frequency that can be processed by the baseband unit. RF MIX board subunit is responsible for controlling the RF front-end parameters such as Tx/Rx frequencies and Tx/Rx gains. In addition, it generates a reference clock for the RF front-end, CTRL unit, and sample + Tx/Rx gating clocks for the baseband unit. RF DSP board implements the interface between the VME and MIX buses. It performs configuration information processing into a suitable format for the MIX board and RF front-end.

The RF unit is similar between the base and mobile station applications. The only difference is on the MS side, where the reference oscillator frequency can be adjusted by means of a voltage-controlled oscillator. The detail properties of RF unit are listed in Table 11.1.

Table 11.1

WTDMA Demonstrator RF Unit Properties

Item	Unit	Value
Frequency ranges	MHz	1920-1980, 2110-2170
RX noise figure	dB	4
Channel spacing	MHz	1.6
RX AGC range	dB	80
Tx power control range	dB	50
Frequency switched speed (Tx and Rx)	μs	1
Tx IF	MHz	14
Rx IF	MHz	30
Reference frequency	MHz	104

Baseband Unit

The baseband unit consists of several submodules, namely transmitter encoders, burst multiplexer, transmitter front-end, receiver front-end, burst demultiplexer, and receiver equalizers/decoders. All of these are physically implemented using multiple Texas Instruments TMS320C40 DSP processors on Pentek's off-the-shelf boards, and a few custom hardware boards. Figure 11.3 shows the functional architecture of the baseband unit.

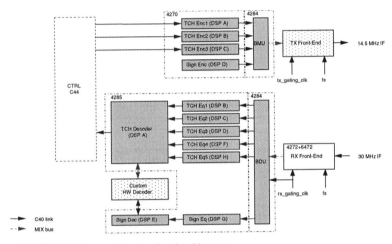

Figure 11.3 WTDMA demonstrator baseband architecture.

Most of the algorithms are written using generic C in COSSAP environment, with some custom coding with assembler/C, and VHDL must be performed. The baseband units both in BS and MS are similar, with the exception on MS side, due to additional measurements for synchronizing the radio link.

On the transmitter side, there are three traffic and one signaling channel encoder responsible for error detection/protection coding, interleaving, and burst building. On the receiver side, the IF signal is first subsampled with A/D converter, after which a receiver board is utilized to perform downconversion (to baseband), and reception filtering. Received bursts are routed to five traffic or one signaling equalizer via burst demultiplexer unit. Equalizers perform multipath effect cancellation by means of a decision feedback equalizer. Following this, the soft bits are deinterleaved and weighted according to burst SIR, and in case of ready packet convolutionally decoded. CRC is finally performed for the packet.

Control Unit

The control unit implements centralized control of physical layer and MAC functionality to allocate resources for circuit-switched and packet-access connections, simple link access control, and simple network layer functionality to manage user data connections. The control of baseband and RF is done on a TDMA-frame basis.

A circuit-switched connection is provided as a comparison system to evaluate performance of the packet-switched connection. Packet-switched service is pure NRT service, so no maximum delay figures are offered. Selective receiver controlled hybrid-II ARQ is used as an error protection scheme for packet-access connections.

Most of the functionality was implemented using SDL to specify the functionality. Figure 11.4 illustrates protocol entities implemented. Acronyms used in Figure 11.4 are RLC, LLC, RBC, and Internet protocol (IP).

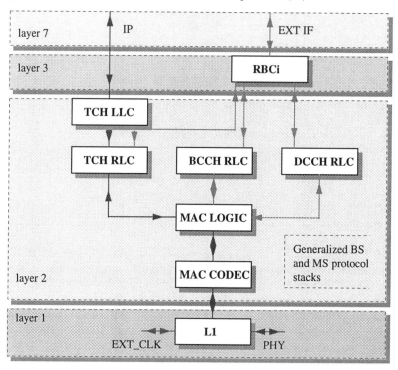

Figure 11.4 WTDMA demonstrator SDL design flow and protocol stack implemented in SDL.

Figure 11.5 presents the design flow for the modules implemented in SDL. As a starting point, message sequence charts (MSCs) are used to identify the flow of events for all the system behavior. After completing MSCs, SDL is used to design the protocol using parallel extended state machines. After the workstation-based simulation validated against initial MSCs is satisfactory, target processor C-language code is generated from the SDL specification.

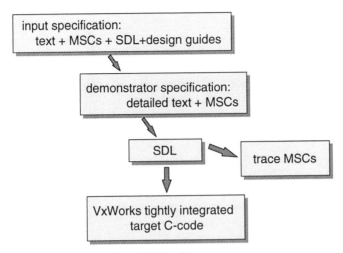

Figure 11.5 WTDMA demonstrator protocol design flow.

The actual PDU processing and physical layer functionality have been implemented with hand-coded C-language routines. User data connection provided is TCP/IP over Ethernet. All common IP services can be used (e.g., web browsing).

Analysis Unit

Support for system analysis is provided by implementing measurement functions directly into the data path and control flow processes. Analysis data is collected from all the protocol layers, from layer 1 to layer 7. All the measurements are first tested in simulation environments and then validated on the real platform. The selected method ensures that adding analysis procedures does not change the behavior of the system. System provides support for storing the main system parameter on a physical layer frame basis to a hard disk for postprocessing and analysis, and for a set of parameters real-time display is available. A UNIX workstation, connected through Ethernet to both MS and BS, is used for centralized control and for off-line and on-line analysis displays. On-line analysis values can be traced either as discrete values or as averaged values over time.

11.1.3   ACTS Summit Demonstrations

Nokia provided the WTDMA demonstrator to the 1998 ACTS Summit. Fully functional WTDMA air interface was already shown in the summer time in Rhodes. Real-time operation was demonstrated using live video application.

11.1.4 Conclusions

The presented prototyping environment is a powerful environment to analyze the performance of WTDMA-based radio links. It facilitates easy modification and seamless design flow for functional prototyping purposes. It is relatively easy to modify for other TDMA-based wireless systems.

11.2 NTT DOCOMO WCDMA EVALUATION SYSTEM

The aim of UMTS/IMT-2000 standardization is to produce a truly global standard for third generation mobile systems. A WCDMA evaluation system facilitates UMTS/IMT-2000 standardization by allowing operators and engineers to demonstrate and test third generation services and technical solutions. It also enables them to learn about and evaluate WCDMA characteristics in ATM environments. Ericsson developed a WCDMA evaluation system to demonstrate the potential of the third generation system, to obtain valuable field experience and input for current commercial development, and to pave the way for the evolution of second generation systems into UMTS. This section describes the WCDMA evaluation system, the technical details of the system, its background, and the wide range of end-user services it offers.

11.2.1 Introduction

A number of manufacturers are fully committed to UMTS/IMT-2000 standardization, the aim of which is to produce a truly global standard for third generation mobile systems. A WCDMA evaluation system facilitates UMTS/IMT-2000 standardization, by allowing operators and engineers to demonstrate and test new, advanced, third generation services and technical solutions. It also enables them to learn about and evaluate WCDMA characteristics in ATM environments. The evaluation system was developed

- To demonstrate the potential of the third generation (WCDMA) system;

- To boost intercontinental standardization processes;

- To provide technical input to proposed new standards;

- To obtain valuable field experience and input for current commercial development;

- To pave the way for the evolution of second generation systems into UMTS.

With the WCDMA evaluation system, the technology has been taken out of the laboratory and put to test in real radio environments. Thus, it is possible to exploit and evaluate the WCDMA-related technology of third generation radio-access systems. Late in 1996, Ericsson received an invitation to tender from NTT

DoCoMo—one of the world's largest mobile phone operators—requesting a WCDMA evaluation system. At the time, however, a standard for WCDMA as a radio access technology did not exist. In 1997, the Japanese standardization body, ARIB, began drafting a standard for WCDMA. This first test system is based on early proposals of the ARIB standard. The standard has since matured and has been globally harmonized as described in Chapters 1 and 12 of this book.

Work on the system began in early 1997, and although it is sometimes called an "experimental" system, the system designers aimed much higher, building what could be termed a precommercial system. In fact, this WCDMA evaluation system has been in small-scale serial production, with close to 30 systems produced to date.

11.2.2 System Overview

The WCDMA evaluation system (Figure 11.6) consists of

- The mobile station simulator (MS-SIM) (Note: mobile stations are also sometimes referred to as mobile terminals);

- The base transceiver station (BTS);

- The RNC;

- The MSC;

- The WCDMA operations system (WOS).

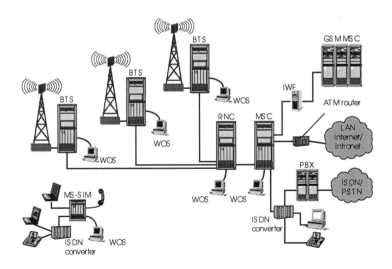

Figure 11.6 System overview of WCDMA evaluation system.

The system was built for experimenting with the WCDMA radio-access network and ATM, and for applying these technologies to different wideband services. Consequently, the architecture of the MS-SIM, BTS, RNC, MSC, and WCDMA operations system is very general and flexible. User data and control information between the RNC and BTS are transported via a preliminary "Iub" interface based on ATM technology ("Iub" is the denomination used in standardization work for the interface between the BTS and the RNC). Similarly, a preliminary "Iu" interface—also based on ATM transport technology—is used between the RNC and MSC. To a great extent, the WCDMA part of the evaluation system is based on previous work conducted at Ericsson research centers, including the wideband test-bed (WBTB, 1995). The ATM part of the system is built on top of a new ATM-based development platform [4].

Obviously, measurement capabilities are essential in an evaluation system. Measurement data is thus collected from appropriate parts of the system and sent to the WCDMA operations system for logging. The data can then be postprocessed for presentation. Two identical WCDMA evaluation systems have been set up in Stockholm, one of which is mainly used for evaluation purposes; the other is used solely for demonstrations. Although not actually part of the WCDMA system, GPS is used in the evaluation system to provide different nodes with accurate timestamps, so that measurement data from the nodes can be correlated. The GPS also provides positioning data that can be used for monitoring the whereabouts of mobile stations.

11.2.3 End-User Applications

Third generation mobile telephony systems will offer end users a wide variety of applications such as high-speed multimedia services, and other services currently only available from fixed networks. The WCDMA evaluation system enables operators to evaluate and demonstrate third generation services. It is difficult to predict what will happen next as the telecommunications, computer, and entertainment industries converge. As mentioned above, the WCDMA evaluation system was developed to meet the demands of wireless communication in a true multimedia environment. The system has been developed independently of third generation end-user applications, providing a wide range of bearer services that permit different end-user applications to be evaluated (Figure 11.7).

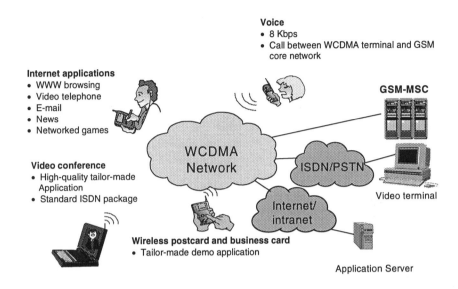

Figure 11.7 End-user applications in the evaluation system.

Voice and Audio

Certainly, standard voice services currently available in second generation systems must also function in third generation systems. The WCDMA evaluation system is thus equipped with 8.8 Kbps voice codecs (G.729). It also provides a means of evaluating and demonstrating interconnection to existing GSM networks.

Video Conferencing

Video conference or video phone applications based on circuit-switched connections (such as ISDN) or packet-switched media (such as TCP/IP) are being tested—for example, Microsoft NetMeeting or Intel Proshare-based products. Other video applications like remote monitoring are also possible. Still and moving-picture codecs are being tested on the WCDMA evaluation system.[1]

---

[1] Microsoft® and NetMeeting® are either registered trademarks or trademarks of Microsoft Corporation in the United States and/or other countries. Intel® and ProShare® are registered trademarks of Intel Corporation in the United States and other countries. PointCast® is a registered trademark of PointCast Incorporated.

Internet Applications

The WCDMA evaluation system supports best-effort, packet-switched data up to approximately 470 Kbps. Access to the Internet and corporate intranets is essential to professional mobile users. "Push applications" (such as PointCast) are becoming increasingly popular on the Internet, allowing users to subscribe to specific kinds of information, for example, weather forecasts, advertisements, news, company bulletins, and updates of company data such as price lists. These applications usually entail a background type of traffic; that is, the information reaches the user even when he or she is occupied with other applications. Packet-switched services in the evaluation system support background traffic. A mobile terminal can be logged onto a server and only pay for updated information sent out and not for the connection time.

Corporate LAN Access

Today, many people remotely access file servers, databases, and GroupWare applications on their company local area network (LAN). For good performance, high-speed connections must be offered that are at least as fast as landline modems operating at 56 Kbps. When working remotely, corporate users want to be "on-line" with their company system, for instance, to see when they receive e-mail. Since cellular calls currently cost more than landline calls, subscribers cannot afford to remain connected eight hours a day. Instead, they must dial in several times a day to check their mail. With WCDMA packet-switched services, however, it will be possible to charge subscribers for volume of data transferred instead of for the duration of the session.

Wireless Postcard and Electronic Business Cards

E-mail has become one of the most common mobile data applications on cellular networks. E-mail often includes more than plain text messages, however. Many messages also carry attachments of additional files—for instance, files of images and video clips. E-mail of this kind requires greater transmission speeds than the present 9.6 Kbps. Electronic business cards are currently being standardized—in addition to name, title, company address, and so on, they may include images and other information. Thus, in the future, instead of seeing the number of a caller, his or her photograph might be displayed on the WCDMA mobile terminal (which also doubles as a digital camera).

Other Applications

Different user groups have different needs as relates to mobile multimedia. For example, the health sector is investigating interactive health and medical applications (telemedicine); the security sector is interested in remote monitoring; and the traffic sector works with traffic telematics and navigation equipment in vehicles. Furthermore, fire brigades and broadcasting companies could benefit from WCDMA systems to assess situations as the first fireman or journalist comes on the scene. Intelligent living involves video-on-demand, on-line entertainment applications, and applications that support working and shopping from home. These areas will also require mobile applications.

11.2.4   System Architecture

Base Transceiver Station

To handle mixed services, the base transceiver station in third generation systems must have a flexible architecture. To meet the requirements for flexibility, designers structured the hardware architecture according to function. That is, instead of using a channel-based architecture, as was done in first and second generation systems (Figure 11.8), the BTS in the WCDMA evaluation system uses a pooled architecture (Figure 11.9). This solution provides greater flexibility for coping with the varied demands of future services.

The antennas, which are passive units for transmission and reception, do not require a direct current (DC) power supply. Two identical antennas are used in each sector: one is used for both transmission and reception; the other is solely used for reception. Instead of two antennas in each sector, a dual-polarized antenna can be used, thereby reducing the number of antennas to one per sector. The antenna for each macrocell sector has a mechanical tilt and a fixed electrical tilt. The architecture of the base station (Figure 11.10) permits the bearer capacity to be freely allocated to users of different types of service:

- Voice service;

- Circuit-data service carried as unrestricted digital information (UDI) up to 384 Kbps;

- Raw packet-data service up to 472 Kbps (packet throughput capacity is dependent on radio interference, and consequently, on the SSCOP retransmission rate).

Limiting factors of the total user data bandwidth are the data capacity of the mobile station, the radio cell size, and radio interference.

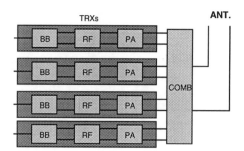

Figure 11.8 Architecture of first and second generation base transceiver stations. BB – baseband, RF – radio frequency, PA – power amplifier.

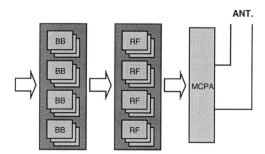

Figure 11.9 Architecture of the third generation base transceiver station. MCPA – multiple carrier power amplifier.

Radio Network Controller

The RNC (which in GSM terminology corresponds to the BSC) is built on a generic ATM infrastructure. The processor and all other devices are connected to the ATM switch, which makes it very easy to extend system capacity. The RNC houses

- The radio resource and macrodiversity combination function (also referred to as the soft handover combination function);

- The speech coding/decoding function according to ITU-T recommendation G.729 (provided the codecs are located in the RNC);

- The "Iub" transport based on 1.5 or 2 Mbps ATM links with ATM adaptation layer 2 (AAL2) link termination;

- The "Iu" interface (carried on a 155 Mbps ATM link);

- Timing and synchronization functions.

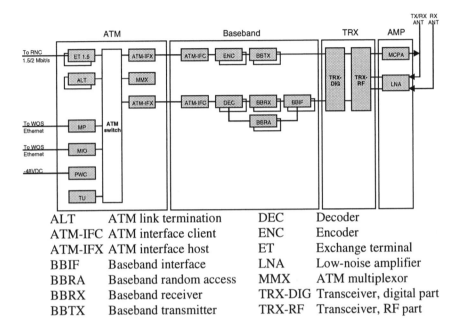

ALT	ATM link termination	DEC	Decoder
ATM-IFC	ATM interface client	ENC	Encoder
ATM-IFX	ATM interface host	ET	Exchange terminal
BBIF	Baseband interface	LNA	Low-noise amplifier
BBRA	Baseband random access	MMX	ATM multiplexor
BBRX	Baseband receiver	TRX-DIG	Transceiver, digital part
BBTX	Baseband transmitter	TRX-RF	Transceiver, RF part

Figure 11.10 Architecture of BTS.

At least three base stations can be connected to one RNC, each with up to two 1.5 or 2 Mbps "Iub" links. The traffic capacity of the example in Figure 11.10 is approximately 160 mobile stations connected to each other or to the fixed network. Every connection is switched in the MSC.

The architecture of the RNC (Figure 11.11) is very flexible, making it ideal for experiments with advanced WCDMA radio-network functions. For instance, to meet new or increased demands, additional printed circuit boards may be installed in empty board positions in RNC subracks. The RNC can also handle large amounts of measurement data in real time, which capability is necessary for experimenting with advanced radio-network functions.

Mobile Services Switching Center

Like the RNC, the MSC is built on a generic ATM switch infrastructure (Figure 11.12) and has the same flexible properties as the RNC. Indeed, certain functionality can be moved between the RNC and MSC. The codecs may be put in either the MSC or the RNC—for example, to comply with "Iu" interface standardization, or for other experimental purposes. The MSC handles the "Iu" interface to the RNC and the fixed network interfaces to ISDN and ATM LAN. The main task of the MSC is to set up calls to and release calls from mobile stations. It includes only a rudimentary level of signaling sequence over the air interface for location updates, authentication, ciphering, and so on.

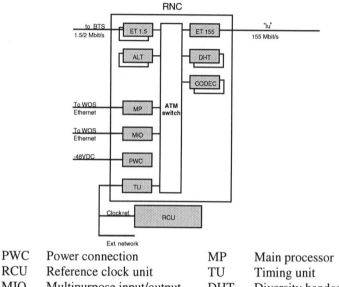

PWC	Power connection	MP	Main processor
RCU	Reference clock unit	TU	Timing unit
MIO	Multipurpose input/output	DHT	Diversity handover

Figure 11.11 Architecture of the RNC.

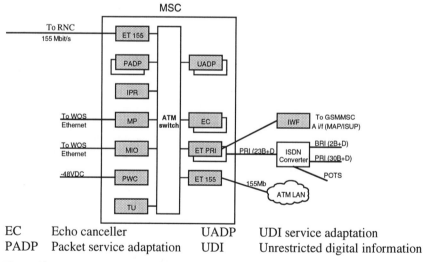

| EC | Echo canceller | UADP | UDI service adaptation |
| PADP | Packet service adaptation | UDI | Unrestricted digital information |

Figure 11.12 Architecture of the MSC.

11.2.5 Evaluations

Numerous tests can be executed within the WCDMA evaluation system. The descriptions that follow highlight some of the capabilities of the system.

Transmitter and Receiver Performance

A large battery of tests is used for evaluating transmitter and receiver performance. The tests cover basic performance under "normal" conditions, quantitative tests made in the laboratory using specified channels, and field trials, which are assumed to be slightly more demonstrative of "real-life" characteristics. Some tests may also be performed under extraordinary and rapidly changing conditions (for instance, when a user enters or exits a building). These tests evaluate the robustness of algorithms and the effects of delay in the system.

Detector Performance

Detector performance is tested to verify searcher/tracker performance in different environments, to find suitable parameter settings for observed and anticipated multipath environments, and to assess performance during soft handover, where synchronous performance can be closely monitored. The detector performance tests

- Verify equipment sensitivity;

- Investigate searcher performance in different dispersive environments;

- Test tracker performance for known dispersion types;

- Verify searcher/tracker performance during soft/softer handover;

- Measure E_b/I_0 detectors for different interference types;

- Investigate the effects of using different numbers of RAKE fingers.

Idle Mode

Tests of the idle mode indicate the general characteristic performance of specific algorithms, radio interface, and mobile-station hardware implemented in the evaluation system. The tests also assess parameter dependency and the performance of sector selection, including time delay while in idle mode. In particular, the tests examine initial and continuous sector selection in idle mode.

Open-Loop Performance

Measurements can be made to estimate the extent to which measured data on one link can be used for predicting the accuracy of another link when setting various parameters. The measurements, which can be made in stationary and variable environments, estimate the quality of the open-loop power control for the RACH/FACH and the initial power setting when setting up the CCH. The tests of open-loop performance predict:

- Power level capability;

- The consequences of RACH/FACH and the initial CCH power setting.

Random Access Performance Characteristics

Certain tests evaluate the performance of the random-access procedure implemented in the WCDMA evaluation system, as well as the effect of different parameter settings of the acquisition procedure. The tests measure the effects of interference levels and the interfering effects that RACH/FACH procedures have on established calls. In particular, the tests

- Assess RACH performance in different power-regulation schemes;

- Examine disturbances from RACH/FACH bursts.

Uplink and Downlink Power Control

Testers evaluate the performance of the fast uplink/downlink power-control loop in terms of speed, stability, performance when E_b/I_0 is low, and operation during soft/softer handover. The practical performance of the control loop, in terms of its ability to maintain a constant E_b/I_0 value in variable radio environments, is assessed for estimating the need for reference value control. A special area of interest focuses on performance at cell fringes and, in this context, the effects of soft/softer handover. The effects of sudden environmental changes (turns in the road, outdoor/indoor usage) can also be studied, as well as step responses during handover and resynchronization. The tests also check the effects of low data rates; for instance, due to poor synchronization or when DTX is used. The downlink tests are similar to tests of the uplink power control. The tests of the uplink and downlink power control assess

- Response times;

- The effects of simultaneous multiple channel types;

- The influence of different dispersion environments.

11.2.6 Conclusions

The architecture of the MS-SIM, BTS, RNC, MSC, and the WCDMA operations system is structured by function and is very flexible. All ATM transport functionality is derived from the Ericsson ATM switch infrastructure in the RNC and MSC. The WCDMA evaluation system was built for experimenting with the

WCDMA radio-access network and ATM, and for applying these technologies to different wideband services.

Following the deployment of the evaluation system at NTT DoCoMo's premises in Japan, there was considerable interest from other operators to get a possibility to test third generation wireless capabilities. As a result, the WCDMA evaluation system has been installed in cooperation with more than 10 operators in eight countries around the world.

References

[1]   J. Eldståhl   and A. Näsman, "WCDMA Evaluation System—Evaluating the Radio Access Technology of Third Generation Systems," *Ericsson Review,* Vol. 76, Nov. 2, 1999, pp. 56-69.

[2]   A. Klein, R. Pirhonen, J. Sköld and R. Suoranta, "FRAMES Multiple Access Mode 1 - Wideband TDMA With and Without Spreading," *Proc. Int. Symposium on Personal, Indoor and Mobile Radio Communications*, Helsinki, Sept. 1997, pp. 37 - 41.

[3]   E. Nikula, A. Lappeteläinen and J. P. Castro, "Performance of FRAMES Multiple Access Mode 1 Without Spreading - WB-TDMA," *Proc. 3rd ACTS Mobile Communication Summit*, Rhodes, Greece, June 1998, pp. 327 - 332.

[4]   J. Reinius, "Cello—An ATM Transport and Control Platform," *Ericsson Review* Vol. 76, Nov. 2, 1999, pp. 48-55.

Chapter 12

Harmonization Activities

The international standardization bodies 3GPP, 3GPP2, and ITU-R are working on the finalization and further development of standards for the ITM-2000/UMTS radio interface. This should ensure the deployment of third generation mobile radio systems in the year 2001 in Japan, and in the year 2002 in other regions of the world. ITU-R is following the goal of a global standard, which fulfills the needs of international operators and end customers. Therefore, the Operators Harmonization Group (OHG) initiated a harmonization process for the CDMA-based radio interface proposals. The achievements of this process have affected the standardization activities.

12.1 ACTIVITIES OF THE OPERATORS HARMONIZATION GROUP

Regional standardization bodies submitted in 1998 different radio interface proposals for IMT-2000/UMTS to ITU-R, which fulfilled the requirements on third generation mobile radio systems. These proposals took into account already deployed second generation mobile radio system technology for backward compatibility reasons, to save investment of network operators as far as possible (see Chapter 1). However, these proposals by different standardization bodies resulted in incompatible solutions in terms of radio parameters and the protocol stack with respect to the requirement on global roaming. Therefore, major international network operators initiated in October 1998 harmonization activities in the OHG in the framework of the ITU-R process, to enable a harmonized approach for the CDMA-based IMT-2000 radio interface. TDMA-based proposals have not been taken into account in the OHG activities. The OHG cooperated with a group of major international manufacturers. This group achieved a consensus for a harmonized radio interface G3G (global third generation) at the end of May 1999 in its last meeting in Toronto [1] and [2]. This concept was submitted to ITU-R and was welcomed by ITU.

The standardization bodies 3PPP and 3GPP2 have endorsed the OHG conclusions as the basis for continued standardization work.

369

The main goals of OHG for this harmonization were

- To develop a single integrated third generation CDMA-based specification, which is based on the proposal WCDMA and cdma2000 being developed by 3GPP and 3GPP2;

- To create a system (G3G) consisting of three modes: DS, MC, and TDD to support the needs of different operators in different regions with respect to different requirements on backward compatibility;

- To ensure a widespread availability of voice and high speed nonvoice services;

- To enable customers to roam with their services across regions, countries, and systems;

- To support a smooth and compatible evolution path from existing infrastructure;

- To enable big markets and economy of scale for the development and deployment of systems.

These general goals resulted in technical requirements.

12.2 TECHNICAL REQUIREMENTS ON HARMONIZATION

The technical requirements are related to radio parameters as well as to the protocol stack to achieve the general goals. The RF parameters of the G3G system should be harmonized between the modes to the greatest extent possible. This would allow the cost-effective development of mobile terminals for global roaming and economy of scale that could significantly reduce the overall costs by minimizing the complexity of dual-mode and multiband terminals and equipment. In addition, the harmonized G3G radio interface should be fully supported by the evolved GSM MAP and ANSI-41-based core networks to support intersystem roaming and seamless handover between the harmonized DS and MC modes, including IS-95 for ANSI-41 and the equivalent to this for UMTS/GSM. Synchronous operation should also be supported. Figure 12.1 shows the generic approach to connect the harmonized G3G radio interface to both evolved core networks GSM MAP and ANSI-41. This approach should ensure on one hand the requirements of different operators in different regions to save their investment, and on the other hand to ensure global roaming.

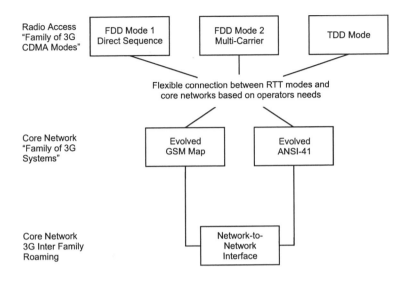

Figure 12.1 Modular 3G harmonized proposal by connection to the two main core networks (Source [2]).

12.3 ACHIEVED CONCENSUS ON THE TECHNICAL APPROACH

The OHG agreed with the Manufacturers Group in the Toronto meeting of the OHG at the end of May 1999 [2] that the harmonized standard for the DS mode will be based on WCDMA, and the harmonized standard for the MC mode will be based on cdma2000. Key technical parameters of the TDD mode will be harmonized with WCDMA as in the original 3GPP proposal for IMT-2000. In addition to this general agreement technical details are part of the consensus.

12.3.1 Interbase Station Synchronization

The original WCDMA proposal of 3GPP does not require a synchronization of base stations, whereas cdma2000 does require synchronous operation. In the G3G radio access scheme, a combined concept was selected with an

- Asynchronous and synchronous approach for WCDMA for the DS mode;

- Synchronous approach for cdma2000 for the MC mode.

12.3.2 Pilot Structure

Pilot signals are needed for channel estimation for signal detection and base station as well as mobile station identification. WCDMA and cdma2000 used in their original proposals provide different concepts. With respect to the harmonization, both pilot structures are combined. For common pilot signals, CDM is applied, where for connection, dedicated pilot signals TDM is used to satisfy the needs of both original concepts for global roaming. 3GPP and 3GPP2 are working on the details of the concept.

12.3.3   Chip Rate and RF Parameters

The chip rate is the key parameter for the design of the RF part of the base station and mobile station equipment. For backward compatibility reasons and the reuse of already deployed equipment, the MC mode does use a chip rate of 3.6864 Mcps, which is three times the chip rate of IS-95. WCDMA and the TDD mode in 3GPP originally used a chip rate of 4.096 Mcps, which can be generated from the GSM clock frequency 13 or 26 MHz. For the design of multimode terminals, practically the same RF part would be required for both modes in the base station and in particular in mobile station equipment. Therefore, the chip rate for the DS mode and the TDD mode is selected as 3.84 Mcps as a trade-off between minimizing the capacity loss due to a reduction of the chip rate from 4.096 Mcps, and to make the chip rate of the DS and MC mode as similar as possible. This will allow for only one RF chain for dual-mode (DS/MC) terminals. Also, the chip rate 3.84 Mcps can be generated similarly from the GSM clock frequency.

12.3.4   Protocol Structure and Interworking

With respect to the situation of different deployed second generation systems in different regions (see Figure 1.12), currently activities are underway by 3GPP and 3GPP2 under the auspices of the OHG to find ways for interworking between the ANSI-41 and GSM MAP-based systems. The OHG proposed a generic protocol stack (Figure 12.2) to enable the connection of the harmonized 3GPP and the 3GPP2 concept either to an evolved GSM core network, as well as to an evolved ANSI-41 core network to ensure terminal mobility and global roaming according to Figure 12.1. As preparation for the necessary extensions to a given protocol stack, so-called hooks will be introduced in both protocol stacks for WCDMA / TDD as well as cdma2000.

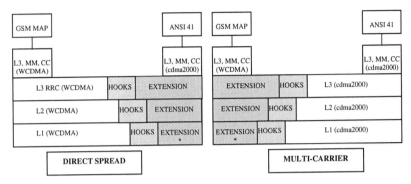

Note: * – Areas for further analysis

Figure 12.2 Protocol structure for implementing the modular concept (Source: OHG [2]).

12.4 FURTHER DEVELOPMENTS

The combination of aligned radio parameters and a combined protocol stack should enable the necessary flexibility for terminal and global roaming to support the needs of international operators and end users. This concept was submitted to ITU-R, 3GPP, and 3GPP2 [2]. Therefore, the international standardization activities are ongoing based on this concept, and they are supported by the community of network operators and manufacturers.

The standardization bodies 3GPP and 3GPP2 are implementing the necessary hooks of this proposed consensus in the release 1999 of the standard. These so-called hooks are the prerequisite for the implementation of the necessary protocol extensions. The changes to the layer 1 will be minimized in these activities, as well as the required changes to the protocol layers 2 and 3 of the core networks. Protocol extensions are part of the release 2000 of the standards that are being standardized by 3GPP and 3GPP2.

References

[1] ITU "Major Progress in Beijing on Standardization of IMT-2000," Press Release ITU/99-7, June 15, 1999.

[2] OHG "Harmonized Global 3G (G3G) Technical Framework for ITU IMT-2000 CDMA Proposal," *Operator Harmonization Group*, May 1999, submitted to ITU-R, Beijing, June 1999.

About the Editors

Ramjee Prasad was born in Babhnaur (Gaya), Bihar, India, on July 1, 1946. He is now a Dutch citizen. He received a B.Sc. (Eng) degree from Bihar Institute of Technology, Sindri, India, and M.Sc. (Eng) and Ph. D. degrees from Birla Institute of Technology (BIT), Ranchi, India, in 1968, 1970, and 1979, respectively.

He joined BIT as a senior research fellow in 1970 and became associate professor in 1980. While he was with BIT, he supervised a number of research projects in the areas of microwave communications and plasma engineering. During 1983 to 1988, he was with the University of Dar es Salaam (UDSM), Tanzania, where he rose to the level of professor of telecommunications at the Department of Electrical Engineering in 1986. At UDSM, he was responsible for the collaborative project "Satellite Communications for Rural Zones" with Eindhoven University of Technology, the Netherlands. From February 1988 to May 1999, he was with the Telecommunications and Traffic-Control Systems Group, Delft University of Technology (DUT), the Netherlands, where he was actively involved in the area of wireless personal and multimedia communications (WPMC). He was the head of the Transmission Research Section of IRCTR (International Research Center for Telecommunications - Transmission and Radar) and also founder program director of the Center for Wireless Personal Communications (CEWPC). Since June 1999, he has been with Aalborg University, Denmark, as codirector of the Center for PersonKommunikation (CPK) and holds the chair of wireless information and multimedia communications. He was involved in the European ACTS project FRAMES (Future Radio Wideband Multiple Access Systems) as a DUT project leader. He is a project leader of several international industrial-funded projects. He has published over 300 technical papers, contributed chapters in several books, and authored and coauthored four books, *CDMA for Wireless Personal Communications*, *Universal Wireless Personal Communications*, *Wideband CDMA for Third Generation Mobile Communications*, and *OFDM for Wireless Multimedia Communications* published by Artech House, Boston. His current research interest lies in wireless networks, packet communications, multiple access protocols, advanced radio access techniques, and multimedia communications.

He has served as a member of advisory and program committees of several IEEE international conferences. He has also presented keynote speeches, invited papers, and tutorials on WPMC at various universities, technical institutions, and IEEE conferences. He was the founder chairman of the IEEE Vehicular Technology/Communications Society Joint Chapter, Benelux Section and is now the honorary chairman. He is also founder of the IEEE Symposium on Communications and Vehicular Technology (SCVT) in the Benelux and he was the symposium chairman of SCVT'93.

He is the founding coordinating editor and editor-in-chief of the Kluwer international journal on *Wireless Personal Communications* and also a member of the editorial board of other international journals, including the *IEEE Communications Magazine* and the *IEE Electronics Communication Engineering Journal*. He was the technical program chairman of the PIMRC'94 International Symposium held in The Hague, the Netherlands, during September 19–23, 1994, and also of the Third Communication Theory Mini-Conference in conjunction with the GLOBECOM'94 held in San Francisco, CA, November 27–30, 1994. He was the conference chairman of IEEE Vehicular Technology Conference, VTC'99 (fall), Amsterdam, the Netherlands held on September 19–22, 1999, and also the steering committee chairman of the Second International Symposium on Wireless Personal Multimedia Communications (WPMC), Amsterdam, the Netherlands held on September 21-23, 1999.

He is listed in the US Who's Who in the World. He is a fellow of the IEE, a fellow of the Institution of Electronics & Telecommunication Engineers, a senior member of IEEE, and a member of NERG (Netherlands Electronics and Radio Society).

Werner Mohr was born in Hann. Münden, Germany, on June 2, 1955. He received a master's degree and a Ph.D. degree both in electrical engineering from the University of Hannover, Germany, in 1981 and 1987, respectively.

He joined BEB Betriebsführungsgesellschaft, an oil and gas company, from 1981 to 1982, where he was responsible for the investigation of a measurement system. In 1982 he returned to the University of Hannover as a member of the research staff of the Institute of High-Frequency Technology. From 1987 to 1990 he was a senior engineer at the same institute. In 1989 to 1990 he was a lecturer at the Fachhochschule Hannover, Germany for telecommunication systems.

Dr. Werner Mohr has been with Siemens AG, Mobile Network Division in Munich, Germany since 1991. He has been responsible for the development of a wideband propagation measurement system, propagation measurements, and channel modeling and he was involved in the European RACE-II Project ATDMA on first investigations for the third generation mobile radio interface. Wideband propagation channel models, which were developed based on extensive measurement campaigns in the ATDMA project, have been internationally standardized by ETSI SMG and ITU TG 8/1 for the evaluation of third generation mobile radio interface proposals. From 1995 to 1996 he was active in ETSI SMG5 for standardization of UMTS. During that time Werner Mohr was also responsible for the evaluation of several mobile radio standards. Since December 1996 he was project manager of the European ACTS Project until the project finished in August 1999. The ETSI SMG decision on the UMTS radio interface (UTRA concept) was based largely on the contributions of the ACTS FRAMES project. He was involved in the decision of ETSI SMG on the UTRA concept in January 1998 in Paris. Werner Mohr was the director of strategic pre-development up to September 1998, and since October 1, 1998, he is vice president of pre-engineering in the chief technical office of the communication on air business area of Siemens, ICN. Currently, he is involved in the 5th framework program of the European Commission. His current research interests are multiple access techniques for wireless communications, and the interworking of systems.

He has published over 60 technical papers in international journals and conferences, including invited papers; he presented tutorials and organized and participated in panel discussions on third generation mobile radio systems. Werner Mohr served as session chairman in several international conferences. He has been a member of several technical program committees of international conferences, e.g., IEEE VTC 1999-Fall, Globecom '99, European Wireless '99, WPMC '98, '99 and 2000.

Werner Mohr is listed in the US Who's Who in the World, the Who's Who in Science and Industry, and other publications. He is a member of VDE, the German Association of Electrical Engineers, and a senior member of IEEE. In 1990 he received the Award of the ITG in VDE.

Walter Konhäuser was born in Ruhpolding, Germany, on November 19, 1949. He studied electrical engineering at the Technical University of Berlin and received a master's degree in 1976.

He was a member of the Institute of Electronics at the Technical University of Berlin from 1977 to 1982, where he became an assistant professor. He was actively involved in the area of decentralized computer networks and the control of industrial processes. He received a Ph.D. degree in electrical engineering from the Technical University of Berlin in 1981.

Since joining Siemens in 1982, he has been involved in a variety of assignments. During the first two years he was responsible for software development of a security system, which was built on a decentralized microcomputer network. He then joined the Siemens Relay Group, where he was responsible for planning and designing full automatic systems for relay assemblies. From 1987 to 1990 Dr. Konhäuser was the plant manager for PCB Assembly with a staff of 600. From 1990 to 1992 he was a quality assurance manager and he has been responsible for the development of an improved operations management concept for the electronic plants.

For the past eight years Dr. Konhäuser has worked in the mobile networks business. In 1992, he took over the responsibility for the application software development for the mobile switching subsystem. A variety of data services were developed under his leadership including the first demonstration of fax transmission via GSM networks. In 1995 he became head of product management for the mobile networks business (Radio Access, Core, and IN). Presently he is senior vice president and chief technical officer within the mobile infrastructure business of Siemens. Dr. Konhäuser is involved in a variety of radio access technologies for wireless communications future network functions (mobile IP networks), and protocols for mobile multimedia communications.

Dr. Konhäuser was PCC chairman of the European ACTS FRAMES Project, which had a significant influence on the ETSI SMG UMTS radio interface decision. He was actively involved in the ETSI SMG decision on the UMTS radio interface in January 1998 in Paris.

He has published over 100 technical papers in international journals and conferences, including invited papers and keynote speeches. He has also authored and coauthored four books.

Walter Konhäuser has been active as a professor in industrial control systems at the Technical University of Berlin since 1996. He is a member of VDE, the German Association of Electrical Engineers.

List of Authors

Author	Affiliation
Acx, A.-G.	France Télécom – CNET, France
Anderson, P.-O.	Ericsson Radio Systems AB, Sweden
Arponen, J.	Nokia Corporation, Finland
Baier, P.W.	University of Kaiserslautern, Germany
Berens, F.	ST Microelectronics, Switzerland Former: University of Kaiserslautern, Germany
Berg, M.	Royal Insitute of Technology (KTH), Sweden
Bing, T.	University of Kaiserslauterrn, Germany
Cedervall, M.	Ericsson Radio Systems AB, Sweden
Cercas, F.	Instituto Superior Técnico, Portugal
Correia, A.	Instituto Superior Técnico, Portugal
Croft, P.	Roke Manor Research Ltd., United Kingdom
Dahlhaus, D.	Swiss Federal Institute of Technology Zurich, Switzerland
Eldståhl, J.	Ericsson Radio Systems AB, Sweden
Erben, H.	Siemens Schweiz AG, Switzerland Former: Centre Suisse d'Electronique et de Microtechnique (CSEM), Switzerland
Frenger, P.	Chalmers University of Technology AB, Sweden
Guerin, N.	France Télécom – CNET, France
Haardt, M.	Siemens AG, Germany
Holma, H.	Nokia Corporation, Finland
Janssen, G.J.M.	Delft University of Technology, The Netherlands
Karlsson, M.	Ericsson Radio Systems AB, Sweden
Klein, A.	Siemens AG, Germany
Konhäuser, W.	Siemens AG, Germany
Latva-aho, M.	University of Oulu, Finland
Lehtinen, O.-A.	Nokia Corporation, Finland

Lindström, M.	Royal Institute of Technology (KTH), Sweden
Lundsjö, J.	Ericsson Radio Systems AB, Sweden
Modonesi, I.	Nokia Corporation, Finland
Mohr, W.	Siemens AG, Germany
Moretti, M.	Delft University of Technology, The Netherlands
Näsman, A.	Ericsson Radio Systems AB, Sweden
Naßhan, M.	Siemens AG, Sweden
Ojanperä, T.	Nokia Corporation, Finland
Olofsson, H.	Ericsson Radio Systems AB, Sweden
Orten, P.	Chalmers University of Technology AB, Sweden
Ottosson, T.	Chalmers University of Technology AB, Sweden
Peterson, S.	Royal Institute of Technology (KTH), Sweden
Prasad, R.	Center for PersonKommunikation, Aalborg University, Denmark
	Former: Delft University of Technology, The Netherlands
Richardson, K.	Roke Manor Research Ltd., United Kingdom
Rinne, M.	Nokia Corporation, Finland
Slanina, P.	Siemens AG Österreich, Austria
Svensson, A.	Chalmers University of Technology, Sweden
Toskala, A.	Nokia Corporation, Finland
Urios, N.	Integracion y Sistema de Medida, Spain
Zander, J.	Royal Institute of Technology (KTH), Sweden

Index